Mary Alice Icee

J. KIERNAN

MEAT HANDBOOK
FOURTH EDITION

MEAT HANDBOOK

FOURTH EDITION

Albert Levie

President
Gulliver's Inc.
Los Angeles

VNR VAN NOSTRAND REINHOLD
New York

Library of Congress Catalog Card Number 79-787

ISBN 0-87055-315-1

Printed in the United States of America

Van Nostrand Reinhold
115 Fifth Avenue
New York, New York 10003

Van Nostrand Reinhold International Company Limited
11 New Fetter Lane
London EC4P 4EE, England

Van Nostrand Reinhold
480 La Trobe Street
Melbourne, Victoria 3000, Australia

Nelson Canada
1120 Birchmount Road
Scarborough, Ontario M1K 5G4, Canada

16 15 14 13 12 11 10 9 8 7 6

Library of Congress Cataloging in Publication Data
Levie, Albert.
Meat handbook.
Includes bibliographies and index.
1. Meat 2. Meat industry and trade. I. Title.
TS1955.L42 1979 664′.92 79-787
ISBN 0-87055-315-1

Contents

Preface

This handbook provides a practical source of information for the food service industry, locker plant, and meat retailers. The material covers the full product cycle from livestock to cut steak. Included are useful background chapters on livestock production and slaughter, including inspection and grading, and the more practical aspects of purchasing, handling, fabricating, storing, and cooking. Since the publication of the first edition in 1963 much new material has been added; photographs of new equipment and current information on preportion cut steaks, along with revised grading standards and new roasting techniques.

The vast funds of research material, the host of people responsible for it, and the willingness of the food service and meat industries to disclose and share, have made this handbook possible. Special mention must be made of the Grading Branch and Meat Inspection Division of the Department of Agriculture, the National Live Stock and Meat Board, to Ben Gutterman, Mike Morganelli, and Cliff Duboff at Elgee Meats, and to the typists who were able to salvage my hieroglyphics.

The author wishes to thank the following for technical assistance in conjunction with the preparation of various segments of the manuscript:

W. O. Caplinger, Chief Staff Officer for Planning and Appraisal, Meat Inspection Division, U.S. Dept. of Agriculture, Washington, D. C.

Max O. Cullen, National Livestock and Meat Board, Chicago, Ill.

Benjamin H. Ershoff, Ph.D., Western Biological Laboratories, Los Angeles, Calif.

Clifford F. Evers, M.S., Manager of Commercial Development-Food, Accent International, Skokie, Ill.

Homer D. Fausch, Department of Animal Husbandry, California State Polytechnic College, Pomona, Calif.

Norman Jaspan, President, Norman Jaspan Associates, Inc., New York City, New York

Professor C. Arthur Jones, Chef Instructor, Culinary Institute of America, New Haven, Conn.

Most gratefully acknowledged is the advice, encouragment, and technical consultation of Professor J. J. Wanderstock of Cornell University, Ithaca, N. Y.

And for the time that has accrued to them the last five years, to my wife La Vonne and to Bradley and Glenn, a deep appreciation.

ALBERT LEVIE

Los Angeles, Calif.
July 1978

1

Livestock

Meat has been used by man for food since the beginning of recorded time. The Bible talks about the fatted calf. Old rabbinical laws dictate which meats are considered sanitary. Pork is recorded as a food as early as 3400 B.C. in Egypt and 2900 B.C. in China.

The first livestock may have been brought to America by the Norsemen. It is possible that about the year 1000 they brought swine and European cattle. Both the colony and the livestock disappeared. On his second voyage in 1493, Columbus carried eight pigs, some cattle, sheep, and goats to Hispaniola (Haiti).

In the sixteenth century, Hernando Cortez landed at Vera Cruz and his military retinue included a few Andalusian cattle with very long horns, and some sheep. Subsequently, the Conquistador became a settler and turned to the business of raising livestock. This was the beginning of the picturesque "Texas Longhorns," the parent stock of later better bred beef cattle, the great "Mexican Herd."

The history of a second herd, the "Atlantic" beef, is controversial. One record reveals that a considerable shipment of beef was received in Jamestown, Virginia about 1610. Other records indicate that three heifers and one bull were shipped from England for breeding purposes and arrived in March, 1624. Whichever may have been the case, the intent was to breed and produce dairy products and meat animals.

Ferdinand De Soto is credited with the first hogs delivered to Florida in 1553. Not too much later, hogs roamed wild in the colonies, were hunted and shot in the fall, and salted for winter consumption and export. The London Company brought the first sheep to Jamestown, Virginia, in 1609.

Today, meat processing is one of America's biggest volume industries. In 1976 reported sales exceeded 38 billion dollars, exceeding 39 billion pounds. On January 1, 1978, cattle population for all breeds was placed at 116.2 million head. (Fig. 1.1.)

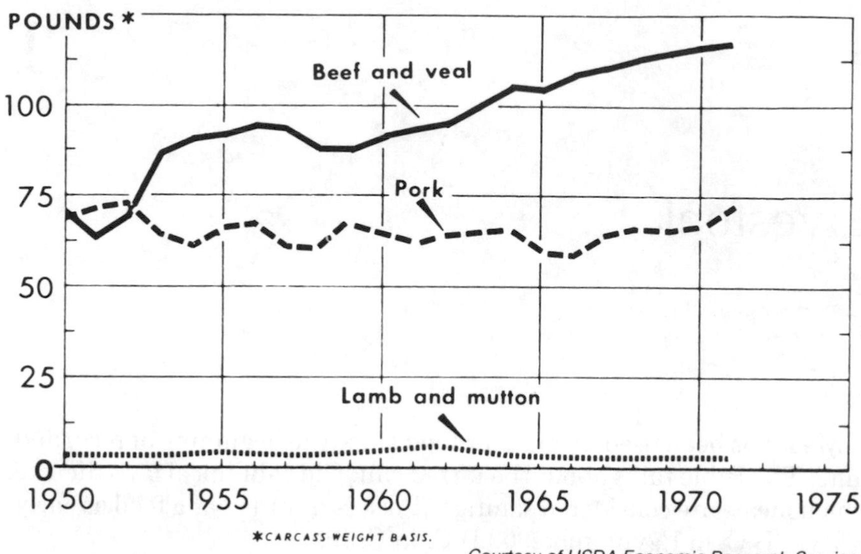

Courtesy of USDA Economic Research Service

FIG. 1.1. MEAT CONSUMPTION PER PERSON

THE CATTLE INDUSTRY

The livestock industry, almost completely revolutionized in the past three centuries, is highly specialized and very scientific. A single firm may be involved not only in livestock, but in slaughter, and distribution as well. On the other hand, there are operations so highly specialized that the total sales projection is but a single product in a single phase of the total industry.

The cattle industry is sharply divided into meat animals and dairy animals. On one hand, dairy breeding has become highly selective with two purposes, maximum fluid yield and maximum butter fat yield, or a desirable ratio of the two. There are additional specialized beef breeds for fighting, draft animals, and animals of worship.

The dairy industry contributes considerably to meat, primarily as a by-product (Table 1.1). Young calves, except those saved for breeding, are slaughtered and account for a large portion of the veal and calf meat produced. Dairy cows and breeding bulls are slaughtered at the end of their productive life. This type of meat is mostly of the lower grades, including USDA Commercial, USDA Utility, USDA Cutter, and USDA Canner. Small quantities of this meat product find their way into retail and institutional channels.

TABLE 1.1. DAIRY TYPE BEEF—BREED AND SEX AS RELATED TO DRESSED CARCASSES INCLUDING HOLSTEIN, JERSEY, GUERNSEY

Male Stock	Female Stock
Calves Greatest number of the male stock is slaughtered for veal.	*Calves* Best stock is saved to replace dairy and breeding stock. The balance are slaughtered for veal.
Steers Some are produced. Considerable research is being conducted on fattening dairy steers	*Heifers* The number raised for commercial meat purposes is increasing under the present grading system.
Bulls Best stock is selected for breeding purposes. They are slaughtered at the end of their economic breeding age, yielding sausage type carcasses.	*Dairy Cows* Best stock is saved for dairy production. They are slaughtered at the end of their economic dairy age, yielding USDA Utility, USDA Cutter, and USDA Canner beef, with some USDA Commercial.

Breeds of Beef Cattle

Beef cattle are usually raised for meat as a food product (Table 1.2). The most widely produced purebreeds are Shorthorn, Hereford, Aberdeen Angus, Brahman, and Santa Gertrudis.

Shorthorn.—The original breed was imported from England in 1783. It was well adapted and consequently widely produced. In color, they range from all red to all white, with combinations of the two as well as red roan. Dark red colors are preferred. Shorthorns make high quality

TABLE 1.2. MEAT TYPE BEEF—BREED AND SEX AS RELATED TO DRESSED CARCASSES INCLUDING SHORTHORN, HEREFORD, ANGUS, BRAHMAN AND SANTA GERTRUDIS

Male Stock	Female Stock
Calves Slaughter is limited. Calves are mostly saved for steers and bulls	*Calves* Slaughter is limited, though probably greater than male stock. Calves are mostly saved for breeding cows and heifers.
Steers Bulk of male stock is castrated and raised for high grade beef. Dressed carcasses will range from about 550 to 1000 lbs, ranging in grade from USDA Prime to USDA Standard, grading mostly USDA Choice.	*Heifers* Depending on range, feed prices and the live market, a large number are fed and marketed as one- to two-year-old fed heifers. They are equal in quality to steers. Heifers are generally smaller retail-market type carcasses.
Bulls Top quality males are saved for breeding. They are slaughtered at the end of their economic life, making a sausage type carcass. Research is being conducted to test marketability of young beef bulls.	*Cows* Best stock is saved for breeding purposes. They are slaughtered at end of their economic breeding life. Cows grade mostly USDA Commercial or lower, and are mostly used for institutional purposes and sausage.

beef, with a very full hindquarter and a thick loin. In many experiments it has been observed that there is some evidence of excessive fat covering and a small eye in the strip loin muscle. The breed is extensively used in the corn belt.

Hereford.—Henry Clay introduced this breed in 1817, importing them from Herefordshire, England. The Hereford caught on after the severe winters of 1881, 1886, and 1887 which demonstrated the superiority of the breed. It is noted for its ability to thrive under the most adverse conditions, such as poor range and short grasses. This stock matures early and fattens readily in feed lots.

The colors are basically red and white. The red color ranges from dark red to a light, almost yellow red. Color bears no relationship to performance or carcass quality. The nickname for the breed, "Whiteface," comes from the universally characteristic all-white face.

Conformation is excellent for beef, with short legs, straight back, compact and blocky profile, well sprung ribs, and a thick broad loin. The rib and loin yield is very high.

Aberdeen Angus.—The first Angus in America were a gift to Sir George Simpson, a Hudson Bay Company official, in Montreal about 1859. Angus were imported in 1873 by George Grant of Victoria, Kansas from Aberdeenshire County and Angus County in Scotland.

The color is solid black. This breed has no horns and is commonly referred to as "Black Angus." In general, these beef are smaller than the Shorthorn and Hereford, and do not thrive on desert range land. However, where good pastures prevail, as in the corn belt and Pacific Northwest, there are some advantages. With optimum pasture conditions, they fatten well and mature early. They have a very desirable quality, showing good marbling and conformation.

Brahman (Zebu or Brahma).—This is an American breed founded on cattle from India imported as early as 1849. They are easily identified by the gray color, a distinct hump over the shoulders, and an excess of loose skin under the throat. The American breed, the results of intermating the several different Indian types, is a more beefy-type animal.

The Brahman sweats and therefore can withstand intense heat; they do very well in the humid Gulf Coast and the semi-arid Southwestern States. The breed is also highly resistant to insect pests.

There is considerable crossbreeding of the English beef breeds, imparting a hybrid vigor. Many feedlot areas in the South and Southwest, as in the Imperial Valley in California, utilize a preponderant number of such crossbreeds.

The distinguishing conformation under previous grading standards prevented the carcass from being accepted for a USDA Choice grade. The meat itself has a desirable pink color. When the carcass is merchandised as wholesale and retail cuts, the identity and the stigma are both lost.

Santa Gertrudis.—This is an American bred purebreed started in 1910 by the famous and vast (940,000 acres) King Ranch near Corpus Christi, Texas, crossing Brahman bulls and beef-type Shorthorn cows. Thirty years of close breeding to "Monkey," his sons and grandsons produced the present breed.

This purebreed will thrive in semi-arid and tropical regions, and has top beef breed conformation characteristics. Rigid selection has produced a cherry red, heat resistant, humpless Santa Gertrudis, which at eight months of age will outweigh the Hereford and Shorthorn by 100 to 200 lbs. This vigorous, quick growing, and fattening breed is now produced in about 24 states and 18 foreign countries.

Beefalo.—The American buffalo has been successfully crossed with the bovine, to coin the name, Beefalo. A typical animal is 3/8 Charolais, 3/8 buffalo, 1/4 Hereford. As yet, it is too early to evaluate the impact of this new breed. The animals are large, lean, and mature early. The assertion is that they are marketable at 1000 lbs at 12 to 14 months of age. There is some question of second generation verility.

Other Breeds.—There are other registered purebreeds, though generally less popular, such as the Devon, Galloway, Red Poll, Scotch Highland, Sussex, Charolaise (French), and the American breeds, Beefmaster, Brangus, and the Charbray. The Charolaise breed is becoming popular in the South and the Southwest. They are also used to breed Charbray cattle.

In commercial practice, large numbers of crossbreeds and mixed breeds are marketed.

Beef Raising

There are three distinct stages in the growing of beef. These frequently occur under different ownerships or at least different phases of management. These are commonly referred to as: (1) cow and calf operations, (2) weaner calves and yearlings, and (3) dry lot feeding. The division is arbitrary and all three phases can occur on a single farm.

Cow and Calf.—A specialized operation occurs mostly in the vast plain areas where the breeding cows and their calves require no more than vast ranges, now fenced, a handful of cowboys to look after all of their needs, a branding iron, plenty of know-how and hard work, and plenty of capital or good banking connections.

Weaner Calves.—Calves are cut out or "weaned" from the cows in late summer or fall at about 6 to 8 months of age. They are sold or moved to more vigorous pastures. Sometimes they are moved hundreds of miles to the natural meadows, sometimes they are put on "permanent pasture" or "rotation pastures" which are fenced, irrigated pastures divided into several separate areas. The cattle are fed in one pasture while the other pastures are irrigated and permitted to grow. The cattle are moved periodically. At about 1 to 1½ years of age, they are generally ready for dry lot feeding.

Dry Lot Feeding.—Fattening in a dry lot is done with the intention of producing USDA Choice and USDA Prime carcasses. The feed lots are confining to restrict physical activity of the animals. The feed is calculated to produce the maximum weight gain per dollar of feed used. Various concentrates are used in the formula, sometimes hard grain, sometimes by-products, such as cotton trash, grape pulp, feather meal, beet pulp, and orange peelings. About three months in the feed lot produces marginal USDA Choice. Hormones are used for growth promotion and feed economy. The oral use of Diethylstilbestrol (DES) is still permitted in prescribed dosages under title 21, 558.225. The Food and Drug Administration is involved in litigation (1978) pertaining to its use. Estrogenic preparations, Synovex S and Synovex H are used as ear implants for the same purpose.

At the conclusion of the dry feeding period, the beef cattle are shipped to one of the major livestock centers, such as Chicago, St. Paul, Kansas City, Omaha, Sioux City, Denver, and Los Angeles, or sold in the "country" to buyers representing the packers. About 60 per cent of the cattle, calves, and lambs are marketed in the "300" rural auctions, as the industry trend continues toward decentralization.

PORK

The development of pork has been much like to that of beef, with careful selection, breeding, and registration of desirable meat-type hogs. There are English breeds, such as the Berkshire, Hampshire, Tamworth, and Yorkshire. There are several basic American registered breeds: the American Landrace, Poland China, Chester White, Duroc, Beltsville No. 1 and No. 2, Maryland No. 1, Minnesota No. 1, No. 2, No. 3, Montana No. 1, the Palouse, and the Hereford. In each instance, the intent was to breed away from the fat-type hog, to the meat-type, which is described as moderately long (30 to 31 in. at 200 lbs), trim, firm, meeting the minimum finish for high quality pork, and yielding over 50 per cent of the carcass weight in trimmed hams, loins, picnics, and Boston butts. These yield less lard and more lean.

LAMB

Production of lamb is reaching for the same stature as beef and hogs, although the demand for this meat product is less. The introduction of synthetic materials since World War II has damaged the wool economy, which is as primary a product as meat. In the Central and Eastern states, lamb is economically produced in small flocks. On the Western plains, much larger flocks are worked.

GLOSSARY OF LIVE MARKET TERMS[1]

This glossary is presented because this language creeps into everyday use and leads to many misconceptions.

Baby Beef.—Originally applied to all yearlings. Today a steer or heifer fattened and marketed by the time it is a year old.

Bangs Cows.—Cows which have reacted to tests for Bang's disease or contagious abortion, and which may only be unloaded in the quarantine division of the market.

Bow Wow (Dog, Rannies).—A small stunted steer or heifer with no quality, unsuited either for beef or feeder purposes. Utilized sometimes as canners or cutters.

Dogey.—Denotes a young animal lacking quality and finish.

Fed Cattle.—Cattle that have been fattened on grain and supplements.

Green Cattle.—Cattle direct from the range and not fed grain. The term is particularly useful in the winter and early spring to distinguish western range cattle from thin young steers that have been warmed up on grain but have to sell as feeders.

Heavy Steer.—One weighing 1300 lbs and up. Those 1500 lbs or above are called "extremely heavy."

Heiferette.—Really a young cow that has had one or more calves, but which, because of straight lines, relative smoothness, and small udder can be used by buyers to fill orders for heavy heifers.

[1] Reprinted by special permission of the Stockman's Journal.

Kosher Cow.—Heavy, fat cow of top quality which, at certain seasons, is very popular among eastern packers catering to the Jewish trade.

Longfed Cattle.—Cattle fed 6 to 13 months or longer, depending on the weight at the start of the feeding period.

Mixed Yearlings.—Usually yearlings of both sexes—steers and heifers—that are sold and weighed together at the same price.

Native Cattle.—Unbranded stock raised and fattened on Corn Belt farms, or under corresponding conditions.

Range Cattle.—Grass fed, shipped straight from the western ranges without being fed grain other than possibly a winter maintenance ration.

Replacement Cattle.—Frequently used interchangeably with stockers and feeders to describe cattle to take back to the country for grazing, wintering, or fattening.

Shortfed Cattle.—Usually cattle fed 30 to 120 days depending on the weight at the start of the feeding period.

Springers.—Cows with calf taken back to the country. Cows are generally later fattened and marketed, calves being fed and sold later.

Stag.—An unsexed male animal not castrated until it has reached maturity. A stag is coarser and displays stronger sex characteristics than a steer.

Warmed-up Cattle.—Originally described practically fat western steers, off grass, and fed a heavy grain ration for a very short time, probably 30 days or less. Now more or less generally used to describe all fed cattle not fat enough to be really desirable from a slaughter standpoint.

HOGS

Barrow.—A male hog castrated when young.

Boar.—A male hog.

Butcher.—A gilt or barrow of any weight, but usually a young animal purchased for slaughter.

Gilt.—A young female hog that has not been bred.

Mixed Hogs.—A group of hogs that includes gilts, barrows, or both, along with a number of sows, or a lot of hogs of uneven weight.

Packers.—Sows that, because of age, weight, or quality, are not suitable for fresh meat demands, and therefore are pickled, canned, made into lard, or cured in other forms.

Shipping Hogs.—Hogs bought on the specific central market to be shipped to other points for slaughter, usually are of selected and sorted grades.

Skips, Busts, Bums, Squeals, Outs, Culls, etc.—Inferior grades, usually short on weight as well as quality and frequently crippled by disease or injury.

SHEEP

Aged Sheep.—All sheep over two years old.

Buck.—A male sheep.

Comeback.—Western-bred animal that has come to market, gone back to the country for feeding, and returned to market in either fat or feeder flesh.

Cornfield Lamb.—A lamb fattened mainly on grain and forage picked up in the cornfield.

Ewe.—A female sheep.

Fat, Slaughter, or Killer Lambs or Ewes.—Suitable for immediate slaughter.

Fed.—An animal that has been fed grain.

Feeder Lamb.—An animal under one year old raised in the Corn Belt, usually in the area served by the market on which the designation is used.

Shearing Lamb.—A heavy lamb, not infrequently in killer flesh, that sells with the wool on to feeder-buyers for shearing purposes and a short feed to regain weight lost in shearing.

Shipping Lambs.—Fat lambs purchased for slaughter in packing plants some distance from the market where sold.

Shorn.—Any animal from which the wool has been sheared. While their grade as meat animals is gauged by the same standards that apply to wooled stock, the length of wool, or fleece, has much to do with its value for either slaughter or feeding purposes.

Spring Lambs.—Young lambs dropped in late winter or early spring; so designated until early summer, when lambs of the previous year become yearlings.

Western.—An animal born on the Western range. Sometimes called "range." Its state of origin is usually indicated in market reports.

Wether (Wether Lamb).—An unsexed male lamb.

Woolskin.—Any animal with wool on—not shorn.

REFERENCES

ANON. 1951. Market classes and grades. A circular, The Stockman's Journal, Omaha, Neb.

ANON. 1957. Facts about beef cattle. A circular, American Meat Institute. Chicago, Ill.

ANON. 1960. Meat Reference Book. American Meat Institute. Chicago, Ill.

BULL, S. 1951. Meat for the Table. McGraw-Hill Book Co., New York.

CLARK, R. T., and BAKER, A. L. 1958. Beef-cattle breeds. Farmers' Bull. *1779,* U. S. Dept. Agr., Washington, D. C.

OPPENHEIMER, H. L. 1961. Cowboy Arithmetic. The Interstate Printers and Publishers, Inc., Danville, Ill.

POTTS, C. G. 1953. Sheep raising on the farm. Farmers' Bull. *2058,* U. S. Dept. Agr., Washington, D. C.

ROMANS, J. R. and ZIEGLER, P. T. 1977. The Meat We Eat, 11th Edition. Interstate Printers and Publishers, Danville, Illinois.

THOMPSON, J. W. 1942. A History of Livestock Raising in U.S., Hist. Series *5*, U. S. Dept. Agr., Washington, D. C.

WOLGAMOT, I. H. 1957. Beef-facts for consumer education, AIB *84,* U. S. Dept. Agr., Washington, D. C.

WOLGAMOT, I. H., and FINCHER, L. J. 1954. Pork-facts for consumer education, AIB *109,* U. S. Dept. Agr., Washington, D. C.

ZELLER, J. H. 1958. Breeds of swine. Farmers' Bull. *1263,* U. S. Dept. Agr., Washington, D. C.

2

Slaughter and Inspection

The first recorded American meat packer was a Captain John Pynchon of Springfield, Mass., who started operations about 1641. He cured with salt and "packed away" salt pork, beef, venison, and bear.

Pynchon led a group of hearty settlers to the lush corn lands of the Connecticut River Valley where the commercial marketing of meat as well as the meat packing industry began. Corn crops and stock production were increased, pork and beef were packed in barrels to be shipped to the colonies.

"Porkopolis," a long forgotten name for Cincinnati, became the first large-scale commercial packing center. The products were prepared for the domestic market and export as well.

Then America moved west. The railroad system developed nationally in its wake. New and improved breeds of cattle grazed over millions of acres. On Christmas Day, 1865, the Chicago stockyard opened. Railroad centers such as Kansas City, Omaha, and St. Paul, soon became major slaughter centers.

In 1867, Philip D. Armour opened his plant and office in Chicago. In 1875, Gustafus F. Swift opened the then modern Swift and Company plant.

Two important technological advances occurred. The refrigerated car was invented by C. H. Hammond. The car was refrigerated initially by ice harvested on the lakes in the winter. In 1880, mechanical refrigeration was perfected and introduced as a means of making ice all year round.

Other changes took place. Large cattle drives had started to the vast ranges of the West, to the lush grazing lands of Kansas, Nebraska, Dakotas, Wyoming, and Montana.

The slaughtering of livestock is complicated, keenly competitive, and ingeniously devised. It is said that the automobile assembly line, set up by Henry Ford, was an adaptation of what he saw in a slaughterhouse. The slaughtered carcass is hung up on a monorail system and moved from worker to worker in regular disassembly line fashion. It is monorailed to the "hot box" for shrouding, shaping and bleaching, and removal of the animal heat; to the cooler for storage; and then to the dock for delivery to truck transportation for ultimate distribution.

ECONOMICS

The economics of the industry are startling. In 1955, total domestic meat production was 27 billion pounds (Fig. 2.1). In 1977 production exceeded 39.1 billion pounds.

Per capita consumption has increased correspondingly (Fig. 2.2).

A most interesting economic phenomenon is that the dressed beef carcass is usually sold for less total dollars than the initial total dollar cost of the live animal. The by-products take care of the difference as well as the profits. It is said that everything is sold except "the squeal." As big as the packers are, relative to their tonnage production or their dollar sales, their profits are far smaller than most any other industry; net profits of about one half per cent of the total packer sales are considered average-to-good returns.

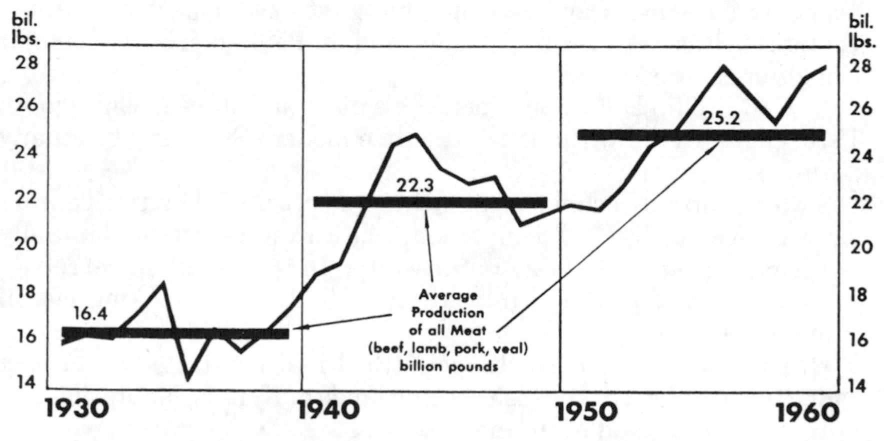

Courtesy of Dr. Herrell De Graff

FIG. 2.1. MEAT PRODUCTION—1930 TO 1960

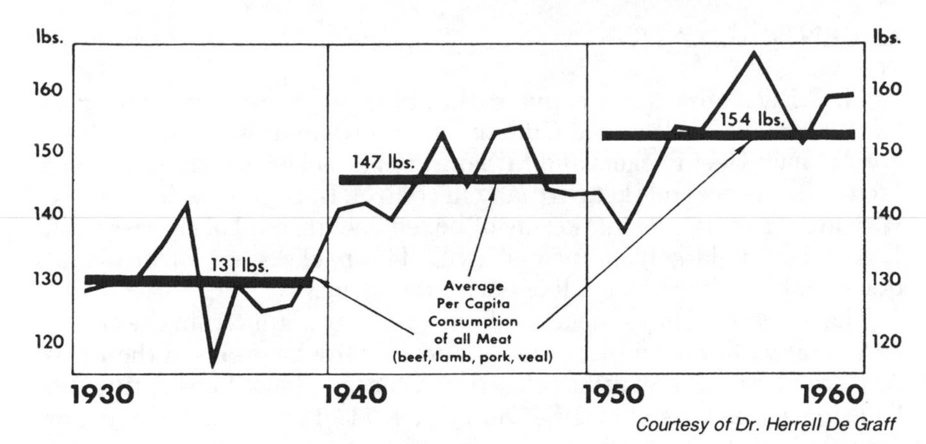

Courtesy of Dr. Herrell De Graff

FIG. 2.2. PER CAPITA MEAT CONSUMPTION—1930 TO 1960

PRODUCTS

The packing house is a complex business organization, engaged in slaughtering and marketing of meat, and the manufacture and distribution of multiple by-products ranging from pharmaceuticals to sporting goods, from paint brush bristles, to fertilizer. For instance, the pancreas of about 100,000 hogs must be saved to produce one pound of dry insulin.

The following is a partial market list of products and by-products:

(1) Meat, fresh and frozen.
(2) Meat, smoked, cured, canned, and cooked.
(3) Sausage and variety meats.
(4) Hides and skins for leather.
(5) Edible fats, and inedible fats for soap.
(6) Wool for clothing and furniture.
(7) Bones for buttons, knife handles, and bone meal.
(8) Blood meal and fertilizer.
(9) Glycerin with literally hundreds of industrial uses.
(10) Pork fat by-product used for truck tires, insecticides, and germicides.
(11) Hair for brushes, felt, rugs, upholstering, and baseball gloves.
(12) Intestines for sausage casing, violin strings, and surgical ligatures.
(13) Gelatin, glue, sandpaper, pet food, and neat's-foot oil.
(14) Pharmaceuticals as pepsin, testosterone, liver extract, thyroxin, epinephrine, albumin, pituitrin, insulin, thromboplastin, thymocrescin, bilirubin, and ACTH.

DECENTRALIZATION

The "Big Four" was a conversation phrase commonly referring to Swift, Armour, Wilson and Cudahy. The statement is currently inaccurate. Iowa Beef Packers and Missouri Beef Packers are the dominant factors in the beef market, virtually in control. Both companies entered with great success the marketing of boxed beef instead of carcass beef. Boxed beef is largely subprimal cuts. The packers are furnishing a disassembly service, much like pork processing.

The industry today is decentralizing, moving plants from the city to rural points. For example, Armour and Cudahy have closed their Los Angeles plants and Swift has closed its Chicago plant. The century old Chicago stockyards closed February 1, 1971. The stockyards in Los Angeles have been shutdown.

What happened is easy to perceive. Old plants became uneconomical to operate. They were voluntarily closed and replaced in rural locations where land was less expensive. Efficient one-story operations were laid out. Labor is usually less costly. In addition, the plants are located close to livestock production, reducing trucking cost to the plant, and reducing shrinkage during transit.

REFRIGERATED TRUCKING

A leading development of the post war era is the extensive use of mechanically refrigerated reefer trucks. These warehouses on wheels ordinarily cost as much as $75,000 and can deliver at least 22 tons of dressed meat on racks, or on rails, in perfect condition, under optimum refrigeration conditions, from the mid-west slaughter plants to large markets as distant as New York, or Los Angeles, in less than three days. Meat shipped from Sioux City on Friday may be marketed on Monday in San Francisco in just the ordinary course of business.

FEDERAL MEAT INSPECTION

History

There are over 70 known diseases that animals may have that can be transmitted to man. In the interest of public health, it became evident years ago that some kind of independent regulation and supervision was necessary.

The Egyptians (2000 B.C.) and the Israelites (1500 B.C.) produced early regulations combining public health and religious concepts. The hog, because it was a scavenger and "full of disease," was forbidden, while certain forequarter cuts of cattle, sheep, and goats were approved. Portions of the carcass which have large blood vessels were forbidden because of the then existing medical concepts of blood-carried diseases. Certain portions of the carcass were to be saved for religious sacrifice.

The Meat Inspection Act of June 30, 1906, inaugurated Federal meat inspection in the United States. The law was effected to control meat and meat products offered in interstate or foreign commerce. Basically, it is a service of ante-mortem and post-mortem examinations conducted by about 3500 inspectors employed by the Federal government. This group is made up of civil service employees who are graduates of a veterinary college and a group of nonprofessional civil service employees under their supervision.

The cost for this vast service is nominal. It was recently estimated (1978) that the cost per animal is $1.24, about 67¢ for each person in the United States.

It may be categorically stated that there is no known disease transmitted by properly cooked, USDA inspected meats, handled in a sanitary manner.

Functions

Inspection occurs in slaughterhouses and other levels for processors, jobbers, canners, or others wishing to handle meat in any form and offer it in interstate or foreign commerce. There are five basic areas of concern for plants under Federal inspection:

(1) Detection and destruction of diseased and unfit meat. The U.S. Department of Agriculture describes this inspection: "Every part of each carcass receives a searching examination, especially the internal organs, because any abnormal condition commonly appears there first." Condemned meats are removed to a special retaining room where they are placed under Federal lock and key, ultimately to be used for fertilizer or destroyed.

Sometimes a portion or the entire carcass of a retained animal may be released for consumption. Where certain infections or bruises are light to moderate, the carcass may be salvaged after a prescribed extensive examination and removal of all visible conditions. Carcasses from which infections are removed must be retained for ten days at temperatures below 15°F. It has been determined that carcasses so processed are wholesome and fit for human consumption.

(2) Clean and sanitary handling and/or preparation of food composed wholly or in part of meat. This is an extensive provision. It means first that the construction of all plants must conform to extremely rigid

standards and specifications. Inspectors oversee the practice of sanitary standards of handling products, and the sanitary practices of plant maintenance.

(3) Prevention of harmful substances in meat food.

(4) Application of an inspection mark, the round purple indicia or inspection number (Fig. 2.3). Shape, size, and language of the stamp is closely controlled. There are various sized stamps for different products. No. 38 is reserved by the U.S. Department of Agriculture for illustrative purposes. This is an ID for each plant under Federal inspection, each having a different number. In some cases, letters will appear with the numbers to indicate a sub-plant of a multiple plant operation.

The shipping of meat in foreign or interstate commerce without proper inspection is a federal offense punishable by a maximum fine of $10,000, or two years in prison, or by both.

A published list of all plants under Federal inspection can be obtained by writing the Superintendent of Documents, Government Printing Office, Washington, D.C. 20402 and requesting the Meat and Poultry Inspection Directory. Enclose the sum of $3.80.

(5) Prevention of false or deceptive labeling or publication of misleading statements. Like all other phases of Federal inspection, labeling is exacting, requiring prior approval of the U.S. Department of Agriculture. All labels submitted must show name of product with an adequate description in acceptable language, inspection indicia, name and address of plant, list of ingredients in order of their importance, and net weight (Fig. 2.4).

For products placed in shipping containers, if not otherwise identified, a domestic meat label must be attached (Fig. 2.5).

Courtesy of U.S. Department of Agriculture

FIG. 2.3. FEDERAL INSPECTION INDICIA

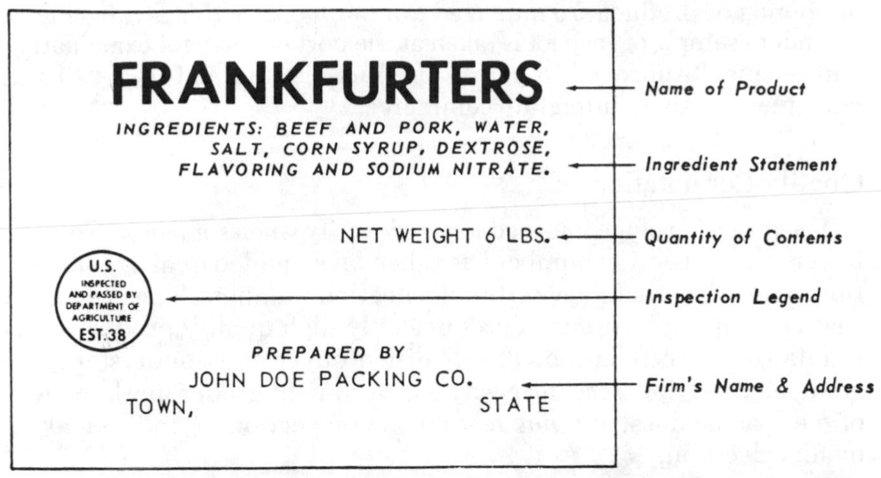

Courtesy of U.S. Department of Agriculture

FIG. 2.4. SAMPLE APPROVED LABEL

Courtesy of U.S. Department of Agriculture

FIG. 2.5. DOMESTIC MEAT LABEL

Imported Products

A foreign nation that maintains a meat inspection system substantially like, or as efficient as the U.S. Department of Agriculture requires, and where experience has demonstrated that reliance can be placed on its export certificate, may offer meat and meat products for import to the United States. An exception to this import regulation is made in the case of certain countries where known communicable livestock diseases exist. These countries may not export to the United States, regardless of the inspection system maintained. A rigid examination by the Department of Agriculture must be passed at the port of entry. Millions of pounds of product are imported annually under this arrangement. A random sample of each lot is taken at the port of entry for examination and testing. "Approved Imported Product" is stamped (Fig. 2.6) and may move freely in interstate commerce.

Quality Connotation

The intention of meat inspection is to certify wholesomeness. To some buyers the inspection number has taken on an added quality connotation. These buyers follow certain inspection numbers. The premise is that certain establishments produce only high quality product. This is a dangerous conclusion. The qualified buyer must understand the grade he is buying, recognize why a government grader marks a piece of meat as he does, and *not* rely on the inspection legend to make a quality decision.

WHOLESOME MEAT ACT OF 1967

The original intent of the act was to assure consumers on a national basis that they will be able to buy only wholesome, inspected, and approved meats. In effect, there were no sweeping changes in the Federal inspection program as it applied to then existing federally inspected plants. Operation "New Broom" brought closer scrutiny of existing federally inspected facilities and operations, and re-evaluation of labeling. A major change was to discontinue the "Certificate of Exemption," which permitted non-federally inspected meats to move in interstate commerce.

The new act places all meat packing and processing under Federal regulation. Under the Federal-State cooperative program meat inspection may be continued as a State program, where State laws are substantially equal to Federal requirements. The Federal government under the Talmadge-Aiken Act would share the direct costs with the states. With few exceptions, only products produced under Federal inspection could move interstate, and be sold to Federal agencies. State inspected products may be sold only within the state produced. The

FIG. 2.6. INSPECTION STAMP FOR APPROVED IMPORTED MEAT PRODUCT

California inspection indicia (Fig. 2.7) is applied officially within the Act. The inspection indicia is similar to the Federal. Other states have their own official inspection indicia.

Slaughterers, processors, wholesalers, and locker plants fall within the Act. Food service operators and retailers, exclusive of their interstate activities, are excluded. All 50 States and Puerto Rico at this time have cooperative agreements.

Full compliance was targeted for December 1969. It was recognized that many of the operations then within the jurisdiction of the law could not meet physical requirements as outlined in USDA Handbook *191*. Under a "rule of reason," if a plant had potable water, good drainage and it was possible to apply principles of good sanitation and produce a wholesome product, it would be qualified for inspection. The target date for full compliance was advanced to December 15, 1970 to avoid creating unfair financial hardships on many operators who found themselves sub-standard.

To protect the public the USDA can settle for nothing less than full compliance. The USDA has actually closed a few sub-standard plants where intention of eventual compliance was not indicated. There are presently (1978) 25 States that have entered into cooperative poultry agreements.

Kosher Inspection

This a religious inspection of wholesomeness. It is an additional inspection, and not a substitute for Federal or local inspection. To get Kosher certification (Fig. 2.8), the animal must be slaughtered according to prescribed rabbinical technique by a specially trained "shohet." The carcass is stamped "Kosher" in script or block letters, sometimes with a tag or metal seal. Literally, the word "Kosher" means "correct," or proper according to the law. It is because the law has sanitary implica-

CAL
INSPECTED
AND PASSED BY
DEPARTMENT OF
AGRICULTURE
EST. 000

FIG. 2.7. CALIFORNIA INSPECTION STAMP

Courtesy of California Department of Agriculture

FIG. 2.8. TYPICAL KOSHER STAMP

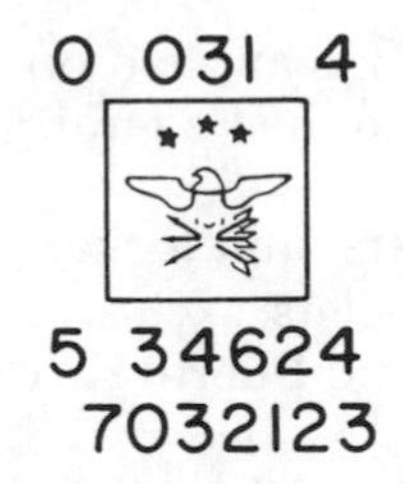

FIG. 2.9. MILITARY INSPECTION STAMP

Courtesy of the U.S. Army Veterinary Corps

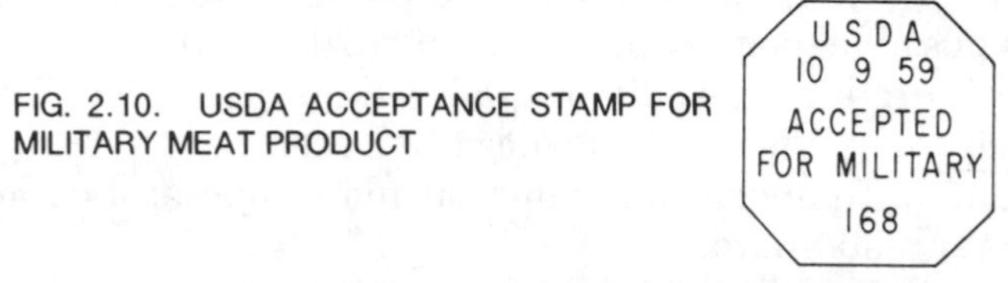

FIG. 2.10. USDA ACCEPTANCE STAMP FOR MILITARY MEAT PRODUCT

Courtesy of U.S. Department of Agriculture

tions, that "Kosher" has taken the connotation of clean or wholesome.

Some buyers look to the Kosher stamp as a mark of higher grade beef. There is no scientific basis to the argument that Kosher slaughtering has anything to do with quality. Generally speaking, the Kosher trade demands a higher type beef because it consumes relatively fresh beef from the tougher forequarter cuts. Some Kosher packing houses, to meet this demand, buy higher than average quality beef. Yet quality will vary from carcass to carcass. Like the Federal inspection stamp, the Kosher stamp is not intended to have anything to do with quality.

Military Inspection

The combined military forces have their own inspection service. The Army Veterinary Corps approves and stamps product for all branches, the Army, Navy, Air Force, and Marine Corps (Fig. 2.9).

In some areas where there is no qualified military personnel, the Grading Service of the U.S. Department of Agriculture inspects and accepts products for the military (Fig. 2.10).

REFERENCES

AGRICULTURE RESEARCH SERVICE. 1952. The inspection stamp as a guide to wholesome meat. Agr. Information Bull. *92,* U.S. Dept. Agr., Washington, D.C.

ANON. 1942. Meat and Meat Cookery. National Live Stock and Meat Board, Chicago, Ill.

ANON. 1950. Ten Lessons on Meat. 7th Edition. National Live Stock and Meat Board, Chicago, Ill.

ANON. 1953. Kosher meats. A circular. American Meat Institute. Chicago, Ill.

ANON. 1960. Marking and labeling program. Agr. Handbook *190,* U.S. Dept. Agr., Washington, D.C.

ANON. 1960. Meat Reference Book. American Meat Institute. Chicago, Ill.

ANON. 1960. Regulations Governing the Meat Inspection of the United States Department of Agriculture. Title 9, Chapter I, Subchapters A and K, Code of Federal Regulations. U.S. Dept. Agr., Washington, D.C.

ANON. 1961. U.S. inspected meat processing plants (no slaughtering). Meat Inspection Division, U.S. Dept. Agr., Washington, D.C.

ANON. 1962. Summary of activities Meat Inspection Division. ARS-*93-2-6,* U.S. Dept. Agr., Washington, D.C.

ANON. 1962. Working reference of livestock regulatory establishments. U.S. Dept. Agr., Washington, D.C.

ANON. 1968. Questions and answers on the Wholesome Meat Act of 1967. American Meat Institute, Chicago, Ill.

ANON. 1969. U.S. Inspected Meat Packing Plants. Handbook No. 191. U.S. Dept. Agr., Washington, D.C.

BULL, S. 1951. Meat for the Table. McGraw-Hill Book Co., New York.

GORDON, I., and GELLER, V. B. 1960. A Handbook on the Jewish Dietary Laws. Union of Orthodox Jewish Congregations of America and Rabbinical Council of America, New York.

ROMANS, J. R. and ZIEGLER, P. T. 1977. The Meat We Eat, 11th Edition. Interstate Printers and Publishers, Danville, Illinois.

SWEM, E. R., and STEVENS, B. 1960. State meat laws. National Provisioner *143,* No. 8, 21–26.

TRESSLER, D. K., and EVERS, C. F. 1957. The Freezing Preservation of Foods. Vol. 1. Fresh Foods. 2. Cooked and Prepared Foods. Avi Publishing Co., Westport, Conn.

WANDERSTOCK, J. J. 1960. Meat purchasing. Cornell Hotel and Restaurant Administration Quarterly *1*, No. 1, 25–29.

WOLGAMOT, I. H. 1957. Beef-facts for consumer education. AIB *84,* U.S. Dept. Agr., Washington, D.C.

WOLGAMOT, I. H., and FINCHER, L. J. 1954. Pork-facts for consumer education. AIB *109,* U.S. Dept. Agr., Washington, D.C.

3

History and Purpose of Grading

On May 1, 1927, the Federal Grading Service was instituted and described by the U.S. Department of Agriculture as "some official yardstick for measuring quality was needed to give the consumer-buyer confidence in his purchase. . . ."

The need for grading is very clear. Like any of nature's products, meat is heterogeneous. Carcasses are of all sizes, many breeds, varying ages, acceptable conformation and malformation, and different classes. The animals have been fed on many different kinds of feeds.

Economically the need for a system is self-evident. How could a Boston jobber buy from a remote point such as Sioux City, Iowa, except on the telephone and with a quality yardstick? How could a modern chain buyer buying for hundreds of units, or a small quality butcher efficiently fill his requirements?

Grading is nothing more than a sorting of meat according to a man-made quality yardstick. It becomes a question of what standard and who is going to apply it.

There are two kinds of grading; that which is done by the Federal government, and that which is done by private individuals or firms. Private grading systems are essentially the branding of merchandise. They attempt to separate the various qualities into rather broad categories. This type of grading carries no certification that any objective standards have been met or maintained.

Government grading is based on objective standards with an attempt to apply them uniformly throughout the United States. This grading system is an exclusive government function. The grades, the graders, and the entire system are under the jurisdiction of the U.S. Department of Agriculture. There is a strong social basis for this independent administration.

FEDERAL GRADING SYSTEM

The meat grader examines the beef carcasses approximately 24 hours or more after slaughter, after the carcass has been transferred from the "hot box" and has had the shrouds removed. He walks down the rail of beef with a coded shield stamp and appraises the carcasses, placing the proper number of stamps on each quarter, one for USDA Good, two for USDA Choice, and three for USDA Prime (Fig. 3.1). After preliminary grade designations are completed, the grader returns and puts a roller (ribbon stamp) on the beef of the same value as the key stamps (Fig. 3.2).

The manner of applying the roller has been carefully worked out so

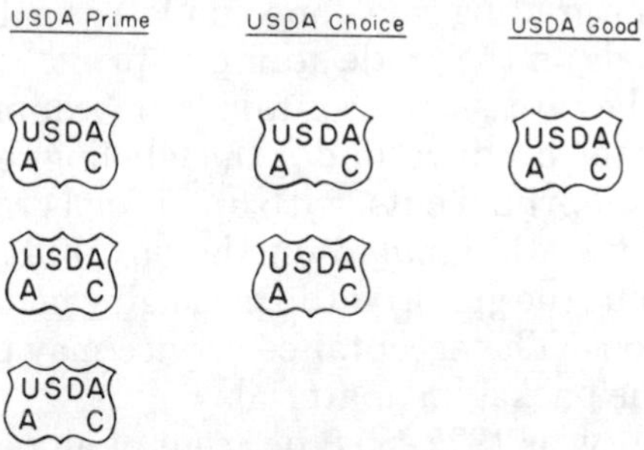

USDA Prime | USDA Choice | USDA Good

Courtesy of U.S. Department of Agriculture

FIG. 3.1. KEY GRADING—CODED SHIELD STAMP

Courtesy of U.S. Department of Agriculture

FIG. 3.2. RIBBON STAMP—USDA CHOICE GRADE

that the grade identification occurs on most of the primal cuts. The grade stamp appears on the chuck, brisket, clod, rib, strip loin, top sirloin, round, and so forth. Cuts from the interior muscles, such as the tenderloin, and those that are produced below the fat surface in the process of fabrication do not generally bear the grade stamp.

The ribbon has one interesting variation. The grade shield will occur five and six times, then the grader's identification initials (Fig. 3.2). These initials clearly identify the grader who performed the grading. The area where the grading was done can later be determined by reference to the initials on the grade stamp imprint.

In addition to the carcass, the container or wrapping may designate the same grade as the product, under the labeling regulations. For example, primal or subprimal cuts, with an official grade, may be boxed and the grade may be designated on the box. Fabricators and steak cutters may continue the grade on the packaging material under closely regulated conditions. The acceptance service may cause the grade to be indicated on the packaging materials.

For the calendar year 1977, 55.8 per cent of all beef, 7.0 per cent of all veal and calf, and 78.3 per cent of all lamb of the commercial production were graded. Take out that portion of product not offered for sale on a fresh basis and the percentages reflected for consumer fresh meat are very high. An extremely high percentage of the beef that is equal in quality to USDA Good or better, is federally graded.

The U.S. Department of Agriculture charges the packer for grading service (the 1978 rate is $20.00 per hour) and the cost is ultimately passed on to the consumer.

Federal grading has produced many benefits. Consumer demands are accurately reflected for the industry. American beef standards are the highest in the world. The system contributes notably to the orderly and efficient marketing of livestock on a national basis. It has encouraged competition through established specifications. During World War II grading was made compulsory in an attempt to make the price control laws meaningful and enforcible.

The use of the Grading Service is voluntary; products may or may not be graded, determined by the packer. Grading is usually performed at the point of slaughter. Either carcasses or primal cuts may be graded.

History of System

Historically, beef grading has undergone a series of changes, the most important of these occurring in the postwar 1940's, 1965, and 1975. There are economics constantly at work for and against the system. There are organized groups that would like to totally undermine, or bring to an end, the federal grading system. The value of the system is attested to by its broad use on a voluntary basis. This, in turn, testifies to its basic pragmatic approach.

Historically, grading was based on three interpolated factors: conformation, finish, and quality. (1) conformation, has to do with the shape of the animal; (2) finish has to do with the quality and distribution of fats; and (3) quality relates to the age of the animal, the firmness and texture of the flesh, the distribution of intramuscular fats (marbling), and the color of the lean.

Conformation

Conformation has to do with shape, the form, or the profile of a carcass. This is closely related to breeding. This characteristic is best observed in the carcass form. Top conformation for beef is described in terms of a short, blocky, compact carcass, short shanked, plump in the rounds, smooth at the hips, and thickly fleshed in the rib and loin area, with a short neck, and smooth shoulders.

Basically, from the food service operator's point of view, conformation is something that will not necessarily affect the product that is put in front of the customer. Conformation is of more concern to the packer and the retail butcher. When the carcass is broken into primal cuts, it will yield a greater or lesser amount of valuable cuts, depending to a great extent, on the conformation of the animal. The question of meat-bone-fat ratio makes conformation somewhat significant.

Recent investigations, however, relating to conformation and breed, have produced some evidence that better bred animals make for higher palatability scores and greater customer satisfaction. If this is the case, conformation is increasingly significant from the operator's point of view. It is impossible to examine portion cuts or primal cuts and come to any valid conclusion relative to conformation. This fact makes the operator dependent on the official grade value on the product.

Finish

Finish relates to the quality and distribution of surface fats. From the point of view of the buyer, finish tells much of the story about quality, and helps the buyer determine that he is getting what he is paying for. Exterior finish itself is nothing that a customer can eat, and may have a negative economic value.

Sometimes an excessive amount of fat will change the fat-lean ratio with the economic effect that the lean becomes more expensive. The buyer must consider that he is paying for the lean relative to the amount of fat present. The higher he reaches for quality, the more he will have to pay both in terms of the basic unit cost, as well as the loss due to the amount of fat that is inherent with the finish. If it is the buyer's intention to buy top quality meat, he must reconcile himself to the fact that some fat will be present.

The color of the fat is not a factor in grading, yet a yellowish cast is objectionable from the merchandising point of view. Although yellow fat is usually associated with older type animals, it is possible for a young, USDA Choice beef to go yellow. The yellow color becomes economically negative. It may be associated with low quality in the customer's eyes.

Occasionally, the fat is "fiery" or red in color. This, too, is of no consequence with respect to grading. There is no decisive opinion on what causes the "fiery" condition. It is sometimes associated with the animal probably being excited or overheated at the time of slaughter, and that there is some blood left in the tissues after slaughter.

Quality

The factors involved in quality are the maturity of the animal; the color, texture, and firmness of the lean; and the intramuscular fats (marbling). Under this system, in order to qualify a piece of beef for USDA Prime, the bones must indicate a youthful animal, the lean must be firm and fine in texture, the color may range from light cherry to deep red, and there must be a liberal amount of intramuscular fat.

Youth of the animal is undoubtedly one of the most significant factors. As the animal matures, the connective tissue increases with a consequent decrease in tenderness. Tenderness, rated very high on the palatability scale, makes youthfulness a most significant factor.

The evaluation of the bone structure and the determination of the youthfulness of the animal are most generally overlooked by the consumer. They are difficult to recognize. Youth is demonstrated best in the bone structure, which changes from a porous, soft, pink-colored tissue to one that is white, hard, flinty, and non-porous with maturity.

Age and tenderness are inversely related. Operators who use boneless primal cuts or portion cuts must rely heavily on the official grade value of the product. The age of the animal is difficult to determine without examining the bone structure. A mature beef, apart from bone structure, may look like USDA Prime or Choice. *No one is truly an expert* in meat quality evaluations when bone structure and conformation cannot be examined.

Three bone structure characteristics should be examined: softness, redness, and porosity. The chine bone is a good guide. In young animals, buttons or soft cartilage will be found at the outer tip of the feather bones. As the animal matures, these buttons disappear, or become hard bone. Comparative photographs show the rib bones of very young veal, youthful heifer, and mature cow (Plate I). The bone cycle is clearly demonstrated. In youthful animals, the rib bones appear red due to an abundant blood supply, which can circulate through this relatively porous bone tissue.

The lean, or the meat muscle, is evaluated from the point of view of texture, firmness, and color. What is meant by "texture" is the fineness of the grain of any given muscle, or the size of the muscle bundles. To the touch, the texture of the muscles will vary from a velvet-light at the extreme end of the scale to that of a turkish towel or coarse, rough structure at the other end. The various muscles of a single carcass will have different texture and tenderness. The best comparison is one of like muscles from different carcasses.

Firmness is a relatively good indication of the type of feed and the amount of feed that the animal has had. As the feed-lot period is extended, the firmness of the muscle structure increases so that it eventually will have that desired set-up appearance. Short fed beef will appear sloppy and flatten concavely when cut.

The color of the lean of beef will vary from light cherry to deep red. As a general rule, the more youthful the animal, the lighter the color. This is not always true.

Marbling, or the intramuscular distribution of fat, is probably the best known, most sought after factor of quality. It is at least the most widely recognized characteristic. It has a special function. As the fats break down during the cooking process, marbling contributes flavor, tenderness, and juiciness.

Marbling alone is not a positive indication that the meat will have a high palatability score. For example, marbling frequently occurs in abundance in very mature beef-type cows. This meat will not have the palatability characteristics of a youthful carcass. Marbling, on the other hand, in combination with a youthful carcass, is a reasonable criterion that the desired results will be achieved. It is most unscientific to examine marbling and from this single characteristic come to a quality conclusion.

The intramuscular fats have an interesting function. They occur between the layers of the connective tissues of the muscle walls. These fats will stretch the connective tissues in much the same manner that a balloon might be stretched when it is blown up. The tissue becomes thin and more easily ruptured, more tender. Fat itself makes the biggest single contribution to flavor. It is not unusual to observe the cholesterol-minded American public eating whole bites of fat. Juiciness is also directly related to fat.

POST WORLD WAR II CHANGES

Shortly after World War II, the Department of Agriculture decided to make a change in grading which did not, however, change the system. Prior to this time, most beef was graded USDA Good. The word "Good" was not consistent with the very high quality or palatability manifested by the grade, so USDA Choice grade was expanded to include top quality beef formerly graded USDA Good. This, in turn, narrowed somewhat the quantity of beef then graded as USDA Good. The transition was made easily and it was logical to give this high quality beef a better name designation.

1965—Revised Beef Grading Standards

The 1965 revisions took new cognizance of youthfulness, and reduced the marbling requirements for certain beef carcasses. Research indicated that the younger animal requires less marbling to achieve the same palatability score. After a one-year trial, the revisions were permanently adopted.

It also became compulsory to "rib down" the side of beef, separating forequarter from hindquarter, so that an actual intramuscular inspection could be made to determine quality. This evaluation was previously estimated by an examination of the degree of finish. Finish was eliminated as one of the three beef grading factors.

Beef Standards 1975

The 1975 Grading standards made the following changes:

1. Conformation was eliminated as a factor in the determination of the grade.
2. When graded, all beef except bulls, will be identified for yield along with quality.
3. The marbling requirements were basically further reduced.

In the elimination of conformation as a criteria, the Department of Agriculture stated that "there was no information which indicates that variations in conformation are related to differences in beef's palatability." However, conformation was a criteria for 50 years, during which time American beef achieved the highest level of quality in the world.

The new yield grade is economically important, especially to the meat packer and meat processor. Yet it has a somewhat negative impact on quality. A premium is placed on leaner beef to achieve the higher yield grade. Lean beef and high palatability are diametrically opposed. The bottom line of the new program is a reduction of the dry feeding period, just long enough to produce relatively lean beef barely meeting the USDA Choice grade, coupled with a high yield grade.

The marbling requirement reduction was strongly opposed by the food service industry. The Department of Agriculture stated that the new standards "will not significantly change the eating characteristics of Prime and Choice grade beef." The fact is, there has been a significant change.

The changes essentially reduce to a single criteria, the standard for beef grading: quality, which has to do with youthfulness and texture. In fact, there appears to be two basic beef grades under the new system, young and old. When a young beef shows a reasonable amount of feed, it will grade USDA Choice; with a little more feed it will grade USDA Prime.

Beef produced today is shorter fed and less palatable than it was prior to the 1975 change. Private grading is increasing. The percentage of beef graded has dropped from a high of 60.4 per cent (1966) to 56% (1978). This drop of 7% might portend a significant trend. At least one national packer has started marketing a private label in significant quantities. This is a trend to watch. An imaginative sales campaign well funded with the many dollars might take a broad share of the market. The consumer will be deprived of his best meat-quality yardstick if grading is substantially reduced or eliminated.

Proposed Changes—1978

The Department of Agriculture has proposed changes having to do with the elimination of fraud and corruption, and to improve efficiency. The proposed changes are as follows:

(1) If the grading service is used, meat must be graded, or marked "U.S. Ungraded."

(2) Restrict grading to the whole carcass or a side.

(3) Require kidneys, and the fat surrounding the kidneys, pelvic region, and heart be removed before grading.

(4) Lesser changes include ribbing down the beef 30 minutes before grading; permitting removal of yield grade under certain circumstances; and redefining "beef carcasses."

The "U.S. Ungraded" change appears as well-intended moral legislation but reckless when one considers the potential side effects. Packers, today, basically grade Prime and Choice and sell the rest of the beef production as "no rolls." This beef is merchandised as "no rolls" or with a packer brand. If the packer is denied this merchandising device, by being forced to grade these carcasses "U.S. Ungraded," he may well elect to have none of his beef graded. The purpose of this amendment is to keep the retail butcher honest, sort of a retail butcher 18th

Amendment. The backlash could well be the discontinuation of beef grading in many packing houses.

This system will not make honest merchants out of dishonest retailers. It does not close the door. As long as the retail butcher has a sharp knife, he can cut off the "U.S. Ungraded" label and represent the meat any way he chooses. In fact, he doesn't have to cut the ink off, just smear it and who can tell it from a smeared USDA Choice grade?

This amendment is a threat to the entire grading system. The trade-off is a poor one. The attempt to make dishonest retail butchers honest will not work. On the other hand, most major packers may find it uneconomical to continue using the grading service.

The proposal to grade sides and carcasses at the packing house level is an important piece of police work. For many years, ungraded ribs and loins have been shipped to remote points and there graded USDA Choice. This has been a widespread, illogical practice, inviting dishonesty. It has never made sense to permit a carcass not making the Choice grade at the packing house, to be cut up and shipped to a remote point, and then have a second chance to meet the USDA choice requirements.

BEEF GRADES

For a practical evaluation of the grades, the USDA Consumer and Marketing Service Home and Garden Bulletin *145* is for sale by the Superintendent of Documents for 25 cents. The latest edition is 1973 and does not include the 1975 grade changes. It is a practical guide but must be looked upon as slightly out of date. Historically, it is an important document for comparative purposes.

It is relatively simple to apply the grading criteria to USDA Prime beef. To arrive at the other grades, it is important to understand how the factors are interpolated. The institutional user must understand each of these grades in terms of functions so that proper selections can be made.

USDA Prime (Fig. 3.3)

As the name implies, beef of this grade is most acceptable and palatable. Prime grade beef is produced from young, fed cattle. The youth of the cattle and the feeding combine to produce quality cuts of beef. Such cuts have liberal quantities of fat interspersed within the lean (marbling). These characteristics contribute greatly to the juiciness, tenderness, and flavor of the meat. Rib roasts and loin steaks of this grade are consistently tender and cuts from the round and chuck should also be highly satisfactory.

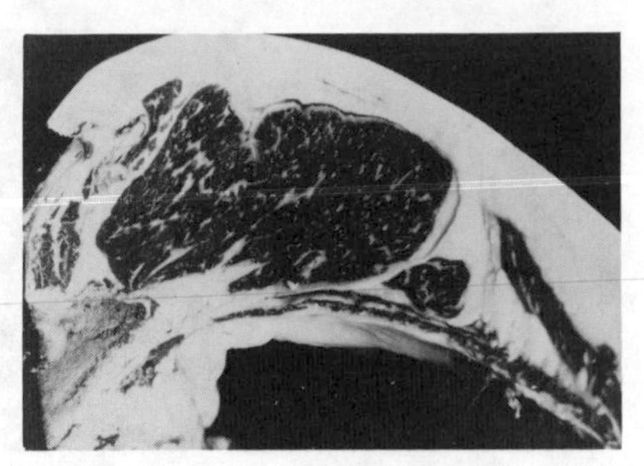

Courtesy U.S. Department of Agriculture

FIG. 3.3

USDA Choice (Fig. 3.4)

This grade is preferred by most consumers because of relatively high quality. More of this grade of beef is produced than of any other grade. Choice beef is usually available the year round in substantial quantity. Roasts and steaks from the loin and rib are somewhat tender and juicy and other cuts, such as those from the round or chuck, which are suitable for braising or roasting, should be tender with a well-developed flavor.

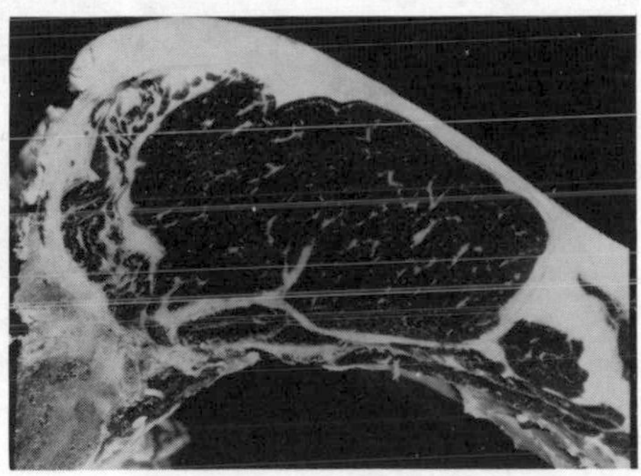

Courtesy U.S. Department of Agriculture

FIG. 3.4.

USDA Good (Fig. 3.5)

This grade pleases thrifty consumers who seek beef with little fat and an acceptable degree of quality. Although this grade lacks the juiciness associated with more marbling, relative tenderness and the high proportion of lean to fat make it the preference of many people. Frequently sold under a private label rather than USDA Good.

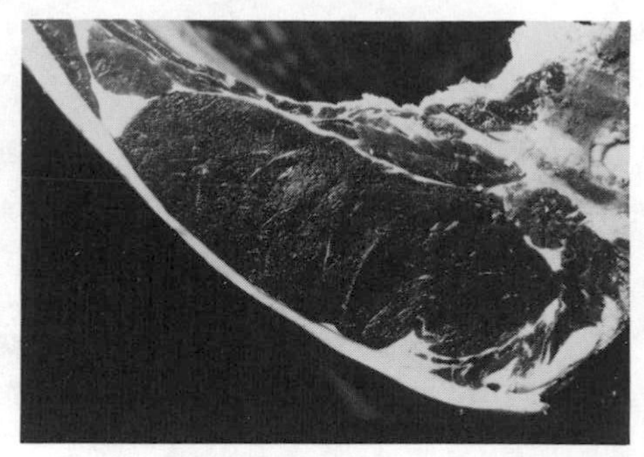

Courtesy U.S. Department of Agriculture

FIG. 3.5.

USDA Standard (Fig. 3.6)

Standard grade beef has a very thin covering of fat, and appeals to consumers whose primary concern is cost. Such beef is lean and usually relatively tender. It is mild in flavor, and lacks the juiciness usually found in beef with more marbling.

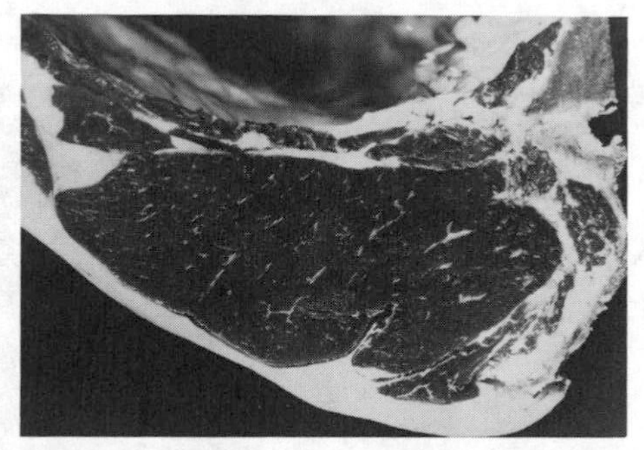

Courtesy U.S. Department of Agriculture

FIG. 3.6.

USDA Commercial (Fig. 3.7)

Beef that is graded Commercial is produced from older cattle and lacks the tenderness of the higher grades. Cuts from this grade, if properly prepared, can be made into satisfactory and economical meat dishes. Most cuts require long, slow cooking with moist heat to make them tender and to develop the rich, full, beef flavor characteristic of mature beef.

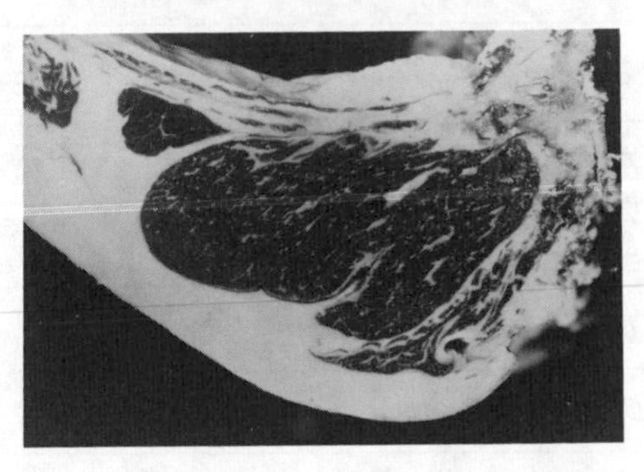

Courtesy U.S. Department of Agriculture

FIG. 3.7.

Other Grades

There are three other grades of beef—Utility, Cutter, and Canner. Utility is produced mostly from cattle somewhat advanced in age and is seldom sold as fresh beef. Cutter and Canner are used, ordinarily, in processed meat products and rarely, if ever, sold in retail stores.

See Fig. 3.3–3.7 for a comparison of the appearance of the different grades. For this purpose a rib cut has been used. Observe these pictures in terms of conformation, finish, and quality.

The relative production of the various grades of beef is illustrated in Fig. 3.8. Each grade of beef may be represented by several classes (Fig. 3.9).

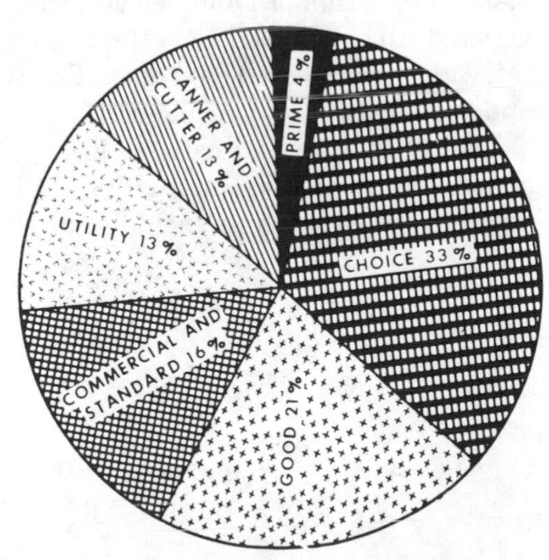

FIG. 3.8. COMMERCIAL CATTLE SLAUGHTER (BY GRADE)

Courtesy of U.S. Department of Agriculture

BEEF TYPE	USDA Prime	USDA Choice	USDA Good	USDA Standard	USDA Commercial	USDA Utility	USDA Cutter-Canner
STEERS	Great percentage	Mostly	Great percentage	Some		Some	Some
HEIFERS	Great percentage	Mostly	Great percentage	Some		Some	Some
BULLS		Some	Some		Great percentage	Great percentage	Great percentage
COWS		Some	Some	Some	Great percentage	Great percentage	Great percentage
DAIRY TYPE							
BULLS		Some	Some		Some	Great percentage	Great percentage
COWS		Some	Some	Some	Great percentage	Mostly	Great percentage

MOSTLY GREAT PERCENTAGE SOME

FIG. 3.9. RELATION OF GRADE AND CLASS OF BOVINES

VEAL AND CALF GRADING

Young bovines are classed as veal and calf. In general, bovines up to three months of age are classed as veal as determined by the grayish pink lean which is smooth and velvety, the soft pliable fat and the narrow, very red rib bones. Calf, over three months, has grayish red lean, harder fat, wider and less red rib bones. The three sex classes of steer, heifer, and bull are of no consequence.

Like beef, veal and calf are graded on a composite evaluation of conformation, finish, and quality. A distinction between veal and calf as well as the grade, is included in the grade ribbon (Fig. 3.10 and 3.11).

FIG. 3.10. GRADE STAMP FACSIMILE—USDA CHOICE CALF

Courtesy of U.S. Department of Agriculture

FIG. 3.11. GRADE STAMP FACSIMILE—
USDA CHOICE VEAL

Courtesy of U.S. Department of Agriculture

OVINE GRADING

Composite evaluation of only two factors, conformation and quality, is used for grading lamb. Conformation is like beef except for the terms leg, loin, and rack.

Fat streaking in the flank muscle, the "overflow" fat and feathering between the ribs and the firmness of the carcass all make points. Lean ranges from a bright red, firm texture, to a dark red, coarse texture.

The ovine is of three classes—lamb, yearling mutton, and mutton. These are basically age differentiations. Four basic characteristics can be observed.

	Lamb	*Yearling*	*Mutton*
Foreshank	Breakjoint (Fig. 3.12)	Spool (Fig. 3.13)	Spool
Rib bones	Narrow	Moderately wide	Wide
	Show red	Traces of red	No red
Lean	Light and fine	Red and slightly coarse	Dark red and coarse

HOG GRADING

There are five classes of hogs:

Barrow—Male	Unsexed as pigs
Gilt—Female	Never having borne pigs
Sow—Female	Having borne pigs
Stag—Male	Unsexed after maturity
Boar—Male	Mature

Grade standards were established in 1952 for barrows and gilts as live animals and dressed pork carcasses. The bases for grading is yield and quality of the meat. Three grades have been established, Choice, Medium, and Cull. Typical of Choice hogs is the grayish pink color, reddish bones, and firm white fat. Three further divisions are set within the Choice grade based on the lean-fat ratio or yield of the ham, loin, picnic, and the Boston butt.

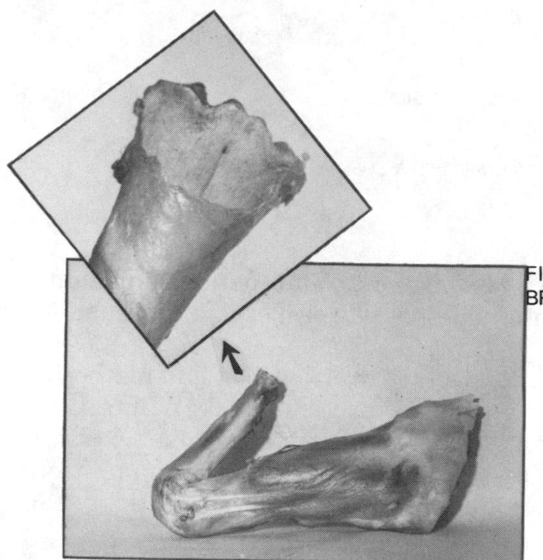

FIG. 3.12. LAMB FORESHANK—BREAK JOINT (INSERT)

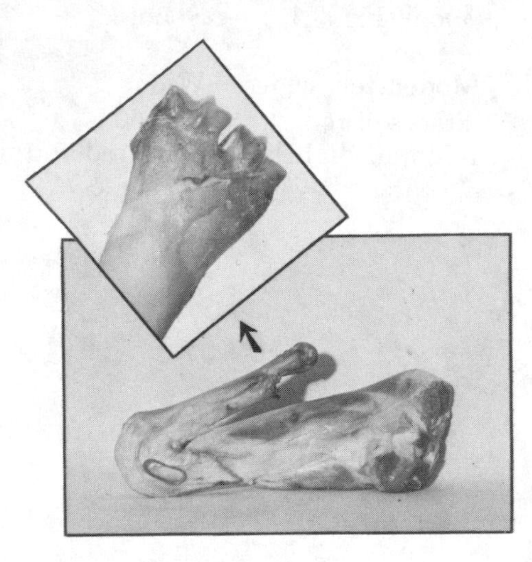

FIG. 3.13. MUTTON FORESHANK—SPOOL JOINT (INSERT)

As yet there is no grade identification on the wholesale cut. The operator must approach grading from the point of view of the weight range of the cut, and the trim in terms of yield of the particular packer. Because most pork marketed is from younger type animals, grading is not critical.

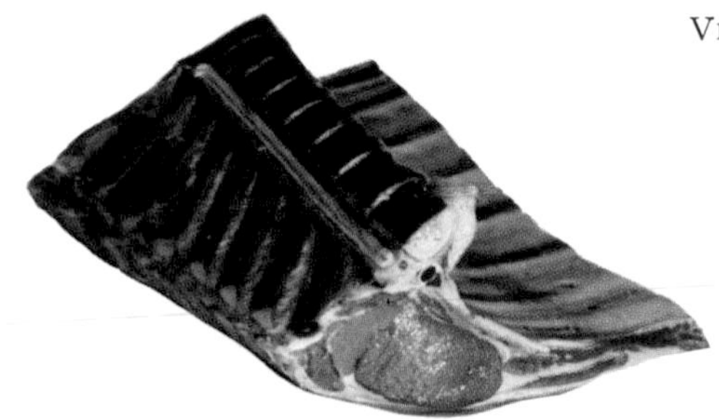

VEAL RIB RACK

Light color lean
Soft cartilage
Soft, porous bones
Redness in bones

HEIFER RIB—USDA CHOICE

Medium red lean
Buttons or soft cartilage disappearing
Harder bones
Less redness in bones

MATURE COW RIB—USDA UTILITY

Dark red lean
No soft cartilage
Hard, flinty bones
Almost no redness in bones

GRADING IS RELATIVE

The grading system is intended to set up an impartial evaluation of meat. Meat is not simple to grade. It is neither manufactured nor molded according to specification. Meat animals are as variable as any other general class that occurs in nature. Consequently, the classifications have to be reasonable. The fact is: the grades are relative. In any particular case, it cannot be assured that they are not ineffective or that they are not arbitrary. Basically the system works and is widely accepted. It performs the function for which it is intended, and is a valuable tool of the entire industry. Yet, there is no denying that borderline situations do exist.

Special Problems

There are a few special problems related to grading which are concerned with the spread within any grade.

A great spread exists within some grades. Frequently, members of the trade speak of two and sometimes three sharply divided areas. This is particularly true in the USDA Choice grade. In the USDA Choice grade, it is not uncommon to talk about and to actually demonstrate low Choice, middle Choice and top Choice. In beef trading, there is a definite division in many instances between the top half and the bottom half of the USDA Choice grade. The language of the grading bears this out. It speaks about the carcass possessing the "*minimum qualification.*" The word "minimum" is important because the grade will then extend up to the point where the top end of the product borders the next higher grade, again becoming a borderline situation. Every carcass within the grade obviously does not have the same value, *only the same minimum value.* This opens a door for romance and merchandising, as far as the distributor is concerned. He can talk about high Choice and high Prime as compared with packinghouse run. Because the assertion of the distributor may or may not be true and because the grades have a spread, it becomes extremely important for the buyer to improve his knowledge. The buyer must understand what he is trying to achieve buying-wise. It is important that the buyer be able to determine if he is paying a premium for "high Choice," and what "high Choice" really is.

Commercial Beef.—Salesmen are not beyond the realm of this exclamation: "Why, this Commercial is just as good as USDA Choice." It is not uncommon for a buyer to be taken in. By carefully observing the curve demonstrating the various grades, the spread of the grade, and how far apart USDA Choice and USDA Commercial are, the buyer should readily note that there is quite a broad differentiation (Fig. 3.14). The salesman is exaggerating and the buyer who is aware of the grading system will suspect that he is only getting a partial truth. There is considerable USDA Commercial beef on the market produced from what is commonly referred to as "red cows," the she-stock for breeding meat-type animals. This beef has an appearance much like that of USDA Choice, and it is not inconceivable for a buyer to mistake it. There is one primary difference. The age of the animal is substantially different in USDA Choice and USDA Commercial. This accounts for a substantially different palatability experience and a substantially different grade.

The commercial grade offers a variety of interesting situations. It is so frequently considered as an economic solution to many cost problems. For example, there have been instances uncovered where beef has been erroneously graded USDA Choice. Not many buyers are able to perceive the difference. Where a buyer stresses the cost of the product, and looks for a tremendous bargain in USDA Choice beef, he is easily duped.

Another interesting variation is the sale of ungraded beef, the equivalent of USDA Utility, USDA Cutter, and USDA Canner, particularly the rib and loin cuts, sometimes represented and sold for USDA Commercial. This is a frequent trade situation, where the buyer is not concerned with seeing the grade. Yet any piece of meat which does not bear an official grade is just exactly that, an ungraded piece of meat. Ungraded meat should not be accepted for anything other than ungraded.

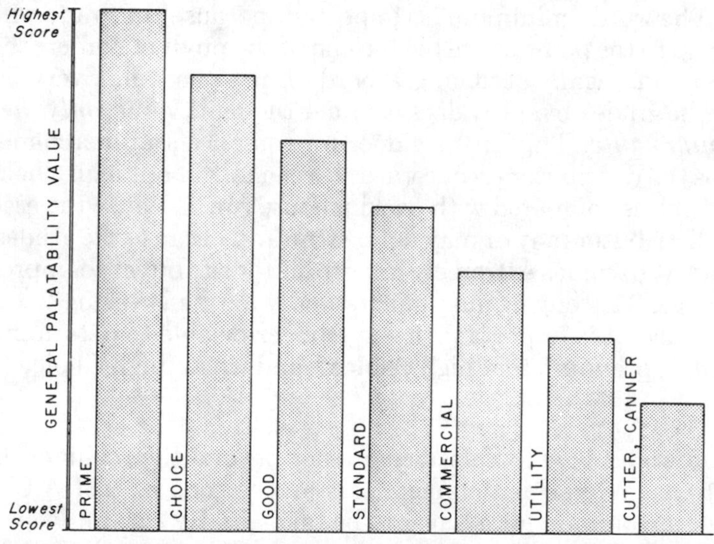

FIG. 3.14. RELATIONSHIP OF PALATABILITY VALUE AND GRADE OF BEEF

Other Quality Factors

There are other quality factors influencing value which should be examined though they do not fall within the province of the grading system. These relate themselves to the condition of the product, the sex of beef carcasses, and regional production.

Condition of Product.—The condition of the meat at the time that it is received by the purchaser is as important a condition of quality as any other, if not the most important. In the first place, if the product is fresh or frozen, it should be cold when received. If the product has been abused, excessively aged, stored without proper refrigeration, improperly handled between the distributor and the ultimate user, shows excessive bacterial growth or mold, or manifests any of the conditions that would make it unfit for consumption, then these conditions are equally important quality considerations. If the meat is to be aged by the distributor, then of course, it is important that this be considered also.

Sex of Beef Carcasses.—A price differentiation often exists between the male and female bovine. The reason for the distinction has little to do with that which might concern the food service industry. Price differentials are primarily concerned with the percentage yields of the primal cuts. Once the carcass has been broken, other than for the weight range, or the foreshank, the udder, and the pelvic area, it is almost impossible to identify whether the meat comes from a heifer or a steer. The main difference is the fact that steers are generally larger than heifers. There is an area between 500 and 700 lbs dressed weight in which they overlap. The standards provide for "the grading and stamping of beef from steers, heifers, and cows, according to its characteristics as beef, without sex identification."

There are six bovine classes:

Steer—Male unsexed while young.
Bull—Fully developed, mature male.
Stag—Unsexed after developing secondary
physical characteristics of the bull.
Heifer—Young female which has not borne a calf.
Cow—Female having borne a calf.
Veal and Calves—Young bovine.

Corn Belt Beef.—No discussion of quality of beef could be complete without some discussion of the controversial "Corn Belt" beef. In the

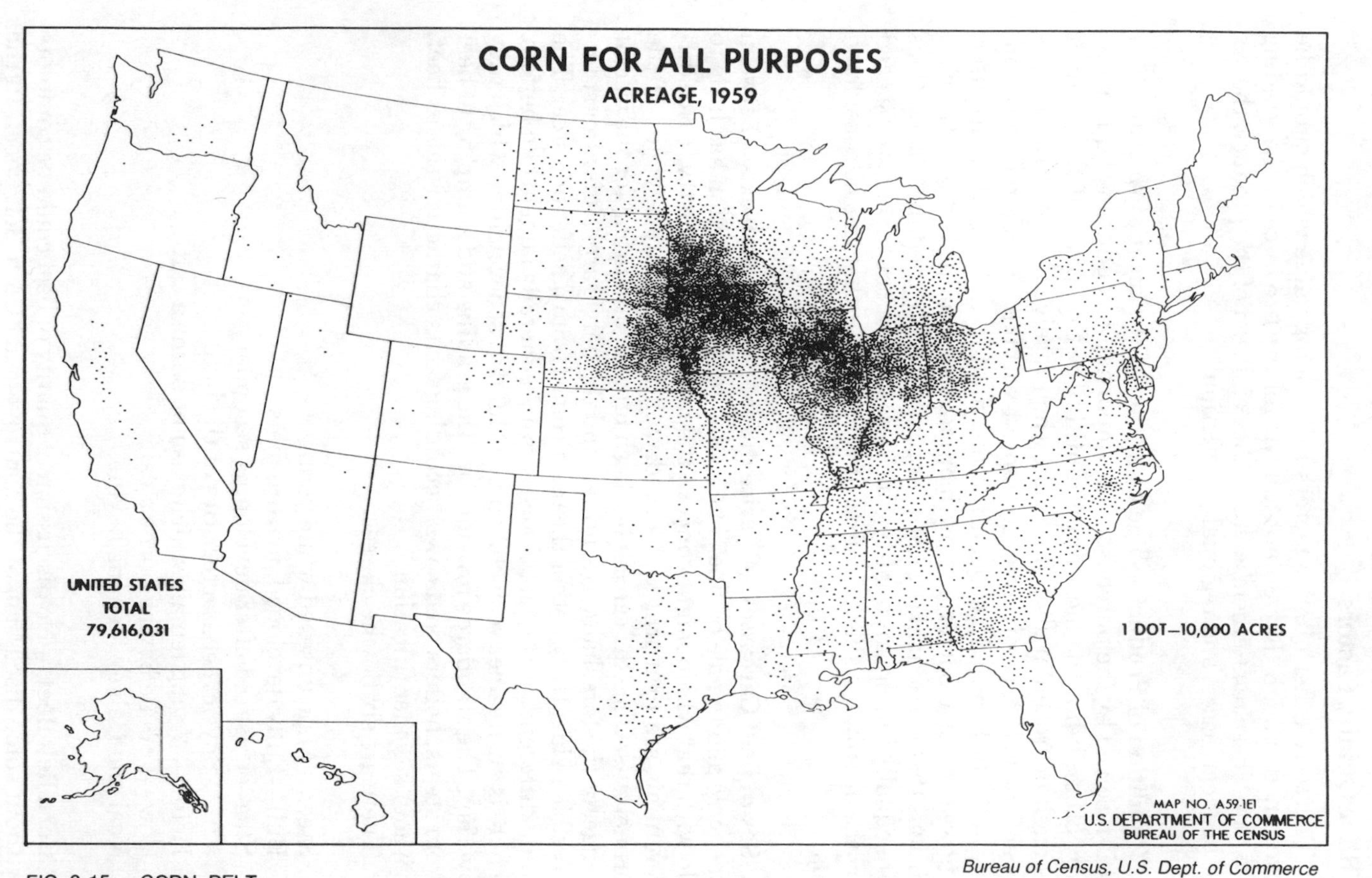

Bureau of Census, U.S. Dept. of Commerce

FIG. 3.15. CORN BELT

Far West this is referred to as "Eastern Beef," and in the East, this is referred to as "Western Beef." The words "Eastern" and "Western" are misnomers. It would be more universal to describe this beef as "Midwestern" or "Corn Belt" beef. Actual reference is to the beef produced in the mid-central western states: Illinois, Minnesota, Iowa, Nebraska, Kansas, and so forth (Fig. 3.15).

As far as the grading service is concerned, when a stamp is placed on a piece of meat with a specific grade, *there is no quality factor related to the regional production.* In order for the buyer to distinguish this "Corn Belt" beef it is necessary to refer to the packing house inspection or identification number. Unfortunately, the packing house number merely tells where the meat was slaughtered, and nothing about quality. For example, beef slaughtered in Chicago or in Sioux City could conceivably be that from the many cars of beef that are shipped in from Arizona, Texas, Nevada, Colorado, or California ranges to a Midwestern point to be marketed.

Primary factors which influence quality in beef are breeding, good pasture, and good feedlot practices. If these are equal, then the final product should be equal without regard to regional production.

No Quality Monopoly.—Some sellers presume that certain packing houses have a monopoly of the top quality beef. This is an impossible situation in a country in which the total cattle population is over 115 million head. It is like suggesting that a dairy has a monopoly on cream. It behooves the purchaser to make a very careful evaluation, and to have a complete understanding of what quality is, apart from a simple purple ink indicia, lest the buyer be trapped into paying a very high price for a *certification of wholesomeness* and the packing house number.

There is a lot of good beef produced outside the "Corn Belt" area. Like many other things in the commercial world in which we live, product is sometimes unfairly depreciated. It is a vicious circle. The consumers' attitude sometimes leads to short feeding because the consumer using this beef is looking for economy in the initial prices as well as an economical lean-fat ratio. This product is frequently aged improperly because it is sold usually at a discount price.

Certain Midwestern packers do attempt to maintain a high quality standard and select better livestock. A degree of success is achieved. They slaughter larger quantities of USDA Prime beef. This is not by accident. There are a few Midwest feeders who select high bred cattle and feed them out for USDA Prime.

Every lot of cattle will show a variable degree of quality with relatively the same breeding, environment, and feed. Some animals will assimilate feed better than others, show a greater weight gain, show finer marbling, have better yielding loins and ribs. Most any lot of USDA Choice live beef will dress out, accidentally yielding a few USDA Prime, sometimes a few USDA Good grade, and a wide range of quality within the USDA Choice grade. An honest understanding of the product must be achieved to make an accurate quality evaluation. The qualified buyer will understand the grade he is buying and recognize why a government grader marks a piece of meat as he does.

ACCEPTANCE SERVICE

This service has existed since 1923. In recent years its use has increased markedly. Under the supervision of a grader, product is prepared and packed according to rigid specifications. The product is literally "accepted" for the food service buyer by a qualified U.S. Department of Agriculture trained grader, at the processing plant. Basically, this service is designed for large-scale users as government agencies, steamship lines, and hotels. It helps create a competitive situation by putting the bidding on an objective basis. Rigid supervision of detailed specifications makes it possible for the buyer to confidently take the lowest price. The cost of the service is nominal at $20.00 per hour for the grader. This cost is reflected in the bid price.

In 1975 and 1976, revised specifications were published. These were the results of the joint efforts of the U.S. Department of Agriculture, the National Association of Hotel and Restaurant Meat Purveyors, and representatives of the food service industry. There are six pamphlets offered by the Superintendent of Documents, U.S. Government Printing Office, Washington 25, D.C.: Series 100—Fresh Beef; Series 200—Fresh Lamb and Mutton; Series 300—Fresh Veal and Calf; Series 400—Fresh Pork; Series 500—Cured, Cured and Smoked, or Fully Cooked Pork; and Series 1000—Portion-cut Meat Products.

The National Association of Meat Purveyors has published a *Meat Buyers Guide* covering all of the above specifications except series 500.

General Requirements.—Initially, the service was complicated, re-

Courtesy of U.S. Department of Agriculture

FIG. 3.16. ACCEPTANCE SERVICE STAMP—MEAT GRADING BRANCH

quiring considerable evaluation. The fact that it is now more widely used demonstrates its practicality. Qualified large-scale users may contact the Meat Grading Branch, Livestock Division, U.S. Department of Agriculture, Washington 20025, D.C.

The buyer has three steps to execute: (1) examine specifications; (2) list bid items by specification number and name along with grade or selection, weight range formula (if applicable), and type of refrigeration; and (3) seek bids, award contracts, and notify nearest USDA meat grading supervisor.

The supplier, upon delivery request of a contract, requests acceptance service, processes and packs product under supervision. Containers are sealed and stamped (Fig. 3.16).

REFERENCES

ANON. 1942. Meat and Meat Cookery. National Live Stock and Meat Board. Chicago, Ill.

ANON. 1945. Meat Handbook of the U.S. Navy. U.S. Navy Dept., Washington, D.C.

ANON. 1950. Ten Lessons on Meat. 7th Edition. National Live Stock and Meat Board, Chicago, Ill.

ANON. 1956. Official United States standards for grades of veal and calf carcasses. S.R.A.-A.M.S. *114,* U.S. Dept. Agr., Washington, D.C.

ANON. 1957. Facts about beef cattle. A circular. American Meat Institute. Chicago, Ill.

ANON. 1958. Official United States standards for grades of pork carcasses. S.R.A.-A.M.S. *171* (Revised). U.S. Dept. Agr., Washington, D.C.

ANON. 1959. Rules and regulations of the U.S. Department of Agriculture governing the grading and certification of meats, etc. S.R.A.-A.M.S. *98* (Revised). U.S. Dept. Agr., Washington, D.C.

ANON. 1960A. Official United States standards for grades of lamb, yearling mutton, and mutton carcasses. S.R.A.-A.M.S. *123,* U.S. Dept. Agr., Washington, D.C.

ANON. 1960B. Institutional meat purchase specifications general requirements for fresh beef (Series 100), for fresh lamb and mutton (Series 200), for fresh veal and calf (Series 300), for fresh pork (Series 400). U.S. Dept. Agr., Washington, D.C.

ANON. 1965. Official United States standards for grades of carcass beef. S.R.A.-A.M.S. *99,* U.S. Dept. Agr., Washington, D.C.

ANON. 1968. How to buy beef steaks. Bull. *145,* U.S. Dept. Agr., Washington, D.C.

ANON. 1969. Meat Evaluation Handbook. National Live Stock and Meat Board, Chicago, Ill.

ANON. 1969. U.S. inspected meat packing plants. Handbook *191,* U.S. Dept. Agr., Washington, D.C.

ANON. 1976. Meat Buyers Guide. National Association of Meat Purveyors, Tucson, Arizona.

BREIMYER, H. F. 1959. Federal grading of meat and its economic measures. Livestock and Meat Situation *103,* 18–22.

BROWNE, A. E., *et al.* 1954. Grades, standards. Yearbook of Agriculture. U.S. Dept. Agr., Washington, D.C.

BULL, S. 1951. Meat for the Table. McGraw-Hill Book Co., New York.

WANDERSTOCK, J. J. 1960. Meat purchasing. Cornell Hotel and Restaurant Administration Quarterly, *1*, No. 1, 25–29.

WOLGAMOT, I. H. 1957. Beef-facts for consumer education. AIB *84,* U.S. Dept. Agr., Washington, D.C.

WOLGAMOT, I. H., and FINCHER, L. J. 1954. Pork-facts for consumer education, AIB *109,* U.S. Dept. Agr., Washington, D.C.

ROMANS, J. R. and ZIEGLER, P. T. 1977. The Meat We Eat, 11th Edition. Interstate Printers and Publishers. Danville, Ill.

4

Structure

Meat may be defined as the carcass of any animal used as food, including domestic mammals raised for food, game animals, shellfish, fish, poultry, and game. It is composed of lean muscles, connective tissue, fat, water, bones, skin, nerves, and blood vessels.

Domestic types include: (1) bovine (veal and beef); (2) ovine (lamb and mutton); (3) swine (hogs); and (4) marginal types (goats, horses, and dogs).

PHYSICAL PROPERTIES OF MEAT ANIMALS

Muscles

The muscles, or edible lean are of two general classes: (1) voluntary or striated; and (2) involuntary or smooth. The heart is the exception, being involuntary but striated. Its fibers have a different structure from the skeletal. The voluntary muscles are the skeletal muscles composed of many thousands of fibers about one inch long ranging from about 0.005 to 0.0004 of an inch in diameter. Appearance of the cells in these fibers will vary with age, sex, and feeding of the animal. These cells are held together by sheaths of connective tissues, and these in turn are bundled together to form muscles.

Muscle bundles will vary in size, length, and thickness. The grain or texture of a piece of meat is determined by the fibers and muscles. For example, small fibers in small bundles make for meat with a fine grain.

The muscles of the alimentary tract, of the blood vessels, of the stomach are examples of involuntary or smooth muscles.

Connective Tissue

The white connective tissues are composed largely of the protein collagen, which is water soluble when heat is applied. Moist heat changes this connective tissue to gelatin, which is more tender and more palatable. When dry heat is applied, the meat generally tends to become tougher. It can be concluded that the amount of connective tissue present, and how the meat is cooked are general factors of tenderness.

Connective Tissues Dissolved During Boiling.—Include the following: (1) endomysium—between walls of the muscle fibers; (2) perimysium—envelope bundles of muscle fibers; and (3) epimysium—sheath of connective tissue surrounding a muscle. Examples—tissue in hip end of strip loin; "silver seam" of a tenderloin.

The quantity of connective tissue is related to the activity of the particular muscle and age of the animal. To illustrate activity, the shank has considerably more connective tissue than does the tenderloin. Illustrating the age factor, the tenderloin of a cow or bull has more connective tissue than the tenderloin of a young steer or heifer.

The protein elastin, another class of connective tissue, is not water soluble. Examples of this are the yellow ligament, commonly called the back strap, and the tendons which connect the muscles to the skeleton.

Fat

All meat contains some degree of fat in and around the cells, between and over the muscles. The deposits may be so fine that they are not visible, or as evident as the cover fat. Total fat in any carcass will vary from about 5 to 30 per cent.

Animals are deliberately placed in feed lots for fattening. The first fat that occurs is stored on the internal organs, the kidneys, stomach, and intestines; then on the surface of the muscles; and finally in the muscles and in the connective tissue. Fat generally can be related to flavor, tenderness, and juiciness, especially when it occurs as "marbling," or intra-muscular deposits.

There is some evidence that fat deposited between the walls of a connective tissue makes them thinner.

These thinner tissues are more easily ruptured when chewed. In this manner fat deposits may contribute to tenderness.

Surface fat, especially on the cover and on the organs, has one negative aspect. It makes the lean relatively expensive. The average chef or housewife indicates a desire for lean beef. In spite of this, the better institutional trade and volume retailers use and profitably sell fat beef.

The feed lot practice to fatten beef actually changes the lean-bone ratio favorably, increasing the net dressed yield of the carcass relative to live weight, and increasing the yield of the more desirable cuts, the ribs and loins (Table 4.1).

Water

All meat contains a great deal of water, which is inversely proportional to the fat percentage. In some lean cows and bulls, water content of the meat will run over 65 per cent and drop to less than 50 per cent in young fat animals. The texture of the meat will change with the water-fat ratio from a wet, flabby, soft, to a velvety, firm and relatively dry texture, as the fat increases.

Organic Extractives

About one to two per cent of organic extractives or nitrogenous extractives are found in lean meat. These may contribute to the flavor of a piece of meat. These extractives are present in proportion to the amount of connective tissue. This may explain why a top sirloin butt is more flavorful than a tenderloin. Because the extracts are water soluble, any boiled piece of meat loses flavor and becomes bland, while the broth or drippings are considerably enriched flavorwise. Argentina produces large quantities of cooked corned beef. It is no coincidence that they export large quantities of meat extract, a natural by-product. These extracts are largely proteins.

Enzymes

Meat contains many different enzymes or natural ferments, which are complex organic substances, and act to hydrolyze certain components of the meat.

The proteolytic enzymes act on or "attack" proteins, breaking them

TABLE 4.1. RIB AND LOIN YIELDS—RELATIVE TO GRADE[1]

Cut[2]	Common, per cent	Medium, per cent	Good, per cent	Choice Prime, per cent
Rib	8.7	9.0	9.2	9.6
Lion	17.2	17.4	17.7	18.0

[1] Source U. S. Bureau of Home Economics.
[2] Cut Chicago style.

down into component parts. For example, enzymatic action makes connective tissues more tender, and contributes to the sensation of juiciness. At freezing temperatures, enzyme actions are extremely slow. As the temperature is increased above freezing, the enzymatic activity increases as a geometric progression. Enzyme activity is inactivated at elevated temperatures as low as 140°F.

Pigments

The pigments contribute color to the lean and fat, but are not important nutrients. The color range of reds found in the lean is largely dependent upon the amount of pigment, myoglobin, present. The quantities are directly proportional to the age of the animal. As the age of the animal increases, progressively the lean grows darker. In the bovine species, the color range runs from light pinkish brown in veal to pinkish brown in calves; to light red in young beef; to bright cherry red in mature bulls and cows. Lamb ranges from bright pink to brick red in mutton, and pork will range from gray-pink to gray-red. In any given animal, different muscles will have different colors. For example, compare a white turkey breast and its dark legs, or the top round and bottom round of a beef.

The "dark cutter" is one exception where a youthful beef will have a very dark lean. The grader may correctly mark it USDA Choice. Yet the lean is as dark in color as a very mature bull. Numerous tests have been made of USDA Choice and USDA Prime beef ranging from pink to "dark cutter" where all other factors are equal, and it has been determined that color is of no significance from the point of view of palatability.

There are reports on experimentation and observation of "dark cutters." It is reported that this phenomenon could be directly correlated with a high pH, the indication of the amount of free acid or a low muscle glycogen. The high pH was associated with animals kept in cold or exposed conditions before slaughter, and/or without adequate food, and/or in an excited condition previous to slaughter. It is generally accepted that dark cutting beef results from a deficiency of animal starch, glycogen, in the tissues of the animal at the time of slaughter.

Other pigment changes will occur in the dressed beef, with exposure to oxygen, enzymes, alcohols, acids, curing preparations, and heat. A piece of beef after it is freshly cut will at first appear very dark, but slowly, as it is exposed to the oxygen in the air, will become a bright red. In the parlance of the butcher shop, this is the "bloom." What actually happens is a chemical phenomenon. The hemoglobin and myoglobin are oxidized and change to oxyhemoglobin and oxymyoglobin.

Heat affects and changes the color pigments. The illustration is for roast beef at the tip of the bulb of the thermometer.

Pigment Changes at Rare 131° to 149°F—Characteristic raw red changes to bright rose red. There is a brown-gray layer next to the crust. Oxyhemoglobin is coagulated.

Pigment Changes at Medium 149° to 158°F—Center is changed to a light pink. Gray perimeter layer is expanded.

Pigment Changes at Well done 158° to 176°F—Heme pigments occur characterized by a brownish gray color. When the product is refrigerated, occasionally the process is reversed, the product taking on a strange red color.

When beef is aged at cooler temperatures, with the aid of enzymes, acids, dehydration, a similar surface effect occurs ranging from dark red at first to the same well done grayish brown. A cut below the surface, however, will reveal a dark red fresh subsurface.

Yellow fat is a limited index of age and breed. In general commercial slaughter, white fat is a characteristic of young, better fed beef, and yellow fat is a characteristic of more mature animals and some dairy breeds. Carotene, the pigment that makes fat yellow generally becomes more predominant as the animal matures, and certain breeds of dairy animals characteristically have yellow fat. Certain types of feed will contribute to yellow fats in all animals. Although yellow fat is commercially undesirable, it has no negative palatability or nutritional characteristics.

MEAT IN THE DIET

Meat plays a most significant role in the diet. The entrée, the important part of the meal, is usually described in terms of the meat dish. Meat has eye appeal; its aroma stimulates the sense of smell; it generally enhances the meal, makes for satiety, staves off hunger. It is almost completely digestible, and high on the nutritional scale.

Proteins, the principal components of living cells, are very complex substances. Both animal and plant proteins consist of amino acids. Much of the value of a protein food is based on its amino acid content. High nutritional value is related to a complete complement of "essential" amino acids. Proteins deficient in one or more "essential" amino acids are relatively low in nutritional value. In general, meat is one of the best sources of high nutritional proteins, essential for life and optimum physiological performance. Lean meat contains from 15 to 20 per cent of protein which varies inversely with the percentage of fat.

Vitamins are certain organic substances found in minute quantities in food, essential to health. There is little vitamin A in fat. It occurs in liver, the richest food source, and variety meats. Meat is a particularly rich source of all of the B complex vitamins. Important ones are thiamin, riboflavin, and niacin.

Minerals or inorganic compounds, are necessary in the diet. Most of the essential minerals except calcium, are found in the lean. Included are phosphorus, iron, copper, and trace elements.

By way of illustration, an average four-ounce portion of meat will provide approximately the following percentages of the daily requirement of nutrients required by a moderately active adult: protein, 24; calories, 14; phosphorus, 14; iron, 25; vitamin A, 6; thiamin, 36; riboflavin, 16; and niacin, 38 per cent.

It should be pointed out that the inexpensive meat cuts are just as nutritious as those which are more expensive.

REFERENCES

AMERICAN MEAT INSTITUTE FOUNDATION. 1960. Science of Meat and Meat Products. W. H. Freeman and Co., San Francisco, Calif..

ANON. 1942. Meat and Meat Cookery. National Live Stock and Meat Board, Chicago, Ill.

ANON. 1950. Ten Lessons on Meat. 7th Edition. National Live Stock and Meat Board. Chicago, Ill.

ANON. 1960. Nutritive value of foods. H. and G. Bull. *72,* U. S. Dept. Agr., Washington, D. C.

AREY, L. B., BURROWS, W., GREENHILL, J. P., *et al.* 1957. Dorland's Illustrated Medical Dictionary. W. B. Saunders Co., Philadelphia, Pa.

BULL, S. 1951. Meat for the Table. McGraw-Hill Book Co., New York.

DOTY, D. M. and PIERCE, J. C. 1961. Beef muscle characteristics as related to carcass grade, carcass weight and degree of aging. Technical Bull *1231,* U. S. Dept Agr., Washington, D. C.

LEVERTON, R. M. and ODELL, G. V. 1959. The nutritive value of cooked meat. *MP-49,* Oklahoma Agr. Expt. Sta., Oklahoma State University, Oklahoma City, Okla.

MORSE, R. E. 1958. What's new in meat technology. Meat *49,* No. 8, 40–42.

ROMANS, J. R. and ZIEGLER, P. T. 1977. The Meat We Eat, 11th Edition. Interstate Printers and Publishers, Danville, Illinois.

SCHWEIGERT, B. S. and PAYNE, B. J. 1956. A summary of the nutrient content of meat. A circular. American Meat Institute Foundation, Chicago, Ill.

WOLGAMOT, I. H. 1957. Beef facts for consumer education, AIB *84,* U. S. Dept. Agr., Washington, D. C.

WOLGAMOT, I. H. and FINCHER, L. J. 1954. Pork-facts for consumer education. AIB *109,* U. S. Dept. Agr., Washington, D. C.

5

Refrigeration of Meat

COOLER STORAGE

To prolong or extend the commercial life of the product in its natural raw state is the fundamental function of the application of cold. This applies to temperatures above the freezing point, and to low temperature storage below the freezing point. Above freezing temperature is commonly referred to as "cooler" and below freezing as "freezer" storage.

Cooler storage is intended to prolong the life of the product for a relatively limited period. The application of refrigeration begins immediately after slaughter when the carcass is removed to the "hot box." This cooler has a very great cooling capacity. The "animal heat" is removed and the internal temperature of the carcass is reduced as quickly as possible to 30 to 38°F depending on the practice of the particular packer. From here, the product, as it is moved through the various trade channels, on and off trucks, in and out of coolers, in good practice, is maintained at a more or less uniform internal temperature. To understand refrigeration properly, it must be pointed out that the internal temperature changes will be very slow, even though the product is exposed to higher temperatures for short periods, and the visible surface temperature may fluctuate widely.

Mechanical Factors

Storing at cooler temperatures involves three primary mechanical problems, temperature, humidity, and air circulation. Although the freezing point of pure water is 32°F, most meat products have a lower freezing point, generally around 28°F. Therefore, satisfactory cooler temperatures will range from a low of 28°F to about 38°F. The upper limit is set by factors which will be discussed later.

Adequate refrigeration should be available to maintain a more or less limited and even temperature range. There are three loads on the cooler, the refrigeration loss through walls and ceiling, the heat given off by workmen, lights and motors in the cooler, and the product load of merchandise moved into the cooler. When a large hot load is moved in, the air temperature will rise quickly, perhaps several degrees. The internal temperature of the product previously in the cooler may change very little; this can be determined by a special thermometer for taking the internal temperature. A good test of the adequacy of the mechanical system itself is to measure the "temperature drop" of the air leaving the blower coils; approximately 10°F or slightly less indicates an adequate system.

Relative humidity should range from about 85 to 90 per cent. The absolute humidity in the cooler temperature range is not very great. Humidity is rarely too high, and most often too low. A proper and adequate mechanical system is relatively good assurance of good humidity. If humidity is too low, various devices to increase it can be employed. The best device, however, is an effective refrigeration system. High humidity will prevent excessive drying and shrinkage.

It is an interesting phenomenon that the desirable white color of fat on high grade beef will tend to become slightly yellow at 80 per cent relative humidity and the color will be restored at 90 to 100 per cent. This has no significance except appearance.

Air circulation is controlled by the coil system, the ceiling height of the box, and the stacking or storing devices. The coils should be large enough to refrigerate the load adequately and rapidly. The ceiling should be high enough so that the air can be circulated above the product and drop gravitationally over the product.

Bacteria

Initial rapid cooling in the "hot box" and a low-holding temperature retard the growth of bacteria, molds, and spores. The antemortem inspection is a certification of a non-contaminated carcass. As the product moves from place to place, the surface is constantly exposed to microorganisms. Meat itself contains all the essential nutrients for the organisms, and consequently they multiply rapidly.

The most important control of bacteria on meat is a low temperature. From 60 to 100°F, most bacteria will grow logarithmically, other conditions being equal, with an optimum growth from 80 to 100°F.

Oxygen, acidity, and moisture are additional factors in the growth of bacteria. There are three different types of microorganism oxygen relationships: those that cannot survive without oxygen (strict aerobes), those that cannot survive with the presence of free oxygen (strict anaerobes), and the facultative types, which can survive either with or without the presence of oxygen. Most bacteria fall within this latter group.

As far as acidity is concerned, as a general rule, bacteria will grow best at a neutral pH, from about pH 6 to 7.5 or 8.0. It has been demonstrated that if the pH of meat, which is normally about six, is reduced, bacterial growth is retarded. In some commercial practices, the pH is reduced by the addition of acetic, lactic, citric, or other acids. In the case of certain femented sausages, the pH is lowered by acid-producing bacteria, which utilize the sugars present for energy, breaking them down into alcohols, producing acid and in some cases gas during the fermentation reaction.

As a general rule, bacteria must have water to perform their physiological functions. Hence, the drying surface of a piece of meat will progressively inhibit the bacterial growth. This can be translated into common practices, such as dried beef, drying by salting, or as in the freezing process, converting the available water into a solid. The water level reduction is illustrated by the curing process, such as sweet pickling, or dry salt and sugar curing, and the subsequent smoking, or ordinary smoking, which achieves maximum dehydration, as in the case of dried beef, and certain dry sausages. This effects a reduction of the bacterial activity.

The factors that affect growth are frequently related in such a manner, that the proper combination of one or more will control most microorganisms, and the best combination of all four will inhibit growth.

For a general understanding of bacteria, it is necessary to investigate the various methods of destroying them. They are considered dead when they cannot multiply under optimum culture conditions. In general, heat, chemicals, and irradiation are good bactericides. In the case of heat, the common temperature used in canning meat is 250°F. Many chemicals destroy bacteria, but are often harmful in human foods. Hence, these chemicals are best used for sanitary purposes. Germicides containing chlorine when used at temperatures above 130°F, are very effective in destroying bacteria if applied to equipment which has been cleaned first.

Radiation (beta rays, gamma rays, or X-rays) is a third effective means of destroying bacteria.

Ordinary ultraviolet light kills bacteria on the surface of meat and helps to reduce the microbiological population of coolers. It is used in some warehouses, maintained at temperatures above those commonly used for aging meat in order to hasten enzyme action and resultant aging.

Molds and Yeasts

The problems encountered in preservation of meats frequently are the same for bacteria, molds, and yeasts, the exception being that yeasts and molds can grow at lower pH and need less moisture. They can frequently use nitrates as a source of nitrogen, and sometimes live on dried, salted, and fermented products; some are able to grow at freezing temperatures. They are destroyed by heat. Molds require oxygen and so often live on the surface of liquids. As molds and yeasts occur principally on the surface of meats, much of the contamination can usually be removed with only a little trim loss. For example, surface molds on hams rarely, if ever, make them unfit to eat. In order to eliminate the mold, simply brush with a stiff brush, wipe with a cloth moistened with acetic acid or salt water, and dry in the air. Heavy mold may be trimmed off and the only real damage remaining is in deep cracks.

Enzymes

Enzymes, which are proteins, are a natural component of the meat tissue. They function in the tissue as a catalyst for virtually every reaction that occurs in the living cell. Enzymes remain functional postmortem.

The proteolytic enzymes play an important role in aging. The degree of activity increases with higher temperatures. This is not a straight line acceleration, but a geometric progression. For approximately every 8°F increase above freezing, the enzymatic activity doubles to 140°F; above this temperature the enzymes are inactivated (Fig. 5.1). The enzymes catalyze the hydrolysis of the various collagens in the meat tissues. These collagens are the connective tissues around each cell, each group

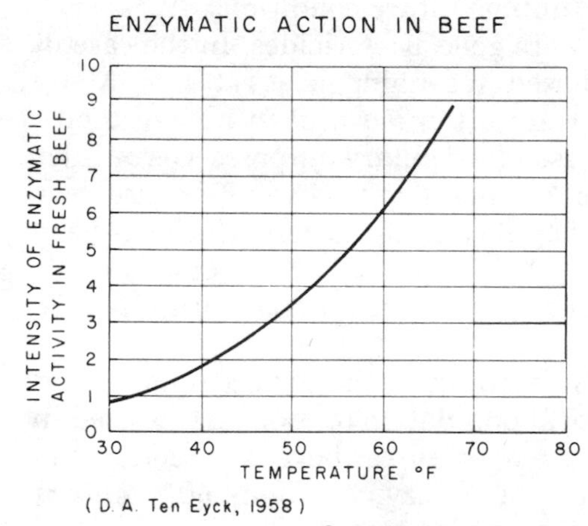

Courtesy of D. A. Ten Eyck

FIG. 5.1. EFFECT OF ACCELERATED TEMPERATURE ON ENZYME ACTIVITY IN BEEF

of cells, and between large muscles. In the process, the collagens, which account in some portion for what is characteristically described as toughness in meat, are hydrolyzed or partially changed from this tough form to simpler proteins and finally to gelatin. A medical suture is a dramatic example of enzymatic action. Strands of sheep gut are frequently used, which consist chiefly of collagen. These stitches, that disappear, are hydrolyzed by the enzymes in the body.

Aging is an enzymatic process intended to partially break down the connective tissues under normal, cooler conditions, at temperatures ranging from 30 to 40°F. This takes two weeks or more. It is sometimes asserted that 10 to 14 days produce maximum results. It must be observed that the enzymatic action continues until whole muscles are "digested." It appears reasonable to conclude that some benefits occur from extended aging up to four or five weeks.

If prolonged aging practices are followed, it is important to use high grade meat well covered with fat, to maintain proper temperature, humidity, and air flow, and keep the cooler clean to minimize growth of bacteria, molds, and yeasts.

A tender cut of tenderloin really requires no aging, and a strip loin requires less than a relatively tough top sirloin. The rib roast, though it has considerable connective tissue, is cooked under much more favorable conditions when it is oven-roasted, than a steak prepared by broiling. The roast does not need as much "age" for equal tenderness.

From the economic point of view, aging has two negative aspects. There is considerable dehydration or shrink, and subsequently reduction of salable weight. With average good cooler conditions, testing exposed primal cuts, it has been demonstrated, though not conclusively, that shrink will run 0.2 to 0.3 per cent per 24 hours. Unlike the common impression, shrink does not occur mostly in the first three or four days, but is nearly a straight line function. Various grades of meat will vary in shrinkage, cuts with a good covering of fat, shrinking the least.

Secondly, molds, yeasts, and some bacteria soon start to grow freely. Fats may turn rancid. These are surface occurrences, a by-product of the aging process, and negative in value. Aging is needed for the enzymatic action that occurs internally. The molds and yeasts on the surface cause what is frequently referred to as the "aged taste." This taste is not desirable. These surfaces have to be trimmed off and consequently the steak cuts become more expensive.

There are several procedures which reduce aging losses and hasten the process. By increasing the temperature, it is roughly estimated that what requires 12 days to achieve at 32°F can be achieved at 42°F in six days, 48°F in four days, and 60°F in two days.

At higher temperatures, the growth of yeasts, molds, and most bacteria is rapid and may cause trouble.

One approach to accelerated aging has been to couple higher cooler temperatures with irradiation or with ultraviolet light sometimes referred to as the "Tenderay process."

The "Cameron" process calls for the product being enclosed in a Cryovac bag, vacuum sealed, and shrunk, exposed to a temperature of 68°F for 24 hours, the equivalent of about nine aging days at 32°F, and then chilled to normal cooler temperature or frozen. The use of this system is very limited.

Cryovac and comparable products offer an economic solution to the long term, 30 to 36°F hold as they reduce shrink, inhibit molds, yeasts, and bacteria on the surface, and the enzymatic tendering process goes on as usual.

FREEZER TEMPERATURES AND STORAGE OF FROZEN MEAT

Temperatures below freezing are intended to prolong for an extended period of time the economic life of the product. All that is required is adequate refrigeration to bring about "sharp freezing," and facilities for holding the meat at a low temperature.

Temperature

Temperatures ranging up to +5°F are acceptable although −10° to 0°F are generally used in commercial practice. It is commonly agreed that quick freezing at −20°F or colder produces smaller ice crystals, whereas larger crystals are produced freezing at temperatures ranging up to 0°F. The location of the crystals will vary with the type of freezing. Sharp freezing crystals form within the muscle fibers, whereas slow freezing crystals form within the connective tissue and the muscle fibers. Sharp freezing of fruits and vegetables is very important. With meat products, on the other hand, quick freezing may extend but a few millimeters into the outer surface and the rest freezes more slowly. What is most significant is that authorities are generally agreed that "there is little or no mechanical damage to meat fibers even during slow freezing." Most ice crystals, even when very small, slowly grow during the storage period and form larger crystals. Temperature cycling, which frequently occurs, will accelerate crystal growth. In the light of these facts, temperatures up to 0°F are now considered acceptable for freezing.

Freezing Meat

Product selected for the freezer should be in good condition and relatively fresh. Freezing is sometimes abused in actual practice. There are instances where a product has been held for a period in high temperature storage, and as a last resort put in the freezer. When defrosted, this product will be no better than it was when frozen. There may be advanced microbiological growth, rancidity of fats, or undesirable odor and flavor changes. This may have started the old wives tale about "cold storage meats."

Today the freezer is an effective commercial device of ever-growing proportion and very significant in the over-all marketing of meat products. There are a few simple steps for a freezer program. The product should be put in the freezer as soon as possible. Small cuts should be properly wrapped in some moisture-proof type of package as foil, wax, or polyethylene wrappings, or dipped or spray coated, or ice glazed, or packed in a vacuum type bag. This will prevent or reduce evaporation of moisture from the surface of the meat and freezer burn.

Especially rapid freezing of the meat is not necessary. The load should be small enough, or the freezer cold enough to freeze the product completely in a relatively short period, lest unnecessary bacterial action occur. It is wise to freeze before boxing. If cases are used, they should not be stacked before freezing. If it is necessary to freeze in cases that are stacked, a suitable low temperature must be maintained to freeze within a reasonable period. Some device should be employed as pallets, or pieces of lumber to create air spaces between cases to insure good air flow and permit heat removal which will help hasten freezing. An impartial test should be made under the actual working conditions to determine if the freezing is being done in a satisfactory manner.

Effects of Freezing

An argument rages between two schools of thought, the one for freezing that contends that a minimum palatability change takes place, and the other claiming vast changes. It is difficult to generalize. In the final analysis, the results will vary from product to product. Flavor, aroma, tenderness, and juiciness must be evaluated. The meat should be tested after it has been cooked.

Tests for juiciness were set up by Child and Baldelli in 1934 and by Tannor, Clark, and Hankins in 1943. The press fluid can be measured by both tests. It has been determined by actual testing that what has been lost in defrosting, as the drip loss, would otherwise be lost in the cooking. If product defrosting methods are carefully followed, or if there is no defrosting, or if careless and indifferent methods are used the total loss of moisture will be about the same after the meat is cooked. Drip loss during defrosting and moisture loss during cooking are inversely proportional. If this is true and drip loss is not related in any manner to the sensation of juiciness, it may be concluded that freezing has little effect on juiciness. It is generally agreed that juiciness is not correlated with press fluid, but relates itself to the concentration and distribution of intramuscular fats.

Ice crystals that form have little if any measurable effect on tenderness. They are basically intracellular. It might be concluded that the rupturing of the cell tissue, and the expansion of the connective tissues might have some mechanical tenderizing effect. Since a very small amount of enzymatic action will continue at freezing or below, protracted storage in the freezer will be accompanied by some degree of hydrolysis and consequent tenderizing effect. There is no evidence that freezing causes beef products to get tougher, except where "freezer burn" occurs.

Slight flavor changes occur which are another negative aspect of freezing. To evaluate this factor properly, flavor tests should be conducted for each class of product. The results also will vary relative to the cooking technique.

Meats stewed and braised take most of their flavor from the drippings or gravy. Dry roasted meats are relatively bland and any loss of flavor may be almost impossible to detect. The flavor of broiled meats reflects the surface seasoning and the surface crust, as well as the amount of chewing required. The flavor of frozen product should not be prejudged, but cooked and tested to determine if the observable flavor changes, if any, are significant for the particular preparation and operation.

Chemical changes are sometimes significant. Pork contains unsaturated fats which tend to oxidize and become rancid. Beef and lamb fats which are saturated are more stable and permit prolonged storage periods. Ground meats tend to turn rancid more quickly due to the intimate contact of the fat with oxygen. Curing and salting considerably accelerates the onset of rancidity. Cured products, if they must be frozen, should not be held for more than a few weeks.

Storage life will vary with the type of meat and the holding temperature (Table 5.1).

Color changes will occur on the surface of products stored in the freezer. This is due to surface dehydration and occurs on unwrapped or loosely wrapped product. The color tends to become lighter, and extreme "freezer burn" is indicated by a straw color and texture. Severe "freezer burn" will result in considerable flavor loss, as well as destruction of nutrients.

TABLE 5.1. STORAGE LIFE OF FROZEN MEATS (IN MONTHS)

	10°F	0°F	−10°F	−20°F
Beef	4	6	12	12+
Lamb	3	6	12	12+
Veal	3	4	8	12
Pork	2	4	8	10

Bacteria will not grow below a given minimum temperature. At about 20°F a few bacteria might grow very slowly. It can be stated that at temperatures used commercially of 0°F or lower there is absolutely no microbial growth, and that some of the microorganisms are killed during the freezing and thawing. It is significant, however, that during the thawing processes, especially as the surface temperature increases above 20°F, those strains of bacteria that are not killed will again grow at the same rate demonstrated by the particular strain at any given temperature. Yeasts and molds are effectively controlled at commercial freezing temperature.

Refreezing is another generally misunderstood problem. There is basically nothing wrong with refreezing. It may occasionally occur in commercial practice. There is one important qualification: the temperature of the defrosted product and the length of time it is defrosted must not be sufficient to permit microbial growth. *Do not refreeze as a last measure to save a product.* At this point, it is already too late! Refreezing will increase some of the undesirable effects of freezing.

FREEZE DRYING

This relatively new process is achieved by freezing under very low pressure so that the ice sublimes, leaving a sponge-like form without moisture. Hence, the product retains its shape, but the total weight is reduced, often more than 50 per cent. Product is generally canned or wrapped in foil to keep out moisture. This product requires no refrigeration, because microorganisms cannot grow at such a low moisture content. Final product moisture ranges from about 1 to 4 per cent.

Reconstitution is achieved by immersion in water before cooking. There is little change because of chemical or enzymatic action. Reconstituted products clearly resemble the original fresh product in flavor and color. Beef appears tougher and dryer. Cooked beef product and freeze dry cooked beef product reconstituted taste alike. The flavor deteriorates after prolonged storage.

The process and its relation to the food service industry is too new to evaluate. New techniques are yet to be worked out. Advantages are apparent for the armed services. Stockpiling can be achieved without freezer facilities, and transportation costs are reduced

REFERENCES

AMERICAN MEAT INSTITUTE FOUNDATION. 1960. Science of Meat and Meat Products. W. H. Freeman and Co., San Francisco, Calif.

ANON. 1945. Meat handbook of the U.S. Navy, U.S. Navy Dept., Washington, D.C.

ANON. 1957. Short-cut in aging beef. Missouri Farmer *49,* No. 2, 16–17.

ANON. 1960. Aged tender beef in 24 hours. Meat *53,* No. 6, 32–33.

ANON. 1960. Cameron process. W. R. Grace and Co., Cambridge, Mass.

BULL, S. 1951. Meat for the Table. McGraw-Hill Book Co., New York.

DIMARCO, R. G., and MORSE, R. E. 1962. Freeze drying: Its relation to meat. MPSM *1*, No. 9, 125–127.

DOTY, D. M. 1955. Meat preservation—past, present, and future. Circ. *13,* American Meat Institute Foundation, Chicago, Ill.

DOTY, D. M., and PIERCE, J. C. 1961. Beef muscle characteristics as related to carcass grade, carcass weight and degree of aging. Tech. Bull. *1231,* U.S. Dept. of Agr., Washington, D.C.

PEARSON, A. M., BURNSIDE, J. E., EDWARDS, H. M., *et al.* 1951. Vitamin losses in drip from frozen meat. Food Research *16,* 85–87.

SLEETH, R. B., HENRICKSON, R.L., and BRADY, D. E. 1956. Effects of controlling environmental conditions during aging on the quality of beef. Food Technol. *11,* 205–208.

SLEETH, R. B., KELLEY, G. G., and BRADY, D. E. 1957. Shrinkage and organoleptic characteristics of beef aged in controlled environments. Food Technol. *12,* 86–90.

TRESSLER, D. K., and EVERS, C. F. 1957. The Freezing Preservation of Foods. Vol. *1*. Fresh Foods. 2: Cooked and Prepared Foods. Avi Publishing Co., Westport, Conn.

TRESSLER, D. K., EVERS, C. F., and EVERS, B. H. 1953. Into the Freezer—and Out. Avi Publishing Co., Westport, Conn.

WEBER, E. S. 1962. Frozen, prepared foods. Cornell Hotel and Restaurant Administration Quarterly *3*, No. 2, 3–8.

WEIR, C. E. 1960. Frozen—dried and dehydrofrozen foods. A circular. American Meat Institute Foundation, Chicago, Ill.

WIESMAN, C. K. 1956. How to handle frozen meats. Refrig. Eng., *10,* No. 10, 60–61, 107–108.

WOLGAMOT, I. H., and FINCHER, L. J. 1954. Pork—facts for consumer education. AIB *109,* U.S. Dept. Agr., Washington, D.C.

ZIEMBA, J. V. 1960. Freeze-drying. Food Eng. *32,* No. 12, 57–64.

6

Distribution

Meat distribution is vastly complicated. Figure 6.1 is an oversimplified representation. The exceptions to the flow of product are many. For example, the boners sell large quantities directly to an institutional user, the Armed Forces. Most jobbers sell some products to other wholesalers. Some wholesalers sell directly to consumers. Figure 6.1 is primarily an attempt to generalize the distribution channels, and focus upon the many specialized segments of distribution.

Branch House

A "branch house" is usually a packer operated sales organization remote from the slaughter house but close to the point of distribution. Usually a full line of product is offered; meat, provisions, and poultry. In some communities, the branch house may function as a hotel supply house.

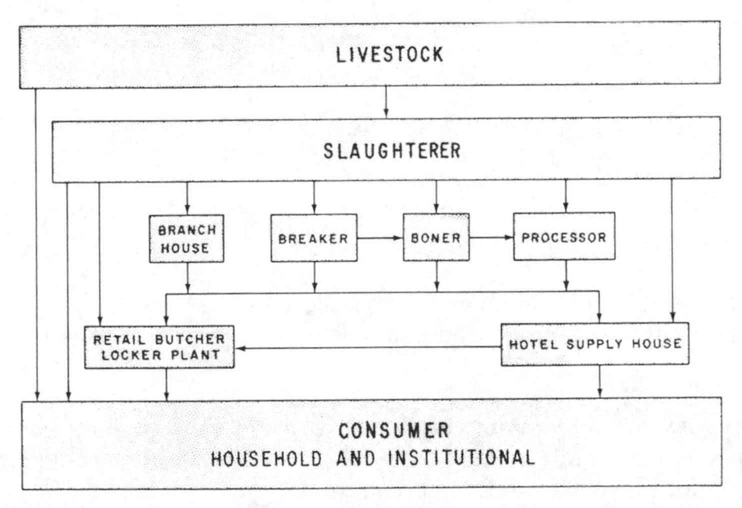

FIG. 6.1. MEAT DISTRIBUTION CHANNELS

Breaker

A "breaker" is a non-slaughterer who buys beef carcasses, and breaks them into primal cuts. Usually the "rough cuts," the rounds and chucks, are sold to chain stores, and the "center cuts," the ribs and loins, are sold to jobbers. Such a breaker in Chicago, for example, could buy from various midwest packers, compete in the Chicago market, or sell to such distant points as Miami or Los Angeles.

The mark-up is usually a very nominal 2 to 2½ cents per pound plus freight. This operation is predicated on very heavy tonnage sales.

Sometimes breakers will include a boning function, as fabricating beef loins, selling tenderloins, strip loins, top sirloins, and bottom sirloins to the hotel supply houses. Usually, the lower end of the grade or "throw outs" are fabricated in this fashion.

Boner

A "boner" is a non-slaughterer who buys beef, veal, or some pork cuts for boning and possibly processing. Some operations are highly specialized, fabricating just beef or veal, usually of the lowest grades. The cuts are sorted, boxed, and usually frozen. Low grade beef cuts are sold partially for sausage, and partially for low grade steaks, usually after being tenderized, sometimes by the boner.

In the case of pork, some firms specialize in boning pork loins for Canadian bacons, back ribs, and pork tenderloins. Some boning is conducted at almost every level of distribution.

Processors

There are a wide variety of processors.

Canners.—Canners pack meat such as hams, meat products as chile con carne, or pet foods, produced under inspection in a sterile or semi-sterile metal container.

Curers.—Products such as hams, bacon, butts, beef briskets, plates, tongues, and rounds are cured in sweet pickle, brine, or corned in some fashion.

Smokers.—Smokers convert cured product into smoked product. Usually curing and smoking take place in the same establishment, though not all cured products are smoked.

Sausage Makers.—Sausage is made by grinding, seasoning, stuffing, and smoking of meat, or a combination of these. For example, fresh pork sausage is made by grinding and seasoning; made into link sausage by stuffing; sometimes it is smoked. There is a large production of sausage, highly specialized, and regionally distinct.

Prepared Foods.—Processors usually cook meat with other ingredients and condiments. This includes the broad field from tamales to TV dinners. Cooked products are usually sold fresh or frozen.

Steak Makers.—Portion-cut steaks are made by cutting, cubing, and/or tenderizing or laminating. Products include veal cutlets, hamburger steaks, tenderized beef steaks, laminated steaks, breaded steaks, etc.

Food Service Supply Houses

The CREST report of the National Restaurant Association states that "One member in 94% of all U.S. families eats out every 2.8 days." In terms of personal consumption expenditures, food service is fourth largest in the nation. In 1939, it was estimated that the public feeder annual volume was 3.5 billion. Fifteen years later, in 1954, 18.0 billion, and for 1976 a reported 49.4 billion.

The food service industry is made up of restaurants, coffee shops, cafeterias, resorts, institutions such as prisons and hospitals, schools, caterers, in-plant feeders, clubs, fraternal organizations, public transportation, government agencies, and the military. They number about one-half million units. About 60 per cent are restaurant, 20 per cent short order, and 20 per cent cafeteria eaters. It is estimated that 75,000 units do 70 per cent of the total volume.

Early in the food service industry, it was customary to buy carcasses, quarters, or primal cuts. The food service industry grew, and became specialized. Economic factors became more significant reflecting special facilities, critical wage scales, and more elaborate decor. Greater service and disassembly of meat became the demand. Where other segments of the meat industry failed to reply, a small segment, the hotel supply houses, responded and emerged as major meat distributors.

It is estimated that there are one thousand firms engaged in servicing the food service industry, most of them doing less than five million annual gross sales. These firms most frequently are specialists within themselves, either in the kind of trade handled or the type and grades of product offered. Special services include selection of product, aging, fabricating, processing, extending credit, frequent delivery, and some marketing functions.

Selection of Product.—The food supply house shows preference for USDA graded products. This has simplified the purchasing from remote points and created dependable outlets for many independent rural packers. Most of the independent packers do not compete with their distributors, encouraging the growth of the hotel supply houses. For the most part, the best merchandise available is channeled through. these distributors. The independent jobber has a great advantage: he can pick and choose each packer's products to achieve the best quality, the most economical, or the most functional.

Aging of Product.—Facilities are usually provided to age product to customer specifications. This requires considerable cooler space, proper air flow, humidity, and temperature, and adequate backlog of inventory.

Fabrication of Product.—This is another step in the meat disassembly line. The extensiveness of the service is relative to the customer's needs or specifications. The ready-to-cook product is the final manifestation of disassembly such as hamburger patties, pre-cut portion steaks, or oven ready roasts. The independent jobber is prepared to fabricate products a number of different ways to meet rigid customer specifications.

Credit and Delivery.—The extension of credit today is a primary function. Whereas the hotel supply house purchases on weekly terms, in many instances monthly billings are extended to their customers. The increased pressures brought on by extension of credit to the restaurants' customers has made longer term credit the rule, rather than the exception. Sometimes special arrangements are made for extended periods.

About half of the hotel supply houses sell to customers within a 50 mile radius. Two-thirds of the firms report less than 200 customers. About 70 per cent of the orders are taken by telephone. Deliveries are mostly made on a daily basis, sometimes on the same day. Refrigerated public carriers enable deliveries to remote areas.

Marketing Function.—Recently, some jobbers have developed skilled marketing departments to aid the food service industry in getting more economically from products, preventing waste, preparing products properly, selecting grades and cuts consistent with the menu plan, devising cutting plans, writing specifications, determining portions and costs, and in some cases, assisting with menu planning.

Pricing and Profits.—One of two pricing formulas is generally followed: (1) cents per pound markup on cost, or (2) percentage markup on cost. Gross profit margins range from about 20 to 25 per cent on sales, and the cost of doing business ranges from 20 per cent up. Net profits based on sales are relatively small. Competition is very keen. A constant effort to maximize volume tends to minimize profits.

TRADE ASSOCIATIONS OF THE MEAT INDUSTRY

National Live Stock and Meat Board

This Board was organized in 1923, a non-profit service organization, to combat propaganda that was reducing total meat consumption. The present Board represents growers, feeders, livestock marketing agencies, meat packers, meat retailers, and restauranteurs.

It became evident that food value information would be a most valuable marketing device. Under the direction of Mr. R. C. Pollock, Secretary-General Manager, research grants to leading universities were instituted. Many important discoveries have resulted. It was determined, for instance, that meat is a rich source of B vitamins, thiamin, riboflavin, and niacin; that liver contributes to the prevention of anemia; and, in general, has many dietary functions.

This initiated the organization of the research project "National Cooperative Meat Investigations" interested in quality and palatability of meat. This has been supplanted by the Reciprocal Meat Conference.

Financing of the Board activities is voluntary and the cost is nominal. Investments are on a per head basis and vary from state to state. In order to expand the Board's activities, it is anticipated that contributions per head will be increased.

A considerable quantity of formal information is available, which is listed in the "Catalog of Publications and Films on Meat." Requests should be directed to: National Live Stock and Meat Board, 444 North Michigan Ave., Chicago, IL, 60611.

American Meat Institute

A trade association, the Institute of American Meat Packers, which later became the American Meat Institute, was founded in 1906. This worldwide organization numbers some 350 members, with headquarters at 1600 Wilson Blvd., Arlington, VA, 22209.

Its functions include services of skilled professionals, the opportunity to form an active industry committee, notices of important legislation, economic reports and analyses, technological and scientific research, media and consumer relations program, educational enrichment, group insurance, and a specialized annual industry convention.

The American Meat Institute Foundation

This was founded in 1944, a direct outgrowth of the research program started at the University of Chicago by the American Meat Institute. The Foundation is an independent, non-profit foundation. The scope is scientific and educational. It has its own laboratory on the campus and the staff exceeds fifty scientists investigating bacteriology, biochemistry, food technology, histology, home economics, nutrition, and the chemistry of meat. In 1960, the Foundation edited the important text report, "The Science of Meat and Meat Products."

National Association of Meat Purveyors

This is the national association of the food supply houses. National offices are maintained at 252 W. Ina Road, Tucson, Ariz., 85704, under the supervision of Raymond F. Thill, the Executive Secretary-treasurer. The program is one of elevating the industry's practices, maintaining representation in Washington, and promoting sounder member firms. This organization has been especially effective in bringing about workable OPA and OPS regulations, lobbying for the federal grading service, and is presently trying to standardize meat cuts throughout the country. It has published a code of ethics to which the member firms are pledged.

REFERENCES

ANON. 1959. Meat distribution in the Los Angeles area. Marketing Research Rept. *347,* U. S. Dept. Agr., Washington, D. C.

ANON. 1959. Membership in the American Meat Institute. American Meat Institute. Chicago, Ill.

BROWNE, A. E., CROW, W. C., LENNARTSON, R. W., *et al.* 1954. Yearbook of Agriculture. U. S. Dept. Agr., Washington, D. C.

ROMANS, J. R. and ZIEGLER, P. T. 1977. The Meat We Eat, 11th Edition. Interstate Printers and Publishers, Danville, Illinois.

ULLENSVANG, L. P. 1959. The Hotel-restaurant meat purveyor. Thesis for Doctorate at Northwestern University, Evanston, Ill.

7

Purchasing

THE BUYER

Purchasing is a job that requires considerable intelligence and character. "Ordering," on the other hand, can be made simple and routine. The responsible buyer must possess good knowledge of the products and be able to evaluate and relate quality and price, be prepared to conduct objective tests, make changes, and cope with an ever-changing economy. The buyer must understand and be able to relate raw product to finished cost, to selling prices, and to profits. The buyer must have great character to face the everyday pressures of the sales world. There is the pressure from friends, from friendly salesmen, from friends of owners, from club members, from the greedy, from the constant offers of personal reward. The buyer can be either an individual or a management team. The person responsible for ordering can be the same person, or one elected to function within a comprehensive plan.

Buying can be reduced to a relatively simple and routine task. The person responsible may be able to complete meat purchasing in a comparatively short time each day, allow a designated time for presentation of ideas and products by salesmen, and fulfill other functions within the organization, depending upon the size and complexity of the operation.

This buyer, if correctly selected, has many obligations that should increase his stature. There is the responsibility to the management team, of which he is clearly a member, for constant improvement. Besides the logistics of product, it is the buyer's duty to be informed about the market, new products, new techniques, seasonally abundant items, current prices, and new ideas. Most of all, he must have unwavering loyalty and motivation that compels him to refuse favors, gratuities, and kickbacks. The buyer's interest and sensitivity to suppliers and his ability to evaluate and present new ideas are cornerstones of progressive management.

BUYER'S QUOTIENT

A self-evaluation or management's examination of the buyer can be made following the format of the "Buyer's Quotient Test." Final scoring is subjective, but some penetrating points can be made.

Buyer's Quotient Test (Self-evaluation)

(1) Do you understand the technical language of the trade?

(2) When an unfamiliar abbreviation is introduced, such as OP, 7 × 10, or any other, do you stop and ask what it means?

(3) Do you encourage competition among the approved purveyors?

(4) Do you listen to and honestly evaluate the story of people with whom you are not doing business?

(5) Are you smugly satisfied with your present sources?

(6) Does "no complaint" indicate that you are getting the best product for every dollar spent, the best value line?

(7) Is quality in line with the price you are paying?

(8) Is the price for the quality in line with the market?

(9) Are your suppliers reliable, "on the ball," energetic, interested, presenting new ideas and new products?

(10) Are your suppliers acting as they should in a wide open competitive market?

(11) Do you buy the same item from the same supplier each time?

(12) Is your buying in a "rut?"

(13) Do you "split the business" arbitrarily and conveniently just to keep everyone happy?

(14) Are there many identical prices from different suppliers?

(15) Could there be a collusive price arrangement among your suppliers?

(16) If you feel a change is in order, is necessary, or would be beneficial, do you bother to recommend it to the management team?

(17) Do you make occasional changes?

(18) Are you friendly with the people with whom you do business? It is almost impossible any other way, but when you get down to brass tacks, to spending company money, can you say: "Sorry, Charlie, but business is business?"

(19) Do you ask questions, make tests, listen, search, and taste, remembering all the time that the company is dedicated to satisfying the public palate and making a profit?

(20) Are inventory counts taken before purchasing?

(21) Are quantities purchased based upon carefully estimated needs?

(22) Are credits received for damaged, substandard, improperly trimmed, and returned merchandise?

Motivation

This psychological aspect of the job has to be carefully examined by both the buyer and management. Is the buyer career minded? Is he buying with other motivations? When the motivations are personal, they are usually innocently misdirected and can be corrected. When personal motivation is deliberate, they are usually intolerable to good management.

ORDERING

A few simple good habits backed up with some formalization will make a giant out of anyone doing the ordering. It is imperative to control and balance inventory. A normal stock must be determined for daily operation, for the week-end operation, and for menus that change daily or operate on a cycle. Demand can be accurately reflected by a daily sales recapitulation. The simplest of devices may be used. Recapitulation on the menu of each sales slip is one way. Keeping a daily butcher cutting report will reflect a daily average usage. Pars must be determined, and constantly revised.

In locker plants where only wholesale cuts are sold, purchasing could be arranged subsequent to the sale. A copy of the sales invoice could be issued to the purchasing agent functioning as a requisition; this makes for inventory control and for matching purchases with sales. Shrinkage can thus be minimized.

Spot Checks

Rough inventories should be taken at regular intervals estimating weights or by counting of pieces. By subtracting the current inventory from the par, the amount needed is easily determined. This can be done daily, every other day, or once or twice weekly, depending on the amount of business. There are advantages in buying daily in some instances, less frequently in others. This inventory check can be reduced to a simple form (Fig. 7.1).

Buying slightly over or under the indicated requirement might be consistent with market, minimum weight shipping orders, or looking ahead to the next order.

MEAT REQUISITION			1/11 1963
	PAR	ON HAND	NEED
STRIP LOIN	10	2	8
TENDERLOIN	20	15	5
TOP BUTTS	10	5	5
RIBS	6	1	5

FIG. 7.1. INVENTORY CHECK LIST AND REQUISITION

Perpetual inventories for each item are sometimes maintained in larger organizations where the purchasing agent is remote from the actual operation. Minimum and maximum inventories must be predetermined and periodically evaluated. While posting, cards indicating minimum inventory levels should be pulled and given to the purchasing agent as a buy sign, to bring inventories up to the maximum again. Periodic reconciliation of the cards and actual inventories should be made.

Written Purchase Orders

Copies should be provided for all departments. They should be issued to Receiving and to the bookkeepers to match delivery receipts. The buyer may require a copy. The purchase order should show exactly what is ordered, the quantities, the date set for delivery, the price, and the purveyor. Quantities, items, and the date will help the receiver. Accounting can determine the correctness of pricing and possibly prevent double billing when there has been no actual delivery. A final function is to help correlate sales with purchases.

The purchase order form should be relatively simple. It can be combined with the requisition. Sometimes a single purchase order can be used for all transactions on any given day (Fig. 7.2).

Price Suppliers

Before ordering, prices should be verified. Do not supply the price in conversation to the supplier! Many times the price will be lower than anticipated. Pricing is often promiscuous in the meat business. When

COMBINED MEAT REQUISITION AND PURCHASE ORDER				1/11 19 63	
ITEM	PAR	ON HAND	BUY	FROM	PRICE
RIBS	6	1	5	B	88
STRIP LOIN	10	2	8	a	1.20
TOP SIRLOIN	10	10	—		
TENDERLOIN	20	15	5	B	1.65

FIG. 7.2. COMBINED MEAT REQUISITION AND PURCHASE ORDER

the buyer "feeds in the price," he is frequently and foolishly putting a floor on a declining market. Where a buyer is aware of an advancing market, by feeding the price to the purveyor, he may get a short deferment on the raise. Buyer pricing is a dangerous practice at best, and a buyer not asking the price is even more foolish. In no case should the buyer permit the purveyor to price the invoice after the sale is made, or after the salesman leaves or hangs up the phone.

File Purchase Orders

Purchase orders are important to price inventories and to establish market trends. Spot checks can be made to determine which items are moving. Pilferage may be spotted by comparing purchases with sales.

Ordering

Ordering should never become burdensome and time consuming. The telephone should be used whenever possible; it enables orders to be given quickly and conveniently. If the daily routine is properly systematized, the ordering should require but a short time.

Cooperate with the Purveyors

Orders should be placed early enough to provide time to fill them properly, to do special fabrications, and to keep delivery schedules. The purveyor is profit motivated and may have overtime to pay. A profitable account gets more consideration and special effort.

RECEIVING AND STORING

Receiving is an important function which must be routine and conscientiously followed for all deliveries. There are honest purveyors, but no infallible ones, and not all of the employees of the honest purveyors are honest. It is entirely possible for an honest purveyor to have a dishonest driver. A good scale, preferably stamp-weight type, should be provided and located in an area where many people are working and not on a remote back dock. This will prevent loitering by the receiver, will encourage him to check carefully all merchandise received, and reduce the possibility of a collusive arrangement with the driver.

There are large quantities of stolen products offered at someone's back door every day. A weak receiving point accounts for most of it. It begins with a scale on a remote receiving dock and requires a single collusive arrangement between driver and receiver. Systematically a portion of the delivery does not leave the truck. The receiver signs for the load, and the driver peddles the stolen merchandise to someone looking for a bargain, who asks no questions. With a willing customer, a low overhead, and a low product cost, some very low prices can be offered with a handsome profit for the "partners."

Scales should be kept in good working condition, checked out at regular intervals by the local Department of Weights and Measures, and used for every item received. No exceptions should be made.

In addition to scales, a ruler is an important receiving device. Many products purchased are specified by length as well as weight. As little as one inch of extra flank on a high priced piece of meat, such as a strip loin, could amount to 30 cents a pound or more.

Receiving procedures should be carefully spelled out and strictly enforced. Careless and indifferent receiving is a certain invitation to dishonesty. A good receiving procedure will help keep most honest people honest.

Merchandise should also be inspected for condition and compliance with specifications. Considerable skill and training is required to accomplish this task.

A device has been developed by the Hotel Corporation of America to measure meat (Fig. 7.3). The device may be used on all measurable products. Simple instructions are engraved on the tool. Another tool called the Spec-Stick is available with a set of instructions (Fig. 7.4). This tool can be adapted to any set of specifications.

Receiving hours should be established with consideration for the purveyor's delivery problems.

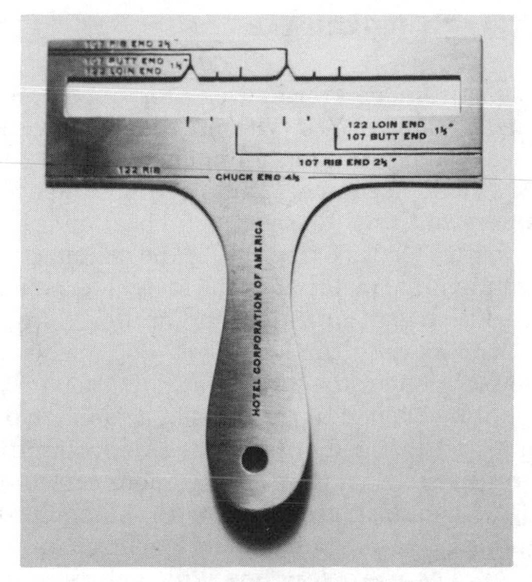

Courtesy of Hotel Corporation of America

FIG. 7.3. RECEIVING CONTROL DEVICE

FIG. 7.4. SPEC-STICK

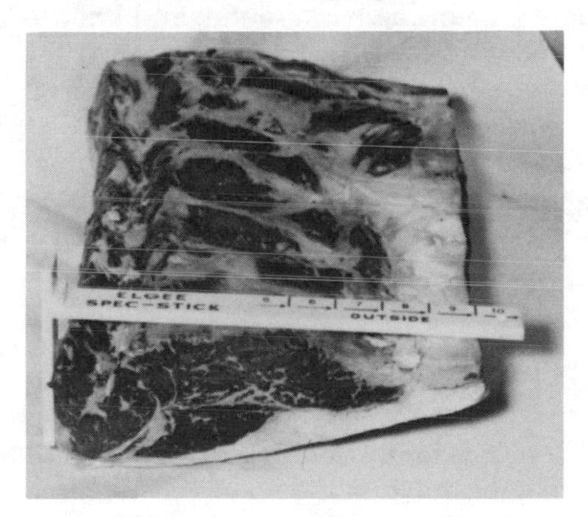

Courtesy of Elgee Meats

Closely Check Weights

Reasonable consideration has to be given for shrinkage. Fair credits should be taken either as an approved notation on the invoice, or on the purveyor's regular credit slip. Consistent credits for shortages should be carefully evaluated. If there is an invoice notation, it should be communicated to the bookkeeper. In the event that merchandise is returned, the same procedure should be followed. These transactions should be reduced to a formal procedure.

THE FORMALIZED PROGRAM

There is no substitute for a formalized program determined by a top level management team with or without the aid of various technical consultants. Help can be obtained from management consultants, accounting firms with management aims, or from skilled marketing personnel of the purveyor firms. The successful beginning of an intelligent operation starts with a plan, a blueprint for the manufacturing operation to which every individual member of the staff is responsible. The plan should be revised, changed, amended, and improved from time to time by the policy team as conditions change. The master plan becomes policy without exception for the day to day operation for all individuals involved. It might be likened to the rules of a game, to be observed by the participants, and changed only by the rules committee.

The most irresponsible act that management can commit is to turn over the buying without instruction and with complete authority. With but a few exceptions, this must ultimately lead to chaos, and sometimes financial disaster. There are very few skilled chefs, stewards, or purchasing agents, who are qualified to direct a complete buying program on the policy level. Too few people combine knowledge of product, adequate accounting background, and knowledge of costs, along with the skill required for their particular job.

Visualize the back of the restaurant as a production line of buying, preparation, cooking, and serving. Visualize the butcher shop or locker plant as a manufacturing business. Buying is the first step upon which the entire program rests. Only an honest evaluation, and constant reevaluation, and the poking and prodding of management insures that the first step is moving in the right direction.

THE PURVEYOR

There are many competent firms in the meat business. There is considerable top quality product available, enough so that any purveyor seeking quality and willing to pay the prevailing price can get it. Conversely, any firm holding their product out as the only quality product on the market is basically dishonest—this is generally a high pressure sales pitch. For example, examine the firm that asserts they have only "high Choice beef." If it is really so high in the grade, why did the graders not grade the beef USDA Prime? Adequate suppliers are available regardless of quality or area. Almost any area is within shipping distance by refrigerated truck.

A good buying plan for most organizations requires more than one supplier—two, three, or four should be used relative to particular requirements. More than one supplier encourages competition—there is nothing healthier for the purchasing organization. Supervised competition can only lead to better prices, better quality, more attention from the supplier, and a greater exchange of ideas. The maximum numbers of suppliers should be set relative to the size of the operation and the amount of time that is planned for purchasing, receiving, and supervising.

There are cases where operators buy exclusively from one firm, sometimes the beef of but one packing house. This policy is rationalized in the following manner: "Buying limited to the product of a single packer (or a single jobber) means a consistent supply.

There are two sides to this story. In either case, the buyer must be professional. A case can be built for buying from a single firm, which should foster better quality control. There is a price problem to contend with.

On the other hand, buying from a single firm might be described as putting "the cart before the horse"—at a high price. The real problem is one of knowledge or lack of the same. The buyer, often professing great product knowledge, yet demonstrating little, depends upon an inspection number, a brand, or a jobber. A little knowledge goes further, opens up the buying, leads to other good suppliers, and generally improves purchasing.

Selecting Purveyors

Purveyors should be carefully scrutinized. Examine the purveyor's company philosophy as well as its physical facilities and its products. Try to determine if the purveyor's business philosophy and concept of "service" is consistent with the success of its customer. The purveyor who is on the team, and not just privately motivated, will present helpful ideas, present products which will achieve maximum customer satisfaction and profits for the house. A visit to the purveyor's plant should be planned. An inspection should be made of the quality of the product and the quantities handled. Examine the quality of some of the purveyor's customers. Is the plant State inspected or Federally inspected? If not, what kind of a sanitation program is maintained? Find out if the firm is physically prepared to handle special requirements of delivery, particular cuts, aging, and grade. Good purveyors will respect the interested operator.

Clubs frequently have a special problem selecting purveyors, especially when a member is also a purveyor. The easy way out is to patronize the member. This may result in an unfair situation to the majority of non-purveyor members, especially if the design of the purveyor-member was to join the club, get the business, and make the most of the captive trade. A logical solution is to approve a nonmember purveyor as one purveyor with the understanding that a "most favored Nation"—or first refusal will be given to the member-purveyors. Some clubs have a policy of doing no business with members. This takes the pressure off the staff and discourages club memberships for trade purposes. It is asserted, that in general, costs are lower and quality is better.

Information from the purveyor should not be overlooked as a management tool. A careful evaluation will lead to dependable firms for helpful and accurate information. The purveyor can appraise the market trends, offer options to buy frozen products on future contracts, present "long or distressed items." The last reference is sometimes a sales gimmick or emotional sales device. Buyer beware! Purveyors have a wealth of knowledge about their product and its use and many of them are prepared to engineer and help improve the operation.

Services.—Many purveyors offer services which will fit into varying operations. Preparation of merchandise at the purveyor level is done in one of two ways. For example, the buying specifications may call for a bone-in strip loin with the specific request that the purveyor bone it out—a good service, but often a dangerous practice. The alternative is to buy a boneless strip loin of a measured length. Because there is a relatively fixed yield relationship from bone-in to boneless, a reciprocal price relationship can be established.

For example:

10 in. bone-in strip loin	20 lbs @ $1.16 = $23.20
IMPS item No. 180 (58%)	11.6 lbs @ $2.00 = $23.20

If the purchasing unit buys a bone-in strip loin and asks the purveyor to bone it out, he is asking the purveyor to destroy the trim measurement. This allows the purveyor to cut a long strip loin. Another pound of fat and bone at strip loin prices makes for a handsome profit for the purveyor. It should be an inviolate rule that all products be delivered as purchased. If fabrication is absolutely necessary, percentage yield checks and reconstruction of the purchased product should be considered.

This might be a typical situation: an order is placed for a beef round, "bone and send, saw bones." Ten extra pounds of bones, sawed with the other bones so that identity is lost, are added to the order for the sake of this example.

80 lbs round @	$0.80 = $64.00	
10 lbs bones @	$0.05 = .50	
Value of product	$64.50	
Billing:		
90 lbs round @	$0.80 = $72.00	
Buyer loses	7.50	
Quoted price	0.80	per lb
Actual cost	0.90	per lb

The solution is to pay the reciprocal price for the boneless round and buy the bones separately, or buy the round "bone and send," checking percentage yield, and reconstructing the bones.

Price

It is not always true that "you get what you pay for." Pricing by meat jobbers frequently is based on what "the traffic will bear." Sometimes prices are set with an ear to the ground for what the competition is doing. Frequently the product cost and market cycles are overlooked in pricing. Some operations are burdened by inefficiency and hidden rebate costs that have to be interpreted into prices.

It behooves the operator to look deeply into the business philosophy of the supplier, as well as the efficiency of the plant. Is the high priced house selling an exceptional product? Or is his operation inefficient? Does he give "kickbacks?" Is the cheap seller selling a cheap product, or is he just competitive? Is the cheap seller confused with the competitive seller?

Buyer types best illustrate the various attitudes that can be taken towards price. There are, generally speaking, four kinds of buyers: (1) quality at any price; (2) quality dominating price; (3) competitive price buying; and (4) bargain hunting.

When an operator insists on quality at any price, he is almost issuing an invitation to disaster. This buyer might remark, "I want the best, and I don't care what it costs." He frequently means it and one of two things happen, either he eventually goes bankrupt or he continues to operate marginally. Anyone operating on such a premise, making no attempt to relate the quality and cost to the menu price, anyone so indifferent to the reasons for being in business, deserves no better fate than to make some purveyor rich. These operators can boast of what a great friend their meat man is—why not?

Quality dominating price operators look beyond price to do an honest job. Where special programs, such as protracted aging or very careful selection or special qualities are required, the buyer emphasizes his special needs. Where the buyer expects special product consideration, he has to provide a constant market to permit the purveyor to develop the program. Fair treatment in price is expected for his loyalty. This buyer opens the door to temptation for the purveyor, whose price practices have to be spot-checked.

Competitive buying requires a program and specifications. It is not unusual to use two to five purveyors, depending on the volume of the buying unit. The price level that is sought and the number of purveyors used to achieve it should be directly proportional to the knowledge of management, and the proper allocation of time for buying, especially for supervision of receiving. When quality is interpolated into the price, the lowest price is not the whole answer. Frequently this type of operator resolves the price situation by approving a given number of purveyors who establish themselves for service and a minimum or better standard. From among the approved purveyors, the low price determines the weekly source for each item.

Some operators are bargain hunters. The door is open for "lowball" prices from anyone. Cheap buyers attract sellers who specialize in products of low quality. The quality purveyors, doing an honest job, pay premiums for their product. These purveyors need a better selling price to offset the higher cost of the product. It must be concluded that price purveyors generally are buying product of marginal quality for the price trade. This often does the job for the buyer, but he should be cognizant of what he is doing.

Buying price frequently opens the door for sharp practices, devices employed to offset cheap invoice prices. Price buyers will attract some purveyors who will offer them prices coupled with short weight, bad trim, low quality, improper aging, substitution of quality, and high prices on items not carefully checked.

The lowest end product cost and the lowest price are frequently not the same thing. Only by objective testing and evaluation can the true cost be determined. In buying meat, price is but a single factor of the end product cost. The lowest price can be presumed to be the best value only when all other things are equal.

Prudent buying will deliver lower prices. These savings could easily be offset by the additional expenses of a constant stream of new purveyors, of very careful receiving, exchanging merchandise, and finally hiring a professional buyer. At a certain point, the expenses incidental to this "buying plan" can exceed the savings. This is the point of diminishing returns, which will vary with and have to be determined for each operation. For example, a fulltime buyer paid 1200 dollars per month plus fringe expenses of perhaps 25 per cent, has to produce a considerable saving in an average operation to just offset wages. There are other considerations too: type of purveyor attracted, the quality standards, and finally the corners that are cut. There is a place for the professional buyer, but too often such a person, with too little knowledge, fails to rise above merely buying price.

"The market" at the level of the food service industry is very nebulous—there are several market services offered to the jobber, some to the buyer. For the meat dealer, the "Yellow Sheet" published by the National Provisioner in Chicago is a daily accurate service. Various U.S. Department of Agriculture reports are distributed on request, without charge. A reliable jobber's weekly price list will reflect the market changes. The buyer's range, relative to the quoted list, will help determine a fair price. Meat prices at wholesale fluctuate daily. Sensitivity to this has to be developed to get "the market," and to realize the savings that should be expressed in lower prices during the market dips. It is not uncommon for a purveyor to freeze a price and take the long profit during the market dips.

Many products exhibit regular price fluctuations, demonstrating rises and dips with seasons and production. Figure 7.5 represents the USDA Choice beef loin market at packer's selling price in Chicago. The cycle may be generalized: two lows occur, one in November, the other in February. The market rises slowly to a summer peak, drops off sharply in October and November, and bounces back for Christmas and New Year's.

Quoting unrealistically low prices to get a new customer is not an uncommon practice among purveyors. This creates an embarrassing situation for the "in purveyors" and sometimes makes the buyer feel foolish. The temptation is often too great to resist. The buyer adds a

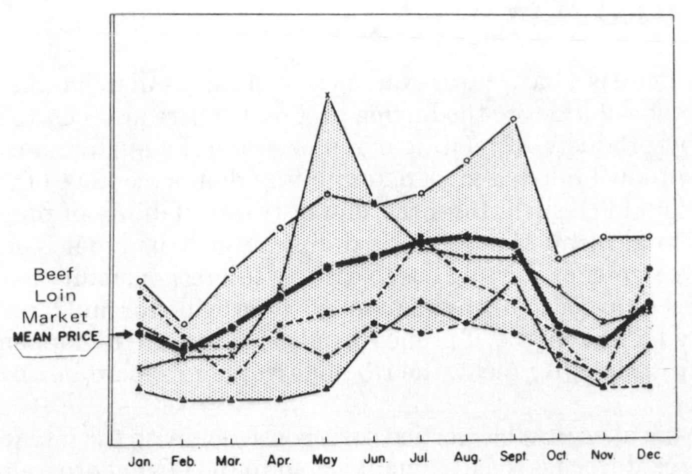

FIG. 7.5. BEEF LOIN MARKET (CHICAGO) REFLECTING A CLASSICAL CYCLE OF FLUCTUATION

new purveyor with the right price. The "low-ball" seller then proceeds to raise prices to the current market. If the program is successful for the purveyor, he has bought a part of the business with a few cheap prices for a short time.

"Name Your Own Price" offers are dangerous. This is a form of "low-balling," wrecking legitimate competition, or offering a "get in" price. If this offer reflects the price policy of the firm, the buyer had better not turn his back, lest the purveyor gain the same prerogative, naming his own price at a later date.

Gathering prices should be the periodic responsibility of the person doing the buying. The optimum situation would be to get firm written weekly quotes from the approved purveyors, to list these on a single comparative sheet, and to buy the best price, quality considered. Written bids are better than oral bids—they are far less time consuming, should be available immediately at the beginning of the business week, and are fair to all bidders, if each has to submit without knowledge of the competitor's prices. Using the successful low bidder on each item will encourage competition. Unfair to the good purveyors is the buyer who has a pet purveyor, lets him look at the prices of competition, and then lets him meet or beat the price on any item. The buyer is letting one firm set the low market, but giving the order to another. This could destroy good competition. On the other hand, it is not uncommon that on the request of a purveyor, the successful prices be disclosed. This encourages competition, and the next week may bring in a new low bidder.

THE MENU PLAN

The menu is a basic portrayal of the restaurant merchandising plan with responsibilities to the buying program. There are the problems of using by-products, of relating menu prices to the quality and portion cost of product purchased, of determining what percentage of the sales' dollar should be spent for meat, and of the availability of products.

The total menu plan includes dining service, banquet menus, and employees' menus. Buying has to assume full responsibility to produce all items required. The menu plan has to provide for any by-products produced. For example, if prime rib of beef is used, what happens to the short ribs, the lifter meat, the rib bones? There should be an over-all plan.

Arriving at acceptable portion costs consistent with the intended food cost percentage, has to be communicated to the buying program, which ultimately reflects itself in the specifications. This is a relatively complicated scheme, as the food cost is a combination of multiple components, frequently having a wide range of mark-ups.

Day to day recapitulation of sales, by item, is often overlooked. Frequently personal prejudice enters the menu plan, reflecting subjective evaluations of the various members of the management team. The ultimate test of the menu should be that which the customer buys from day to day. A near fatal error is committed where cuts are purchased with by-products that are then forced into the menu. These by-product items are frequently "sluffed off" on the menu and sold at distress prices, downgrading the entire operation. This type of thinking frequently contributes to relatively high percentages for both labor and overhead.

Small institutions, such as small hospitals, boarding schools, college fraternities, and other relatively small feeding units have a food problem that seems insurmountable. The usual approach is an attempt to "buy price" as a cure all. The real purchasing power of the unit is relatively small, and price buying invites sharp price practices. It is not uncommon to end with a poor food cost and a bad product.

The only apparent solution is an over-all program, stemming from the premise that the purchasing power itself is an inadequate approach. A simple program can be evolved: (1) get reliable purveyors, and pay a fair price; (2) control costs by practicing careful portion control and eliminating waste; (3) set up a menu projection for several weeks striving for an average daily meat portion cost; check by dividing total cost of the cycle by the total number of days. The result represents the average daily cost.

A working knowledge of portion sizes and finished portion costs is needed. Almost any intended average cost can be achieved by the proper combination of items for the cycle.

The cycle plan can be concealed by altering the sequence of the weeks, or by using an odd number of weeks. Seasonal or low priced items should be substituted for variety and cost advantages.

Control of portions and waste is aided through buying. The daily purchases have to be carefully calculated in terms of daily census or portions required. In most small institutions, where one or two entrées are offered, overbuying is almost a complete loss. Buying requirements can be easily translated to raw weights for meat. The number of portions required, divided by the servings per pound, equals the number of pounds required. For example, if a meat loaf portion raw weight is four per pound, and the census is 150, the meat required is about 38 lbs.

The reliable purveyor plays an important role by filling the order according to specifications and as close to the required weight as is practical. Prudent buying, following the concept outlined, can achieve

A SMALL INSTITUTION
MEAT REQUISITION AND PURCHASE ORDER 1/25 1963

MEAL PLAN				ORDER		
DAY	ITEM	EST CENSUS	SERVINGS PER LB RAW	BUY	FROM	PRICE
1/26	Meat Loaf	150	4/1	38#	B.Co.	.49
1/27	Lamb Leg	180	15/leg	12pcs	B.Co.	.69
1/28	Beef Liver	120	4/1	30#	B.Co.	.55

FIG. 7.6. A SMALL INSTITUTION MEAT REQUISITION AND PURCHASE ORDER

remarkable results. Figure 7.6 is one approach to formalizing this program.

A FORMAL RESTAURANT BUYING PROGRAM

The food cost of any operation is actually predetermined by the menu prices, the quality specified, and the size of the portions. The kitchen should be responsible for making the program work. The role of the kitchen is to conserve the program, to control waste, to practice portion control, to manufacture products according to specification, and to put the total policy into effect.

Steps for a Formal Program

(1) Plan menu and revise as needed
(2) Write specifications
(3) Provide accounting controls, projections, unit sales analysis, current profit and loss information
(4) Provide forms for purchasing, receiving, butchering, yield and cost tests, perpetual inventories
(5) Formalize a kitchen conservation program:
 (a) For by-product
 (b) To control waste
 (c) For employee meals
 (d) With reports
  (1) Cost and yield tests
  (2) Butcher reports
  (3) Products made, sold, leftovers
 (e) Control shrink
  (1) Inventory controls
  (2) Pilferage control
  (3) Proper product handling

(f) Improve butchering techniques
(g) Provide accounting reports
  (1) Breakdown purchases by classes
  (2) Reflect purchases by classes as a percentage of sales, percentage of food cost; prepare on a comparative basis
  (3) Recapitulation of items sold daily
(h) Provide a kitchen manual of rules and procedures

CARE OF PRODUCT

A frozen product should be received in temperature, and placed in the freezer immediately. First in, first out should be an absolute rule. Currently received items should be placed under or behind products already in inventory. Dating or lot numbering may help to insure rotations where large quantities are handled. Maximum optimum storage periods, as well as packers' special instructions, should be observed. Any product that is to be frozen should be wrapped.

Fresh products should not be wrapped except in Cryovac or similar packaging or loosely covered, dated and lot numbered for rotation control. It should be placed, not piled, in the cooler, to be withdrawn in proper sequence. The cooler should be fresh. Regular, periodic checks of temperature and humidity should be made. Cryovac bags should not be punctured. Cured meats should be kept wrapped. Fresh steaks should be plattered or wrapped.

Cooked products should be loosely covered.

Considerable savings can be achieved by using refrigerators for multiple purposes. Because the restaurant cooler is generally intended for very short holding periods, it is possible to store different kinds of products together in spite of different optimum storage conditions. As an example, for a short period, fresh fruits and vegetables can be stored in the meat box with no serious effects. Other items to be considered are fish, poultry, dairy, cheese, vegetables, fruits, prepared foods, and sauces.

Fresh fish can be successfully stored in the same cooler as meat with excellent results, if placed in a chest of ice which may be connected in some fashion to empty into the cooler floor drain.

When a number of items are stored in a cooler, it is important to evaluate compatibility of items, not only in their relation to meat, but to each other. For example, what is the odor transfer problem of butter and onions?

Tagging.—It is important to date and rotate products. This can be controlled by a date on the wrapper or a simple date tag. A well developed functional set of specifications provides all other information with reasonable accuracy.

REFERENCES

ANON. 1954. Hospital Food Service Manual. Am. Hospital Assn., Chicago, Ill.

ANON. 1960. Meat Manual. 6th Edition. National Live Stock and Meat Board, Chicago, Ill.

ANON. 1962. Food buyers guide. Food Publications Inc., Los Angeles, Calif.

BRODNER, J., MASCHAL, H. T., and CARLSON, H. M. 1962. Profitable Food and Beverage Operation. 4th Edition. Ahrens Publishing Co., New York.

DUKAS, P., and LUNDBERG, D. E. 1960. How to Operate a Restaurant. Ahrens Publishing Co., New York.

FROOMAN, A. A. 1953. Five Steps to Effective Institutional Food Buying. 2nd Edition. Institutions Publications, Chicago, Ill.

GREEN, E. F., DRAKE, G. G., and SWEENEY, F. J., 1978. Profitable Food and Beverage Management: Planning. Hayden., Rochelle Park, N.J.

HARWELL, E. M., ANDERSON, D. L., SHAFFER, P. E., and KNOWLES, R. H. 1953. Receiving, blocking, and cutting meats in retail food stores. Marketing Research Rept. *41,* U.S. Dept. Agr., Washington, D.C.

KEISTER, D. C. 1957. How to Increase Profits with Portion Control. Ahrens Publishing Co., New York.

KNIGHT, G. E. 1956. Quantity food purchasing. Circ. *R-500,* Mich. Agr. Expt. Sta., Michigan State Univ., East Lansing, Mich.

LIFQUIST, R. C., and TATE, E. B. 1951. Planning food for institutions. Agr. Handbook *16,* Agr. Research Administration, Washington, D.C.

LUKOWSKI, R. F. 1960. Receiving practices in food service establishments. A circular, Coop. Exten. Service, Univ. of Mass., Amherst, Mass.

MACFARLANE, A. M. 1958. Key to success in food purchasing. Institutions Magazine *42,* No. 1, 81.

STOKES, J. W. 1960. Food Service in Industry and Institutions. Wm. C. Brown Co., Inc., Dubuque, Iowa.

WANDERSTOCK, J. J. 1960. Meat purchasing. Cornell Hotel and Restaurant Administration Quarterly *1,* No. 1, 25–29.

WRIGHT, C. E. 1962. Food Buying. Macmillan Co., New York.

8

Specifications

Specifications should be written in such a way as to prevent inadvertent policy changes and to communicate accurately with purveyors. Two general forms are suggested, one which can be presented as an over-all columnar work sheet, or a second as an individual specification sheet for each cut. On the work sheet, columns may be set up for each pertinent specification as cut, grade, weight, age, pack, and trim. This is a simple and effective form. A specification sheet for each cut is illustrated in Fig. 8.1. Individual sheets lend themselves to interim changes with a minimum of retyping. They can be exact and comprehensive. Kitchen plans, including cutting, cooking, use of cuts, garnish, and possibly a color photograph of the finished product can be appended to the individual specification.

Formalized specifications, which require considerable research and development, are the cornerstone of a business-like operation. First of all, in order to be reduced to writing, they must be carefully thought out. Then, they should be submitted to the constructive criticism of the entire staff. Consideration should be given to the idea of calling in key personnel to support the project, to counter dissenters and personal prejudices, to assemble the best ideas from the total experience of the staff, and to coordinate the suggestions of the sales or menu department and the production or kitchen departments. The chef, purchasing agent, butcher, steward, receiving clerk, manager, comptroller, and maitre d'hotel may be included in a round table discussion.

Written specifications will not be forgotten or overlooked, subject to inadvertent minor changes, or aborted by changes of chefs or purchasing agents. Top management will have to police the program. The customer will benefit by a consistent product. Written specifications contribute stature to the parties responsible for receiving. Changes may occur from time to time, but only as they are presented and approved by management. Specifications are not designed to eliminate change, but to control it.

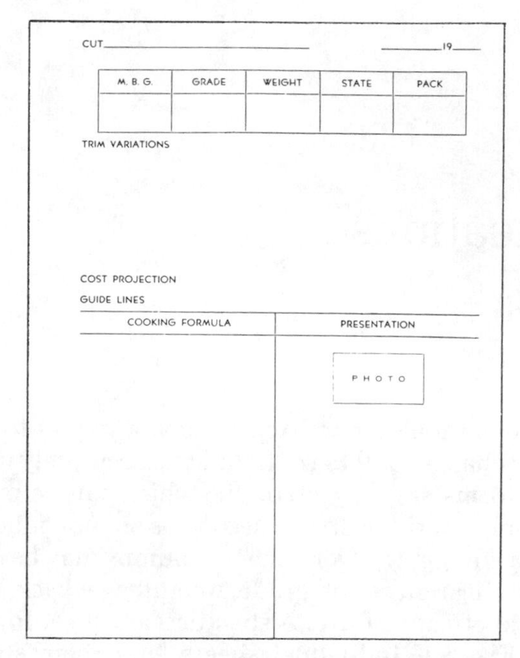

CUT________________________ ________19____

M. B. G.	GRADE	WEIGHT	STATE	PACK

TRIM VARIATIONS

COST PROJECTION

GUIDE LINES

COOKING FORMULA	PRESENTATION
	PHOTO

FIG. 8.1 TYPICAL SPECIFICATION SHEET

CUT

This is the title of the sample specification sheet. It should describe in acceptable terminology the particular meat item for which the specification is drawn, as, for example, beef loin, strip loin, or strip steak.

An important economic decision has to be made, which relates to the following: (1) a sound knowledge of the carcass breakdown and components; (2) an understanding of what is available on the market; (3) familiarity with the by-products of the cuts selected; (4) an understanding of the "average cost" price concept; and (5) for the restaurant operator, a sound approach to the practice of portion control.

A technical description of the cut is not necessary if the name of the cut is correct. Make the definition so simple that it communicates to the entire staff, as well as to the purveyors, exactly what is intended. If a "bone-in strip loin" tells the story, why complicate it?

To get maximum exposure to the range of cuts that are available, professional help should be enlisted. A meat jobber, familiar with restaurant problems, may be consulted. The possibility of a slanted point of view or a sales "pitch" is an inherent problem of such advice. Food consultants or food management accounting firms may be employed.

"Buy-Make" Decision

Determining the cut, whether it is a packing house cut, a primal cut, or a pre-cut portion, must be reduced to a "buy-make" decision. Five primary factors have to be evaluated:

(1) What are the relative costs of the cuts?

(2) What degree of quality control can be achieved? Is there adequate supervision to control production and waste?

(3) Will labor costs be incorporated in price? If so, is there still a price advantage? Will a fabricated item reduce the total labor load? Will overtime be required? Can more cuts be fabricated with no increase in labor costs?

(4) Will the by-product be used or sluffed? Does the by-product fit the operation or will it result in low priced distressed selling?

(5) Are accounting techniques advanced sufficiently to provide defined unit costs to make menu costing and planning accurate? Will portion buying improve the general pricing on the menu?

The first step, the determination of the cut, obviously requires good judgment and planning integrated with the menu and the available labor. Points one and four are most important for the retail butcher.

Replacement cost of the by-product should be determined to make a sound arithmetical comparison between different cuts. The yields of the primary cuts must be known as well. "Guesstimation" is inadequate. To demonstrate the many variables involved, compare a packing house style rib (IMPS 103) with a roast ready rib (IMPS 109). Assume yields to be roast ready rib 70, short ribs 20, and bones 10 per cent.

IMPS	103	7 × 10 Ribs	@ $1.10/lb vs.
IMPS	109	R.R. Ribs	@ $1.70/lb

7 × 10 Ribs—100 lbs @ $1.10 = $110.00

Yielding			
R.R. Ribs	60 lbs	@ $1.71	$102.60
Short Ribs	20 lbs	@ .45	9.00
			$111.60

By examining the hypothetical illustration, it can be stated in this particular case:

(1) The 7 × 10 rib is a better buy, but by only $1.60 per cwt.

(2) Additional labor must be applied to the 7 × 10 rib as compared with the R.R. rib, and if labor cost is added to the cost of the raw material, in this case the 7 × 10 ribs are not a better buy.

(3) If the short ribs are not used, it would be even more economical to use the R.R. rib. Sixty pounds of R.R. ribs would cost $102.60. 7 × 10 ribs needed to yield 60 lbs of R.R. ribs would cost $110.00.

(4) An infinite number of price combinations can be set up. An objective evaluation can only be achieved with knowledge of all the percentage yields, and application of replacement cost to primary cut as well as by-product.

Restaurant Portion Control

Every operator must practice portion control to achieve any degree of success. Sound management must take the attitude that their establishment is basically a factory, where labor is applied to raw materials, and uniform finished goods, uniform as to quality and quantity, are offered to the public at a price presupposing a profit for that portion of finished goods. Quantity and quality of the finished product will vary with each operation, but the most important single objective is that they be constant, uniform, and consistent within the establishment. This is the practice of portion control and must apply through the full range of the industry from lunch wagon to the most expensive and elegant continental cuisine.

To achieve food cost control, a program of portion control must be instrumented. This may be expressed in four steps: (1) cost control (buying); (2) standardization of product (specifications); (3) standard kitchen procedures (formalization); and (4) merchandising plan (menu).

Reduce this to a formal, communicable plan. Add intelligent supervision, willing personnel and a new establishment has a good chance of escaping the high food service industry casualty list.

Portion cut meat is a natural evolution of the meat disassembly line. It has been a short history of change from hindquarter, to loin, to boneless strip loin, to strip loin steak. In this era of specialization, the question of how much sub-assembly should be purchased has to be answered. What are the economic considerations? What are the quality considerations? Though the practice of portion control is a must, the purchase of portion cuts is optional and should be carefully evaluated in terms of the particular operation. Portion control can be achieved with either prime cuts or pre-cuts.

Five questions must be answered to determine overall policy.

(1) **Will labor cost be reduced?** Will there be less work for the present staff? Will a shift be reduced? Can related portion prepared food items be purchased to develop an over-all program which will reduce total labor force?

(2) **How is the by-product being used?** Is it evaluated at replacement cost and producing its share of income? Is it integrated in the menu, the employees' meals? Does it create unpopular items that end up as wasted left-overs? Can better, more economical functions be created? By-products treated with care and imagination can conceivably become the most profitable items used.

(3) **Are most products and portions uniform?** How much supervision is required? Do kitchen practices conform to the plan? Is there any check on portions? Is the entire program of cutting subjective, the "guesstimation" of a single person in the kitchen? Have real cost tests been made? Are tests current? Are periodic tests conducted, recorded, and filed?

(4) **Is food cost being controlled?** Is there any control of pilferage? Of waste? Are preplanned yields being achieved? Are sales slips recapitulated? Can sales be related to purchases? Are they? Has a perpetual inventory of finished goods been set up to control cutting and to relate purchases to sales?

(5) **What does the customer really want to eat?** Is he attracted by the by-product? Does he want low priced by-product entrees? Would customer satisfaction and sales go up if the menu plan more closely resembled the customer wants? Does merchandising of the by-product at attractively low prices create increased sales?

Buying portion control offers management part of the answers. In our competitive economy there is little room for error. Poor estimation of portions could spell the difference between profit and loss. Buying portions makes for a skillful and simple relationship of cost and selling price.

The buyer's major price problem is to make the transition cents-per-pound from raw material to a finished product which includes labor. To achieve this stature, it is necessary for the buyer to orient himself in terms of unit cost per portion and relate this to menu selling price and food cost. The cost per pound of portion cuts is awesome at first. In fact, a simple exposure to pre-cut items, should create a new awareness of cost.

Whereas pre-cut items offered currently generally put too little stress on quality, they do assure a consistency at a given minimum quality level. The quality available has to be evaluated and applied to the individual operation. By using pre-cut items, control can be improved for the following: (1) quality; (2) portion sizes; (3) unit cost; (4) cost-selling price relationship; (5) waste; (6) by-product; (7) inventory; (8) pilferage; (9) shortages; (10) employees' indifference and inefficiency; (11) standards inadvertently changing; (12) specialized menu plan variety; (13) slow moving items; and (14) irresponsible purchasing and receiving.

The average buyer concludes that cost is the single biggest argument against pre-cut. Careful testing and evaluation would prove this conclusively incorrect. The biggest problem facing the meat industry today in the field of pre-cutting is the generally low quality standard, leaving much to be desired for many operators.

Ready-to-cut, as opposed to pre-cuts, is a compromise plan attempting to achieve a maximum quality control with most of the pre-cut advantages. Examples of ready-to-use or ready-to-cook items are the roast-ready rib and boneless short cut strip loin, items 109 and 180 of the U. S. Department of Agriculture acceptance specifications. Many other items are offered by various jobbers.

The ready cut program offers the elimination of skilled butchering, good portion control foundation, easy to determine portion costs, simple inventory and receiving procedures, and the elimination of by-products.

Retail Butcher.—The buy-make decision requires a searching examination. Almost every shop in any given neighborhood finds it easy to merchandise some cuts, difficult and almost impossible to move others. Where multiple units are involved, they will sometimes complement each other. The independent operator generally finds himself with a difficult "buy-make" decision.

By buying carcasses instead of cuts there are sometimes savings from one to two cents per pound. In some areas, a better quality may be achieved through cuts, and sometimes through carcasses. In spite of price and quality, a primary consideration should be given to the cuts hard to merchandise. They may require so much time, promotion, and distress pricing as to create a loss in excess of the gain from carcass buying. A dedication to carcass buying sometimes leads to a general lowering of gross profit margins and net operating losses.

In a "carriage trade market" it does not appear economical to buy carcasses which yield 25 per cent desirable rib and loin cuts and 75 per cent rough cuts. On the other hand in a more competitive neighborhood rib and loin cuts sometimes have to be sold actually at or below replacement cost.

In general, it can be concluded that the retail butcher is better off economically buying and selling the cuts best adapted to the particular neighborhood. A trial and error approach will generally demonstrate this. Whereas the total sales volume may be reduced, the controlled volume probably will lead to greater profits.

Locker plant.—Where only primal cuts are offered to the patrons the decision is simple: buy the cuts sold. Where a retail service department is combined with the locker plant, the "buy-make" decision is as complicated as it is for the retail butcher. To avoid an inadvertent subsidy of retail sales by the locker operation, department accounting of sales and purchases should be maintained.

"AVERAGE COST" PRICE CONCEPT

There is little knowledge generally of the relationship between the cost of a primal cut and the primary cuts produced from it. Consequently, there is a great fear of making any changes. The same "argument" is frequently used by buyers supporting exactly opposite positions.

Buyer A uses 7 × 10 ribs. Jobber suggests that short ribs end in the freezer and are finally discarded. The jobber points out that he makes good use of the short ribs and "credits" the selling price of a roast ready rib for this. Thus the roast ready rib would be more economical to use. The buyer's standard reply is: "If I now pay $1.10/lb for 7 × 10 ribs, how can you suggest I pay $1.71/lb for the same thing?"

Buyer B uses roast ready ribs. Jobber suggests plan of merchandising short ribs in the restaurant. Jobber points out that buyer will enjoy a lower cost, and can make use of short ribs. Buyer's standard reply: "If I now pay $1.71/lb for an R.R. rib, how can you suggest that I pay you $1.10/lb for short ribs, fat, and bones?"

Both buyers ignore the "average cost" price concept. Every piece of meat is made up of components of different value. The price quoted is nothing more than an "average price." This price reflects the value and percentage yield of each component. The name of the primal cut is only a generalization, a name given to include a group of items in the same carcass proximity. The name is a misnomer. For example, a strip loin bone-in, intended for strip steaks is made up of fat, bone, backstrap, flank trim and boneless strip loin. To make the transition in cost from bone-in strip loin to boneless strip loin, it is necessary to relate the two items price-wise and eliminate the guesswork. Only testing can lead to an intelligent decision on the question of "what cut." (See Buy-Make Decision in Chap. 9.)

GRADE

The food service industry and the retail trade overwhelmingly indicated preference for quality based on federal grading, especially for beef and lamb.

Federal Grades Are Reliable

The grading service, although not perfect, is the most objective and the best available. Constant efforts are being made to improve this system. This approach excludes brands and private grading, and excludes the inspection indices which are a certification of wholesomeness, not a quality grading device.

The grade selected should be specifically related to each item. There is no rule that each item should be of the same grade. For example, it may be the decision of the operator to use USDA Choice steak meat, and boneless USDA Good chucks for ground beef. Different grades for different items are consistent.

The proliferation of steak operations, cutting USDA Choice cuts, lead to a modification of the grading service which permits an operator to identify the grade of the product on the box in which it is packed. If USDA Choice tenderloins are cut into steaks, the jobber may label the box accordingly. In the case of tenderloins, which are interior muscles, and are not grade identified with a stamp, a grading certificate must accompany them in order to continue the grade. In the case of exterior muscles like a top butt or strip loin, the grade on the meat itself is sufficient to permit the identification of the grade on the box.

Ungraded products and private brands pose a significant problem. This includes those beef items produced from internal muscles which are not "rolled" with a grade, and portion cut items with no grade identity. The grade on the invoice is no better than the integrity of the seller.

All unidentified products must be associated with the jobber producing it, must be purchased by brand, not grade. Unidentified product will have to be tested and evaluated on the basis of a private brand. If the "brand" qualifies quality-wise it then can be compared with other approved brands costwise. One fact cannot be avoided—where the product is unidentified, it must be tested. Quality of ungraded items, as represented, and as they exist in fact, may be two vastly different entities. Invoicing a tenderloin as "USDA Choice" is no guarantee of quality beyond the integrity of the company. The tenderloin is not rolled or identified. Let the buyer beware!

Sometimes grades are invented, such as "fancy" or "AA." There are no such Federal meat grades. Specifications could take the place of a grade. Instead of "fancy" sweetbreads, it would be more accurate to describe them as veal sweetbreads, 12 oz and up, in pairs, and individually wrapped.

Private Grades.—The independent retail butcher and locker plant operator can sometimes use a private beef brand profitably. It is important that the packer producing the brand provide constant quality control. Frequently a small cost premium is associated with the private grade, if produced by an independent packer. This is no quality assurance, but this type of marketing effectively limits the number of outlets using the product. If the retail patrons are willing to pay a small premium, then an exclusive and profitable trade may be established.

Selling Prices.—This should determine the grade specified. It should be no higher than the patron is willing to pay, no higher than the patron can afford. The quality, on the other hand, should be as high as that implied by the price and language of the operation. The quality must be consistent with a good value for the customer. It behooves the operator to give the patron good value, yet to act consistent with achieving a profit. A simple statement of policy is to buy for the patron's pleasure and pocketbook, yet buy a quality consistent with a profit for the operator.

WEIGHT

Portion control begins with the specification of size for many items. This is especially true when a steaking or carving plan has been determined to achieve a portion that is satisfactory both for appearance and cost.

Steaking Plan

Assume that strip steaks for the sake of appearance as well as satisfactory preparation have to be a minimum of 1¼ in. thick (the figure is arbitrary). Strip loins within a five pound range, on the average, will yield more or less the same number of steaks of any given thickness. From a 10 lb, strip loin and a 13 lb strip loin assume a yield of 10 steaks, 1¼ in. thick. Both strip loins cost $2.00 per lb.

Strip Loin Cost	*Yield*	*Portion Cost*
10 lb strip loin @ $2.00 = $20.00	10 steaks	= $2.00 each
13 lb strip loin @ $2.00 = $26.00	10 steaks	= $2.60 each

Of course, the ten steaks from the 13 lb strip loin have a larger eye and weigh more. The primary points to be settled are how thick, and at what cost relative to the menu. In the restaurant, the weight of the steak is just a by-product. The correct strip loin size specification for any given menu price can be arrived at experimentally. In the retail meat operation a smaller cut makes for a lower take-a-way price, the usual concern of the housewife.

Carving Plan

Prime rib portions are usually cut relative to the rib bones in terms of one, two, or three servings to the bone. Servings per bone is a must to achieve portion control, to regulate thickness and size. Assume two roast ready ribs of 17 and 21 lbs, both at $1.80 per pound, with a 14-portion or two servings per rib carving plan.

Rib Cost	*Yield*	*Portion Cost*
17 lb rib @ $1.80 = $30.60	14 portions	= $2.19 each
21 lb rib @ $1.80 = $37.80	14 portions	= $2.70 each

Thickness should be the primary appearance qualification. The cost per serving as related to the menu price is the primary economic consideration. The weight of the portion, cooked or raw, is merely a by-product.

Yield is a correlative of weight in many instances. For example, a 40 lb rib will yield better percentage-wise than a 30 lb rib when making a Spencer eye. The lean of a bone-in strip loin will generally yield higher as the strip loin gets larger. Yield, sometimes will have little of anything to do with size, as in the case of boneless chucks used for grinding, top rounds for roasting, etc. Frequently a high yield will have to be evaluated in terms of other factors. Using the higher yielding size may actually result in a higher restaurant unit cost. This might be true in the case of a strip loin, where the advantage of a high yield is offset by a high portion cost. On the other hand, the retail butcher, selling weight and net portions, is benefited by a higher yield.

Quality is sometimes a correlative of weight. Young hogs of premium quality usually fall in given weight ranges. For example, sheet spare ribs, three pounds and down and 8 to 12 pork loins are premium items. Given pork weight ranges relate product to youth, to the qualities of lean, to tenderness and flavor.

Trim is frequently related to weight. For example, an 18 lb beef rib has probably been oven prepared in some manner. A 32 lb rib is probably 7 × 10. A rib invoiced as oven-prepared, weighing 32 lbs might indicate improper trim. Usually a heavy lamb rack is associated with a desirable large eye. A lamb rack "can be made" heavy by poor trim, by leaving extra shoulder chops on the rack. It behooves the buyer to correlate weight with trim.

AGE

The specification for age should generally cover the post-mortem period during which the product is to be held in a cooler. There is no single rule that will be all-inclusive to cover every single product. A portion of the product may be specified with considerable age on it, some fresh, some even fresh frozen, depending on the particular product, and the use for which it is intended.

The palatability score of beef cuts to be roasted or broiled will improve with age. High quality beef tenderloins may be eliminated from any aging program as they are intrinsically very tender. Other beef cuts, used for low priced entrées, generally prepared by braising, stewing, or grinding should not be aged. Surface bacterial growth will cause undesirable flavor changes and necessary surface trimming will be uneconomical. High quality lamb, pork, and veal cuts are generally not aged.

What age is, what the benefits and the negative aspects are, should be determined before the specification for age is written. Only a few people are aware of how age manifests itself. The operator should know what 1, 2, 3, or 4 weeks age looks like. After the specification is determined, it should be policed. It is easy for the purveyor to deviate.

Age will manifest itself first, on the rib end of the strip loin, and the loin end of the rib. This is where the forequarter and hindquarter are first separated. Next, it will show up on the ball tip and top sirloin on the loin, and the chuck end of the rib. Exposed surfaces will dry, discolor, and finally the growth of molds, bacteria, and yeasts will be visible. The fat will ultimately show signs of breaking down.

The changes will vary with the temperature at which the product is stored, and the conditions of humidity, sanitation, and air flow that prevail. An aging program requires extreme cooperation between the purveyor and the user, with some system of tagging to keep a record of the age. Occasional visits to the purveyor's cooler will determine if the program is being conscientiously followed. On the other hand, the user must be a consistent customer to keep the program moving.

Aging is a very expensive process. During the aging period, considerable shrink occurs, storage space and refrigeration is required, money is tied up in inventory, and finally a portion of the meat is lost in the final trimming. There is no question that a reasonable aging period increases the palatability of the product. The operator must reconcile the economic losses with the palatability gains. It is a question of how much quality can be afforded. Sometimes a compromise of some kind must be made in order to keep the cost of the aged product consistent with the menu selling price. Above all, the operator must be able to judge an aged product, particularly if he is paying for it. Through a process of trial and error, it seems reasonable that age can be recognized, and the characteristics can be pointed out to the people responsible for receiving. The purveyors can be held responsible for the promises that they have made with respect to the delivery of the product aged according to specifications.

Meat, successfully aged in Cryovac, bears no physical evidence of how long it has been in the bag. It is very tempting and often the practice of salesmen, and sometimes the purveyors they represent, to "give" the user any age requested. Product "aged three weeks" or more may be offered in any quantity. At the same time, the seller may be working from an "empty wagon" selling fresh product. The buyer has no way to evaluate accurately the age of the product in Cryovac.

The responsible buyer using Cryovac has some additional responsibilities. The product must be periodically taste-tested. Visits should be made to the purveyor's plant to evaluate aging facilities, aging programs, and inventories progressively aged. When protracted aging is held important, the buyer should develop his own program with the purveyor.

FROZEN MEAT PRODUCTS

Frozen meat products are finding an ever expanding market in the institutional trade. The finest restaurants, where there is no compromise of quality, find use for frozen products. For example, frozen veal sweetbreads are very popular. There still exists, however, a general negative attitude toward frozen products. Though this is not intended as an argument for a change to frozen meat, there is sufficient reason to test objectively frozen meat items to determine if the final product is consistent with the standards of palatability set by the restaurant.

Pricewise, frozen products offer some advantages. Meat that is frozen is usually stored when supply is excessive or when the market is depressed. Frozen products often reflect the lower market. There are some distributors specializing in frozen meat who pass on these lower market costs to the consumer. Some frozen food jobbers establish their price lists for 90 days, with only four seasonal changes a year.

Quality-wise, frozen product has some advantages. Where the local markets are unable to produce a product or the quality equal to remote regional production, the advantages of a frozen shipped-in product are very clear. Sometimes a product frozen at the peak of freshness is fresher than "fresh." For example, halibut taken from the Alaskan waters and frozen immediately on the boat is fresher than halibut that is iced and held on the boats, trans-shipped to a distributor and finally reshipped to a consumer. Highly perishable halibut starts to go downhill as soon as it is caught. This same fish is preserved almost in perfect condition by immediate freezing. It is not implied that frozen fish is better than fresh fish. Commercially, however, frozen fish is often more palatable than fresh fish that has run the full course and time of distribution.

Frozen meats require less attention than fresh products. They can be stored easily, do not require as careful rotation, and do not place the operator in the position of being overstocked with a highly perishable commodity.

It is most important that the operator does not categorically set aside frozen product because it is frozen. Usually fresh meat cannot be distinguished from a frozen product after seasoning and cooking.

Many frozen meats are available which are frozen as a part of normal commercial production. This includes pork loins, spare ribs, butts, and sausage items. There is some lamb at times of surplus production. Much veal is marketed frozen as cutlets, legs, and roasts. Beef, mostly sold fresh, is appearing in considerable quantity as portion cut items. Fish, for the most part is sold frozen. Frozen poultry is now available in large quantities. More than 19 per cent (1977) left federally inspected slaughter plants in a frozen state. Frozen portion control chicken parts are well received.

Locker plant operators who offer carcasses to their customers and process on the band saw may purchase economically frozen measled beef carcases that have been released.

Seasonal Purchasing.—Frequently overlooked is the opportunity to buy some cuts at the low end of a price cycle to be used later after prices start to climb. Frozen products can be purchased and stored, or taken on future contracts. Simple future contract agreements are made usually for a given quantity of merchandise to be delivered during a future period at the current low market price, plus charges for handling, storage, taxes, extra processing, and interest. These charges are usually nominal, compared to the wide market fluctuation. Strip loins, normally depressed in January and February, are usually considerably lower than strip loins in the summer months. Some summer reports have already discovered the value of the "future contract plan."

TRIM

This is the most misunderstood of the specifications. Almost uniformly, when specifications are attempted, there is a definition of the cut rather than a description of the trim of the cut. For example, in the case of a strip loin, there is a long description of what a strip loin is, usually overlooking how long from the chine (inside or middle) or how far from the eye the flank should be cut.

Uniformity of trim does not exist in the industry. Yet it is almost universally assumed that what is intended as an OP rib (oven-prepared rib) by one jobber is the same for another jobber. The buyer, in order to achieve a uniform product from different purveyors, must describe the trim he expects. It might be wise to show the jobber what is desired, and have the jobber put this in written form. His language should be descriptive and communicable. The following three examples show the economic significance of trim. In each case the oven-prepared rib is cut from a packing house 7 × 10 rib assumed to weigh 30 lbs. The "OP rib" price is assumed to be $1.40 per lb. The cutting plan calls for 14 servings per rib.

Example 1 (Fig. 8.2)

(1) Chine bone left on.
(2) Short ribs cut off in a straight line from a point two inches from edge of short ribs at both ends of rib.
(3) Yield 85 per cent.
(4) OP rib weighs 25.5 lbs.
(5) Portion cost = $2.55 (14 servings).

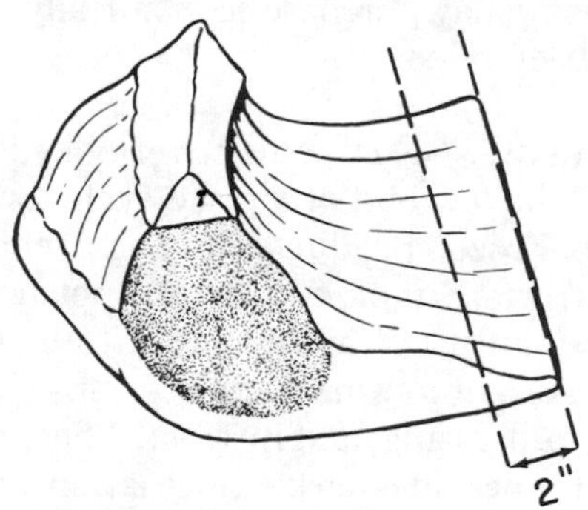

FIG. 8.2. OVEN-PREPARED RIB—85 PER CENT YIELD

Example 2 (Fig. 8.3)

(1) Chine bone cut off.
(2) Short ribs cut off in a straight line three inches from the eye on the rib end cutting parallel with the edge of the short ribs.
(3) Yield 75 per cent.
(4) OP rib weighs 22.5 lbs.

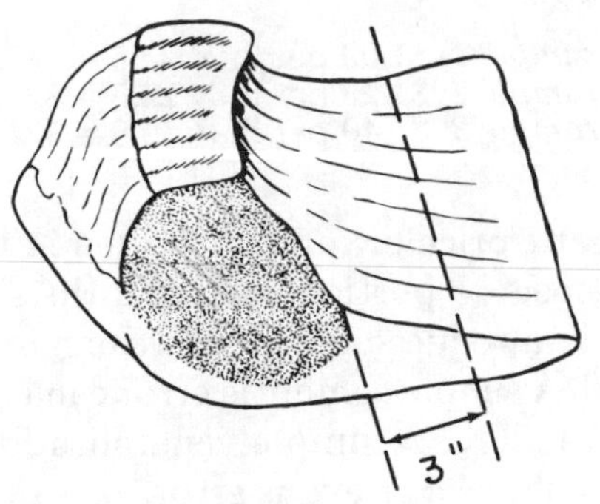

FIG. 8.3. OVEN-PREPARED RIB—75 PER CENT YIELD

(5) Portion cost = $2.25 (14 servings).

Example 3 (Fig. 8.4)

(1) Chine bone cut off.
(2) Short ribs cut off in a straight line from a point three inches from eye on the rib end to a point four inches from eye on chuck end.
(3) Yield 66 per cent.
(4) OP rib weighs 19.8 lbs.
(5) Portion cost = $1.98 (14 servings).

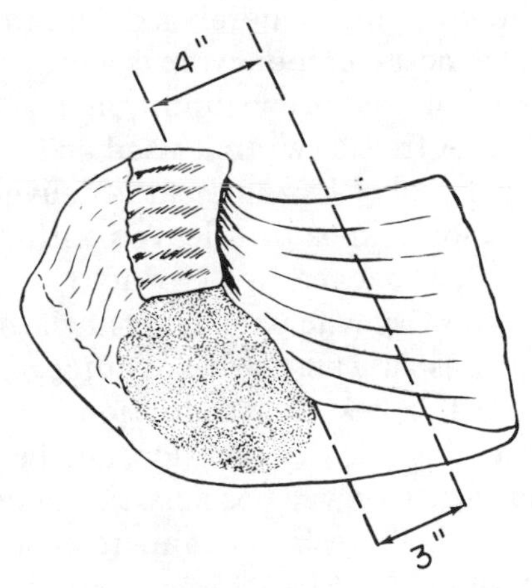

FIG. 8.4. OVEN-PREPARED RIB—66 PER CENT YIELD

In each case the size of the eye or usable portion is the same. The ribs in Example 1 and 2 would have to be priced as follows to result in the same portion cost as Example 3.

Example 1: \$1.09 per lb × 25.5 = \$27.80
Example 2: \$1.23 per lb × 22.5 = \$27.68
Example 3: \$1.40 per lb × 19.8 = \$27.72

There is a dramatic price difference hidden in trim. The role that trim takes with respect to portion cost, and the significance it takes where competition is introduced between two or more purveyors, cannot be overemphasized. A simple semantic error can be the difference between success and failure. A simple generalization by management about all OP ribs could be a grievcus error.

The three different rib preparations sometimes actually occur in a single jobbing house. The "OP rib" delivered to any particular restaurant might vary with the prudence exercised, the competition, the independence and interest of the buyer, and the general knowledge of management.

PACK

This specification is important with respect to a few items in most operations, very important in some cases as ocean ship stores or preparation for protracted frozen storage.

In most cases, the wrapping is merely a device to insure sanitary delivery of product. The most common wrap is a heavy butcher wrap, with the product delivered in a sanitary aluminum lug or plastic box, or in a new cardboard box. A freezer wrap is used on request or on product that is frozen. A great deal of frozen meat is delivered in paper boxes. Fresh cut steaks are sometimes interleaved with peach platter wrap, wrapped in MSAD 80, or packed in Cryovac.

Most purveyors offer Cryovac packaging at the option of the buyer. It has some advantages as well as disadvantages, depending on the particular operation. The best determination is to weigh the various factors, relate them to the specific operation, and be objective in judging the functions of the pack. Cryovac has a particular advantage where age trim is undesirable, or where fresh steaks are to be stored for a prolonged period without freezing.

Special containers are imperative for some operations like ocean vessels taking on provisions. The containers have to conform to space, handling equipment, and rough handling. Special sealing with wire or strapping, special wrap, and detailed identification stenciled on particular places may be required.

MENU PLAN

The specifications at this point relate to product purchased for the menu. If the item is portion cut, there is no problem except that this cut may have multiple uses. For example, a four-ounce hamburger patty might be used for hamburger sandwich, beef patty and eggs, and "low-cal" patty, all on the same menu. Non-portion cuts usually have a primary cut as well as by-product. At this point the specifications create an awareness of the by-product, the intended use of the by-product, a possible determination of portion size, and an estimated average annual cost of the portion. For example, a choice strip loin may be used as follows: New York Steaks for dinner menu (1¼ in.); New York Steaks for extra cut (1¾ in.); flank lean for braised sirloin tips; fat trim for hamburger steak (ten ounces). This specification makes for an integrated purchasing-menu plan, with a specific use of by-product.

CUTTING

A plan should be worked out to get maximum yield from each cut. This could be worked out on paper like an architect's plan, subject to modification through trial and error testing. How involved the cutting plan is would depend upon the basic items selected. For example—the cutting plan of a beef loin would be more complicated than that for boneless primal cuts. The chapters on cutting present many basic approaches.

COOKING FORMULA AND GARNISH

This specification is really an addendum to the general specifications. It is not necessary, but worth considering. Such a specification seems to be the logical conclusion to the writing of specifications, and certainly an integral step where a complete formal kitchen plan is intended. This idea is not original. It is practiced in some highly specialized operations such as Gulliver's in Los Angeles, and in complicated operations such as the cuisine offered by Trader Vic's in a multiple operation. A color photograph of the complete entrée would finalize the specification.

Such a complete set of specifications would be the architecture for an operation that should perform consistently, display attractively, and maintain quality standards as high as the operator intends.

While the benefits of specifications are obvious, there are some pitfalls. They carry no guarantee for success. Specifications must be communicated, interpreted, and put into effect. Specifications grow old, useless, and outmoded.

A number of items are bought with a grade stipulation, yet by their very nature, must be delivered with no identification. When the bidding becomes competitive the purveyors are encouraged to turn corners, especially where it takes the lowest bid to make the sale. For example, a large institutional buyer buys pre-cut fillet steaks. The specification is USDA Choice. The lowest successful bidder might come in with a price lower than the steaks can be produced by the buyer from whole USDA Choice tenderloin. The buyer can see only the price tags and has no doubts because he believes he is protected with "rigid" specifications. In this hypothetical case, the key might be that the purveyor fabricates large quantities of USDA Good, USDA Standard, and USDA Commercial loins. All tenderloins without regard for grade are used. This buyer is buying a brand and a price, not a grade. In this case other purveyors, unwilling to violate the grade specification, price themselves out of the sale, or resort to meeting competition with a similar devious device.

For the most part, specifications create competition and open doors on an objective basis. When properly set up and intelligently administered, specifications are a big step in the right direction.

PALATABILITY TEST

"The proof of the puddin' is in the eating." In the final analysis, the determination of the grade, the option to be exercised among brands, the evaluation of new products and new recipes, and any other decision relative to finalizing on a product, should be made in terms of the customer's palate, or a palatability test. Too frequently this test, if made, is made by one person who tastes and declares that he does or does not like, reflecting a highly personalized palate, without a control sample, and frequently with no concern about the customer's reaction or desires. A few basic principles, outlined below, should be followed to achieve an objective palatability evaluation:

(1) Use the same cut, same amount of age, same trim, same cooking procedure on all samples.

(2) Code pieces with a letter or a number so panel cannot be influenced.

(3) Test your panel. Use two samples from one piece. The decision of the panel should be questioned if they rate identical pieces differently.

(4) Select the panel carefully. A good panel will represent your cus-

tomers' palates. Exclude management and unusual individuals. Attempt to achieve a good cross section.

(5) Two conclusions should be sought: (a) a general reaction to the product, and (b) a particular reaction to each of the palatability characteristics: appearance, aroma, flavor, tenderness, and juiciness.

(6) "Appearance" generally reflects the preparation in the kitchen rather than the product. Particular attention should be given to this evaluation.

(7) Use a control item in the test. For example, if testing frozen product, include a fresh sample; when testing a frozen item, include a fresh one. If a new quality is being tested, include the regular quality.

(8) A single sample or a single test is never adequate. Tests must be repeated on different occasions to be relatively conclusive. Do not inform the purveyor of a quality test. It is most important that the sample is selected at random.

(9) To assure objectivity, disinterested parties should supervise all phases of test. Exclude the chef, the purchasing agent, etc.

(10) Various grading scales may be employed and different systems may be developed for mathematical evaluation. For example:

Excellent = 5	Fair = 2
Very Good = 4	Poor = 1
Good = 3	Bad = 0

(11) For the purposes of scoring, some of the test factors may be excluded such as appearance, which is kitchen controlled and not a part of the product tested. Extra points may be allocated for key characteristics. For example:

Product A	*Panel Score*	×	*Factor*	=	*Total*
Appearance	4		0		0
Aroma	4		0		0
Flavor	5		1		5
Tenderness	4		2		8
Juiciness	5		1		5
	22				18

In this hypothetical test tenderness is rated most important. Appearance and aroma may be considered kitchen controlled and carry no value. Every product in a given test must carry the same factor value.

Instructions:

1. Rate each sample tested according to the following scale: Excellent—5; Very Good—4; Good—3; Fair—2; Poor—1; Bad—0
2. An evaluation of every characteristic must be made. Your test can be scored only if you grade every item.
3. Personal comments and criticism will be appreciated.

Sample A	*Score*	*Sample B*	*Score*
Appearance	____	Appearance	____
Aroma	____	Aroma	____
Flavor	____	Flavor	____
Tenderness	____	Tenderness	____
Juiciness	____	Juiciness	____
Texture	____	Texture	____
Comment on Acceptability:		*Comment on Acceptability:*	

FIG. 8.5. SAMPLE PALATABILITY TEST

(12) Comments should be carefully examined although they are subjective and cannot be "scored." It is possible that they indicate more than the score. Testing is primarily for subjective reactions. (See Fig. 8.5 for sample test).

REFERENCES

ANON. 1960. Meat Manual. 6th Edition. National Live Stock and Meat Board, Chicago, Ill.

ANON. 1953. Food purchasing guide. A circular. Am. Hospital Assn., Chicago, Ill.

ANON. 1962. Food buyers guide. Food Publications Inc., Los Angeles, Calif.

BRODNER, J., MASCHAL, H. T., and CARLSON, H. M. 1962. Profitable Food and Beverage Operation. 4th Edition. Ahrens Publishing Co., New York.

KEISTER, D. C. 1957. How to Increase Profits with Portion Control. Ahrens Publishing Co., New York.

KNIGHT, G. E. 1956. Quantity food purchasing. Circ. *R-500,* Mich. Agr. Expt. Sta., East Lansing, Mich.

LEVIE, A. 1968. Convenience meats. Cornell Hotel Restaurant Admin. Quart., *9,* No. 3, 85–91.

McDANIEL, R. 1955. Profit in portions. Restaurant Management *77,* No. 6, 46.

STOKES, J. W. 1960. Food Service in Industry and Institutions. Wm. C. Brown Co., Dubuque, Iowa.

WOOD, A. 1957. Quantity Buying Guides. Revised. Ahrens Publishing Co., New York.

9

Arithmetic of Meat Merchandising

The arithmetic of meat merchandising starts with testing. Testing is nothing more than a factual report systematically prepared to supply at the source information for comparative purposes, to help examine and redirect as necessary, that which is being tested. Testing requires a report form, and the collection of data under actual or simulated working conditions to establish standards for evaluation. The primary function of any test is to supply the management team with information about any segment of the operation, to replace guesses with facts, and to produce data to help make intelligent decisions. Testing cannot be overemphasized.

SHRINK TESTS

There are three steps in conducting a shrink test: (1) weigh the finished product and determine the percentage of the original weight; (2) convert percentage into a reciprocal (see Applications at end of chapter); and (3) multiply cost of original product by reciprocal to get cost of product tested.

Example 1—A canned ham weighs ten pounds @ $1.62 per pound. To test shrink take ham out of can:

Out of can weight—8 lbs	
Yield 80 per cent	
Reciprocal	1.25
Cost (in can)	$1.62 per pound
Cost out of can	$2.025 per pound

Example 2—Same ham after cooking:

Cooked weight 6 lbs	
Yield 60 per cent	
Reciprocal	1.67

Cost (in can)	\$1.62 per pound
Cost of cooked product	\$2.71 per pound
4 oz cooked portion cost	\$0.68

The in-can weight needed for a 4-oz cooked portion in this case is 4×1.67 or 6.68 ozs. This demonstrates another use of reciprocals.

Shrink tests are especially important in comparing brands of pumped products such as ham, bacon, and corned meats in terms of the lowest end product cost.

Shrink tests should be made to compare cooked portion yields with raw portions provided. To illustrate, assume the portion size and cost of a pot roast entrée is based on five ounces raw. A recapitulation of the actual sales will accurately reflect the cooked portions realized, the actual portion cost, and the correctness of the pricing. Also from this test, the correctness of the amounts prepared and the effectiveness of dish-up procedures can be evaluated.

Prudent retail meat operators will make a simple over-all shrink analysis. A physical inventory will have to be taken periodically, once a month, or once a week for optimum control. An analysis should be made as follows by pounds:

Start	Beginning inventory
Plus	Purchases
Less	Sale of by-product
Less	Sales
Equals	What inventory should be
Less	What inventory is
Equals	Shrink

Reduce shrink to a percentage and determine a pattern. Deviations should be investigated immediately.

The locker operator, in order to achieve the mark-up anticipated, must either sell products at the arrival weight, or make shrink tests.

To illustrate, assume a beef hindquarter arrives at 150 lbs, is hung and sold to the patron at 140 lbs. The preshrunk cost is 96¢ per pound and the operator wants a 20¢ per pound gross profit. The shrink is 6.67 per cent. The yield is 93.3 per cent with a reciprocal of 1.07.

96¢ (cost) × 1.07 (Reciprocal) = \$1.03 True cost
0.20 Markup
\$1.23 Selling price

YIELD TESTS

This is similar to the shrink test except that it is generally used for raw product fabrication. This test is especially useful to compare the trim of product from the various competitors and the comparative end product costs.

Assume that USDA Choice tenderloins are to be defatted before cutting. By determining the percentage yield, a comparison of trims can be made. Convert percentage yield into a reciprocal, multiply by price per pound and the net cost of a defatted tenderloin can be determined.

Extensive yield tests are a must for retail meat operators who fabricate carcasses and primal cuts. What the return is on any particular item can only be determined by testing.

The test in Fig. 9.1 is based on 7×10 ribs weighing 93 lbs for three pieces at $1.10 per pound.

BUY-MAKE DECISION

The decision to buy or to make, an option that is exercised almost daily, can be aided by testing.

A simple procedure may be set up:

(1) Cut product and record weight of primary and secondary products. Total should approximate weight of beginning product.

9/15 1978

ITEM 7×10 Ribs

UNITS 3 PACKER XYZ

WEIGHT 93 COST $1.10

CONDITION BY S.D.V.

YIELD	WEIGHT	%	SP	RETURN
First Cut Rib Roast	27	29	2.10	60.90
Chuck End Rib Roast	35	38	1.80	68.40
Beef Stew	2	2	1.75	3.50
Trim	1	1	.10	.70
Short Ribs	10	11	1.39	15.29

GROSS RETURN 148.79

COST 110.00

CENTS PER POUND YIELD 38.79

GROSS PROFIT PERCENTAGE 26.03

FIG. 9.1. TYPICAL YIELD TEST—7×10 RIB

(2) Convert weights into percentages. These should approximate 100 per cent.

(3) Apply replacement costs and determine total return for primary and secondary products.

(4) Compare return with cost of primary cut.

The test in Fig. 9.2 is based on 7×10 ribs weighing 93 lbs for 3 pieces @ $1.10 per pound.

The gain of $5.60 is per cwt or about $1.87 per rib. The buy-make decision has to be made by comparing the economic gain with consideration for the additional shrink, the use of the secondary cuts, additional administrative problems, and additional labor. This test can be used to compare yields from different suppliers. Assume that the standard expected yield for 7×10 ribs based on a standard cutting procedure with values of $1.71 on the R.R. rib and 30¢ on the short ribs, is $112.00 per cwt.

Assume that a lot of five ribs is tested. In this hypothetical case, the R.R. ribs weigh 91 lbs and the short ribs weigh 41 lbs (Fig. 9.3).

9/15 19 78

ITEM 7 x 10 Rib

UNITS 3 PACKER B

WEIGHT 93 COST $1.10/lb.

CONDITION OK BY S.D.V.

YIELD	WEIGHT	%	SP	RETURN
O.P. Ribs	65	70	1.57	109.90
Short Ribs	17½	19	.30	5.70
Fat and Bones	9½	10		
	92	99		
				115.60
		Cost		110.00
		Gain		5.60

GROSS RETURN ______

COST ______

CENTS PER POUND YIELD ______

GROSS PROFIT PERCENTAGE ______

FIG. 9.2. SHORT TEST—7×10 RIB

9/15 1978

ITEM 7X10 RIB

UNITS 5 PACKER A

WEIGHT 163 COST $1.10

CONDITION OK BY S.D.V.

YIELD	WEIGHT	%	SP	RETURN
R.R. Ribs	91	56	171	95.76
Short Ribs	41	25	30	7.50
Fat and Bones	31	19		
	163	100		
				103.26

GROSS RETURN ______

COST ______

CENTS PER POUND YIELD ______

GROSS PROFIT PERCENTAGE ______

FIG. 9.3. TEST FOR STANDARD EXPECTED YIELD—7 X 10 RIB

The return of $103.26 is $8.74 per cwt less than the anticipated return. The trim on the ribs from jobber A is substandard. Additional tests should be made before a final decision is reached, as the test could be incorrect.

The buy-make decision is a classical argument among meat retailers and locker plant operators. Where there is no simple answer, there are three factors that should help make the decision.

Is the economic gain or loss significant?

Can the by-product be used? Would excessive waste, shrink, or distressed selling occur?

Does the sale of the secondary product hurt the total operation? Does it attract an undesirable type of trade? Does it produce very low income sales? Require more time and skill than the primary product? Is new volume generated? Good volume?

The difference in cost between carcass and primal cuts can usually be stated somewhere between one to two cents per pound. Yield tests should be made to determine the difference. Many operators, recognizing this small difference, will specialize in the cuts that move best.

The specialists believe they can do a better job, and show a greater profit. These operators believe that they would do a bigger volume handling straight carcasses, but because of low mark up and forced selling, would end up with a smaller net profit.

STANDARD EXPECTED YIELDS

Statistical information can be developed to reduce the buy-make decision to a simple economic evaluation. Formulas can be worked out where there is a primary product with a relatively uniform yield, and secondary products of small, relatively stable value. For example, the comparative values of a bone-in and boneless strip loin may be formulated based on standard expected yield tests: (1) cut, weigh, and figure percentage yields; (2) evaluate total by-product at replacement cost; repeat test to develop a pattern for by-product value; (3) subtract value of by-product from total cost, to get net cost of primary product; and (4) multiply reciprocal of yield of primary product times net cost to get price per pound of trimmed primary product.

Step 1

Assume a 20 lb bone-in strip loin @ $1.48 per pound to be cut into a 10 in. boneless strip loin.

Item	*Yield, lbs*	*Percentage*
10 in. boneless strip loin	14.4	72
Lean trim	1.0	5
Fat, bone, shrink	4.6	23
	20 lbs	100%

Step 2

Trim 5 per cent @ $0.80 lb	= $ 4.00
Fat and bone 23 per cent	= —0—
By-product value per cwt	$ 2.00

Step 3

100 lbs @ $1.48	= $148.00
Less by-product value	− 4.00
Net cost of primary product	$144.00

Step 4

Reciprocal of 72 per cent yield	= 1.39
Times net cost	$1.44
Equivalent price of 10 in. boneless strip loin	$2.00

Test this calculation as follows:

Boneless

$2.00 (price) × 72 (lbs)	= $144.00
Plus by-product	$ 4.00
	$148.00

Bone-in

100 lbs @ $1.44	= $144.00

From this working formula involving reciprocal and by-product value a calculation can be made at any price level. Conversion tables can be set up for quick calculations.

Conversion Tables 9.1 and 9.2 for steaks and ribs are based on yields for a particular cutting specification. These are presented only as illustrations.

TABLE 9.1. STEAKS—CONVERSION TABLE

Primal Cut	Strip Loin	Fillet	Top Sirloin
1.35	2.86	3.18	3.11
1.40	2.96	3.27	3.20
1.45	3.05	3.36	3.29
1.50	3.14	3.45	3.37
1.55	3.23	3.54	3.46
1.60	3.33	3.64	3.55
1.65	3.42	3.73	3.64
1.70	3.51	3.82	3.72
1.75	3.60	3.91	3.81
1.80	3.70	4.00	3.90
1.85	3.79	4.09	3.99
1.90	3.88	4.18	4.07
1.95	3.97	4.27	4.16
2.00	4.07	4.36	4.25
2.05	4.16	4.45	4.34
2.10	4.25	4.55	4.42
2.15	4.34	4.64	4.51
2.20	4.44	4.73	4.60
2.25	4.53	4.82	4.69
2.30	4.62	4.91	4.77
2.35	4.71	5.00	4.86
2.40	4.81	5.09	4.95
2.45	4.90	5.18	5.03
2.50	4.99	5.27	5.12
2.55	5.08	5.36	5.21
2.60	5.18	5.46	5.30
2.65	5.27	5.55	5.39
2.70	5.32	5.64	5.47
2.75	5.45	5.73	5.56
2.80	5.55	5.82	5.65
2.85	5.64	5.91	5.74
2.90	5.73	6.00	5.82
2.95	5.82	6.09	5.91

TABLE 9.2. CONVERSION TABLE—RIBS

7 × 10	Roast Ready	Export	Banquet	Rib Eye
86	133	166	202	273
87	135	168	204	277
88	136	170	207	280
89	138	172	209	283
90	140	174	212	286
91	141	176	214	289
92	143	178	216	293
93	144	179	219	296
94	146	181	221	299
95	147	183	223	302
96	149	185	226	305
97	150	187	228	308
98	152	189	230	312
99	153	191	233	315
100	155	193	235	318
101	157	195	237	321
102	158	197	240	324
103	160	199	242	328
104	161	201	244	331
105	163	203	247	334
106	164	205	249	337
107	166	207	251	340
108	167	208	254	343
109	169	210	256	347
110	171	212	259	350
111	172	214	261	353
112	174	216	263	356
113	175	218	266	359
114	177	220	268	363
115	178	222	270	366
116	180	224	273	369
117	181	226	275	372
118	183	228	277	375
119	184	230	280	378
120	186	232	282	382

PORTION COST TESTS—SHORT FORM

The simple test procedure is to count the units cut from any given piece or lot of meat and divide this into the total cost. This can be expressed as a simple formula:

Pounds × Price ÷ Units Yielded = Unit Cost

Example 1

12 lb strip loin @ $2.00 per pound yielding 10 steaks

12 × $2.00 = $24.00 ÷ 10 = $2.40 each

When two or more different size steaks are cut, for example 8 and 12 ozs, a unit value system may be employed where the 8-oz steak is 1 unit and the 12-oz is 1½ units. An inventory of "units" can be taken and the cost of each size thus determined.

Example 2

Same strip loin as in Example 1, yielding 9 pieces of 8 oz steak (1 unit value) and 4 pieces of 12 oz steak (1½ unit value) for a total of 15 units.

12 × \$2.00 = \$24.00 ÷ 15	= \$1.60 unit value
8 oz (1 unit) costs	= \$1.60 each
12 oz (1½ units) costs	= \$2.40 each

The cost per ounce or pound and the weight yield of the steaks are ignored. Portions are prepared and portions are sold, which might be a reason for disregarding weight and focusing on actual portion cost.

Where a daily cutting report is prepared, and the products fall within a weight range specification, it is possible from the briefest report to reasonably calculate the unit cost. The butcher only has to report the number of steaks cut and the number of strip loins cut.

Example 3

Five strip loins are cut up. The specification is 12 to 14 lbs, cost \$2.50, and the yield is 51 steaks.

5 × 13 × \$2.50 = \$162.50 ÷ 51 = \$3.19 each

PORTION COST TESTS—LONG FORM

By-product at replacement cost (R/C) is included in the complete test.
The formula can be expressed as follows:
Pounds × price − by-product credit = Cost ÷ units yielded = unit cost

Example 4

Assume the same conditions as in Example 1, with the following by-product:

Item	*Pounds*	*R/C*	*Total*
Lean trim	3.0	\$0.80	\$2.40
Fat	6.5	—0—	—0—
Bone	3.0	—0—	—0—
By-product			\$2.40

Fat and bone are usually assigned no value for testing.

13 × \$2.50 = \$32.50 − \$2.40 = \$30.10 ÷ 10 = \$3.01 each

A daily butcher cutting report can be set up along with a daily inventory to be submitted to the bookkeeping department. Some basic controls can be achieved if a sales reconciliation is made. Daily unit costs can be determined. Cutting deviations can be quickly detected. Kitchen pilferage, shortages, and shrink can be checked.

Figures 9.4 and 9.5 illustrate a simple daily cutting report and the auditor's inventory reconciliation. The illustration is complicated by two sizes of strip steaks being cut. The auditor's analysis is nothing more than a perpetual inventory system: (1) unit cost can be tested by

DATE	AM INVENTORY			CUT TODAY		
	STRIP LOINS	STRIP STEAKS 8 oz	STRIP STEAKS 12 oz	STRIP LOINS	8 oz STEAKS	12 oz STEAKS
1/20	6	10	10	4	4	38
1/21	12	4	12	10	40	80
1/22	2	20	40			

FIG. 9.4. DAILY CUTTING REPORT

PERPETUAL INVENTORY

STRIP LOINS						
DATE	BEGINNING INVENTORY	+ PURCHASES	− CUT UP	= BALANCE (should be)	ACTUAL INVENTORY	+/−
11/20	6	10	4	12	12	0
11/21	12	0	10	2	2	0
11/22	2					

8 OUNCE STRIP LOIN STEAKS						
DATE	BEGINNING INVENTORY	+ CUT UP	− SALES	= BALANCE (should be)	ACTUAL INVENTORY	+/−
11/20	10	4	10	4	4	0
11/21	4	40	23	21	20	−1
11/22	20					

12 OUNCE STRIP LOIN STEAKS						
DATE	BEGINNING INVENTORY	+ CUT UP	− SALES	= BALANCE (should be)	INVENTORY	+/−
11/20	10	38	36	12	12	0
11/21	12	80	52	40	40	0
11/22	40					

FIG. 9.5. PERPETUAL INVENTORY

applying average weight, current cost prices, and the short form unit cost formula; (2) the auditors use the next A.M. cutting inventory as the current day ending inventory; and (3) auditor must determine daily purchases and recapitulate items sold daily.

A simple control can be set up on key items combined with a daily, weekly, or monthly physical inventory. A running cost of the items can be maintained by taking the starting inventory, adding purchases, and deducting ending inventory. A running account of the sales of the key inventory items must be maintained as well. The difference between the sales and the cost is the gross profit. In a restaurant this might be converted into a percentage figure. In a locker plant or retail butcher shop, the cents per pound yield would be a more constant index.

Figure 9.6 is a simple inventory form. The inventory must be taken

MEAT INVENTORY 8 AM			
DATE	STRIP LOIN	TOP SIRLOIN	TENDERLOIN
1/11	8	4	5
1/12	6	4	2

FIG. 9.6. DAILY INVENTORY FORM—WHOLE PIECE COUNT

at the same time, either the close of business, or the beginning of the day.

The person responsible in accounting might use a separate control card for each item (Fig. 9.7).

Assume the following:

(1) Beginning inventory (at 8 A.M. on 1/11) was eight strip loins. Assume a cost of \$2.40 per lb, and a specification of 12 to 14 lbs, use 13 lbs as the average.

$$8 \times 13 \times \$2.40 = \$249.60$$

(2) Purchases (1/11) taken directly from delivery invoices—assumed \$160.00.

(3) Ending inventory (at 8 A.M. on 1/12) is six strip loins. Using the same value as in (1):

$$6 \times 13 \times \$2.40 = \$187.20$$

(4) The cost of the strip loins

Begin Inventory + Purchases − End Inventory = Cost of Sales

\$249.60 + \$160.00 − \$187.20 = \$222.40

(5) Sales are recapitulated from the dining room sales tickets. Hypothetical sales for the day of all kinds of strip loin steaks are \$530.00

STRIP LOIN CONTROL CARD						
DATE	BEGIN INV	+ PURCHASES	-END INV	*FOOD COST	SALES	FOOD COST
1/11	249.60	160.00	187.20	222.40	530.00	42%

FIG. 9.7. INVENTORY CONTROL CARD—STRIP LOINS

(6) Food cost ÷ sales = food cost percentage

$$\$222.00 \div \$530.00 = 42\%$$

This statistical analysis is valuable for control as it portrays trends. Once the pattern is established and broken, then the cause must be determined. The figures may reflect changes in cost of product, some deviation from specification as the trim or steaking, waste, theft, or simple bookkeeping errors. When the trouble sign goes up, the operator is on notice to start looking.

Pounds as well as dollars may be analyzed. Pounds being constant, provide a better statistical basis than cost dollars which fluctuate with the market.

The locker plant operator buying product solely to process in bulk form can profitably use this accounting procedure to determine gross profit: (1) subtract cost from sales for gross profit; and (2) divide gross profit by sales for percentage of gross profit.

Where the locker plant combines retail sales and bulk sales, a two key cash register may be used, recording the two distinct types of sales. Both may be tested.

The retail meat dealer may use this test to spot check a particular item. Spot checked items should be rung separately on the cash register. All product may be divided into classes as beef, veal, pork, lamb, and provisions, keying the register to the divisions, testing each class.

TIME AND MOTION TESTS

The labor applied to process raw material must be considered to evaluate properly any products tested. Simple records can be maintained with an application of the proper labor rate. Included in the rate must be the fringe expenses, social security taxes, workmen's compensation insurance, pensions, sick leave, vacations, paid holidays, rest periods, overtime, etc. These may run as high as 50 per cent of the base rate.

Where large quantities of meat are used, and one or more butchers are employed, a careful evaluation of butcher shop efficiency should be made. The labor cost per pound should be determined. Many food service operations run a very high butcher cost per pound. Use of jobber fabrication might result in substantial labor savings.

DAILY MENU RECAP

Meat entrées should be recapitulated daily. This practice should have no exception. A daily recap will reveal poor selling and unprofitable items. The recap will function as a pricing guide, pointing out the area where obvious price increases can be made.

PRICING

Markup on Cost

The most popular method of pricing meat entrées on the menu is a markup on cost. Conversely, cost of the meat is reflected as a percentage of the total selling price. This is nothing more than the use of a reciprocal. For example, to figure a 40 per cent food cost, the markup is two and a half times the raw material. The reciprocal of 40 per cent is 2.5.

In actual practice, no markup formula is simply followed. There are items that bear a greater markup, some that require an adjustment downward. In these cases the markup is really not based on cost. The operator wishes to achieve a given food cost and hopes somehow out of the "mix" and a shrewd understanding of the customer's sense of value that a given cost will be produced. By trial and error, by adjusting quantities and portions, the all-important percentage is obtained. This, of course, does not guarantee the necessary volume to produce enough gross profit (sales less food cost) to take care of labor, overhead, administrative expenses, and profit.

Food cost markup has one inherent merchandising error. Low priced menu items even with a low food cost are sometimes unprofitable, and high priced menu items, even adjusted downward, are often so high that they drive trade away.

To illustrate: a restaurant serves dinners only. An accurate unit cost analysis for several months reveals:

Labor cost per dinner served	$1.80
Overhead per dinner served	0.50
Total	$2.30

It is reasonable to assume that the average entrée costs about the same to produce and serve. In fact, low priced entrées usually involve costly preparation while high priced broiler items are usually inexpensive to prepare.

Two typical items on the menu are a fish entrée and a steak entrée. The operator feels that fish is more profitable because of the low food cost. Assume the following: fish entrée 30 per cent cost on $3.00 item and steak entrée 40 per cent cost on $5.00 item. On the unit-cost basis, how much does each entrée contribute to the cost of doing business?

	Fish	*Steak*
Unit selling price	$6.00	$9.00
Unit food cost	(30%) 1.80	(40%) 3.60
Unit gross profit	$4.20	$5.40
Unit cost per meal (labor and overhead)	3.50	3.50
Profit on meal	$.70	$1.90

Unit-cost Pricing

Unit-cost accounting is not new. Basic manufacturing accounting uses three factors: material (food), labor, and overhead. The total of these is multiplied times a reciprocal to cover administrative expenses and profits.

A restaurant could approach the fish and steak price in this manner: To the total cost add 20 per cent of the selling price to cover administrative expense and intended profit margin. To state this conversely, food, labor, and overhead should be 83 per cent of the selling price. The reciprocal of 83 per cent is 1.20.

	Fish	*Steak*
Unit food cost	$1.80	$3.60
Unit labor cost	2.50	2.50
Unit overhead cost	1.00	1.00
Cost of finished goods	5.30	7.10
Mark-up reciprocal	1.20%	1.20%
Selling price	6.36	8.52
Round out price to	6.35	8.50
Food cost	28%	42%

The unit-cost approach reveals unprofitable items, helps determine prices that will redirect the customer's purchasing to more profitable items, and leads to more attractive pricing of "red meat" items which invariably produce more customers. There may be a wide variance of percentage food costs from item to item, but the total cost of "finished goods" (food + labor + overhead) should be 83 per cent, leaving 17 per cent of sales for administrative expense and profit. The average food cost will reveal itself as an operational pattern, a product of the sales "mix."

A unit cost on administrative expenses plus a unit mark-up for profits is a variation on unit cost pricing.

	Fish	*Steak*
Cost of finished goods	$5.30	$7.10
Unit administrative cost	.50	.50
Unit of profit	1.00	1.00
Selling price	6.80	8.60
Food cost percentage	26%	42%

Whichever pricing method is selected, it is most important that it be consistently applied.

The locker plant operator can use unit pricing as a simple approach with excellent results. Total labor and overhead applicable to processing, divided by pounds processed, will portray the average cost per pound. Testing will demonstrate that the processing cost from one item to the next will not vary widely. A percentage markup on the other hand could result in unprofitable pricing.

To illustrate: assume a beef forequarter at 68¢ and a hindquarter at 96¢ per pound; a desired 20 per cent gross profit or 25 per cent mark-up.

Forequarter		*Hindquarter*	
	68¢		96¢
+25%	17¢	+25%	24¢

This hypothetical locker plant with an average gross profit of 21¢ per pound and an operating cost of 18¢ per pound may be profitable. The forequarter sale is unprofitable yielding 17¢ per pound against an average 18¢ cost. The low priced forequarter may actually cost more to fabricate per pound than the higher priced hindquarter.

The retail meat operator should give serious consideration to such a mark-up system, especially for an assembly line self service operation. At least the markup should be tempered by the unit cost concept. The sale of unprofitable items might be slowed down or possibly eliminated where unit cost pricing is applied.

APPLICATIONS

Reciprocals

This is the quantity derived by the division of the dividend 100 by any divisor. Because the division process is the reverse of multiplication, it follows that the reciprocal times the divisor used equals 100.

Reciprocals are commonly used, though infrequently associated by name. For example, to achieve a 40 per cent food cost, multiply by 2.50, the reciprocal of 40 per cent. To illustrate the arithmetic:

```
        2.5
.40)1.000
        80
        200
        200
```

Gross Profit

The difference between percentage markup and gross profit is frequently confused. Gross profit is the difference between sales and cost of sales. A 20 per cent markup does not yield a 20 per cent gross profit. The use of reciprocals offers a very simple solution, especially for meat retailers.

In the following example, a 20 per cent gross profit is desired. Hence cost represents 80 per cent of the selling price. The reciprocal for 80 per cent is 1.25.

Formula—Cost × Reciprocal = Selling price

The example is for a 60¢ cost and a 20 per cent gross profit.

60¢ (cost) × 1.25 (reciprocal) = 0.75¢ (selling price)
Cost = 0.60¢
Gross profit = 0.15¢
Gross profit = 20 per cent

Any gross profit can be quickly and accurately determined.

REFERENCES

ANON. 1948. Food Cost Accounting. Am. Hospital Assn., Chicago, Ill.

BRODNER, J., MASCHAL, H. T., and CARLSON, H. M. 1962. Profitable Food and Beverage Operation. 4th Edition. Ahrens Publishing Co., New York.

DUKAS, P., and LUNDBERG, D. E. 1960. How to Operate a Restaurant. Ahrens Publishing Co., New York.

STOKES, J. W. 1960. Food Service in Industry and Institutions. Wm. C. Brown Co., Dubuque, Iowa.

10

Commercial Bribery

DEFINITION OF "KICKBACK"

In our competitive business world the attitude must be taken that the kickback is nothing more than a system of organized crime. Kickbacks are nothing more than devices to dupe management, devices to take cash from the operation, devices to put cash in the pockets of purveyors and employees involved. Kickbacks are a form of bribery, not to be tolerated as the customary way of doing business.

No one has more astutely described the kickback than Harry Rudnick (N.A.M.P.). "A kickback is a word generally ascribed to the surreptitious giving of money (or a thing of value) to an employee of a customer in consideration of (or appreciation for) purchases made by the customer from the firm making the payment. Although a distinct line cannot be made which would clearly separate "giving" which is a kickback, from "giving" which is not, it may be stated that infrequent gifts at times when gifts are generally given (Christmas, Thanksgiving, births, and marriages) as a token of appreciation, to an employee of a customer, do not constitute a kickback. On the other hand, large and regular gifts to an employee of a customer, especially where the amount bears a relationship to the amount of business done, are definitely a kickback."

Kickbacks can be simply defined in a three-fold manner: first, that regular payments be made which bear some relationship to the amount of business involved; secondly, there is no disclosure to the owner or management of this arrangement; and third, payments are made in product, cash, payroll, premiums, prizes, gifts, or some other devices of value.

The payoff is made in many different ways. For example, it may be cash picked up at the purveyor's plant, or it may be distributed by salesmen on their regular calls. It may be a payroll check on a monthly basis, based on a percentage of purchases. It may be meat product, which is sometimes delivered to the employer's address, to the home of the employee, or to another establishment, in which the employee may have an interest. It may be relatively large gifts, as a new set of tires, or an automobile, or furnishing of an apartment, or enough premium stamps for a new TV set. It may be a pair of season box seats for the theater, race track, baseball park, or an all expense paid vacation to Las Vegas.

One thing is certain: whatever the medium of payoff, and whatever the interval, the payoff is always a dishonest arrangement. The code of ethics of the National Association of Meat Purveyors clearly states the attitude of this association (Fig. 10.1).

THE PROFILE OF A THIEF

It is estimated that billions of dollars change hands annually in kickbacks, payoffs, and bribes. Among the male population, with respect to the distribution of embezzlers, over 35 per cent are among the executives in managerial and professional positions. They have been placed in this particular position of trust by management, based on the faith and judgment of management.

The profile of a successful thief reveals that he generally has four common characteristics: (1) he dislikes change; (2) he has some special need; (3) he is highly skilled; and (4) he has overcome the problem of his conscience.

Change

Change is feared because it could create new situations. Status quo creates confidence. Change, whether it be for the better of the organization or not, could ultimately lead to exposure.

National Association of Meat Purveyors
CODE OF ETHICS

III To pay a commission prohibited by law directly or indirectly, to an employee of a customer for goods sold to such a customer.

IV To make false or deceptive statements to a customer concerning time of delivery, grade, quality or trim of merchandise, or ability to supply.

Courtesy of National Association of Meat Purveyors

FIG. 10.1. CODE OF ETHICS

Special Needs

Perhaps there is an addiction to gambling, or women, or a debt, or a sickness such as being addicted to drugs or alcohol. There may be a great need for foreign cars, or a similar passion. There may be a wife who is extremely demanding, as Veblen so aptly put forth in his concept of "conspicuous consumption," in keeping up with the Joneses. These special needs and others could cause the person involved to require extra income. A case history of a special need is an owner who fails in business and then takes a job as manager, buyer, chef, steward, or some other position where he is responsible for, and involved in meat buying. This buyer permits regular overcharges to the present employer by the purveyor, which somehow pays off the old debt of the previously bankrupt operation.

Skill

Skill is required to conceive of how to make a plan work, or to put it in effect, and to sustain it. A skilled person is usually the only one who can find the right opportunity.

Management, with a preconceived idea of what a good gross profit or a good food cost is, is the perfect dupe for the kickback. All that has to be achieved is the "right" percentage. This can be achieved by controlling waste, controlling stealing by other employees, using low cost items, careful menu planning, cutting quality and portions, "taking weight on the scales." The fact is that percentages can be run up or down by anyone in control by changing the value given the customer.

Conscience

The thief must overcome his conscience. This can be done, perhaps, by one of several rationalizations: management is getting more than it is paying; the side income is a relatively small amount of that which is really due; there is nothing more involved than a "commission"; everybody does it. Whatever the rationalization may be, and however small the amount may have been at the beginning, a point can be stretched, and is finally limited only by the magnitude of the operation.

Case Types

The man involved, his character and his attitude toward the people around him, are most interesting. For example, the surly type, who can barely tolerate himself, has assumed this role to run off all of the salesmen, to maintain his control and power, and in general, to have people generally avoid him so that he cannot be trapped. His usual excuse for not buying from a new purveyor is that they do not have the quality of merchandise that he is looking for.

There is the case of the buyer who is very aloof, very soft-spoken, very willing to cooperate, and very much a gentleman, professional and polished. He takes his place in society working for charitable organizations; he is frequently a leader in his professional organizations. He listens to all salesmen, he is sympathetic to those who work with him and under him, and he is particularly cooperative with management, *the perfect confidence man.* He knows that as long as he is friendly to a salesman, when he says "no," it is likely that the salesman will not go over his head to management. He knows that as long as he is cooperative with management, the subtleties of his actions will not be questioned.

There is the case of the chef who takes charge although he is not the appointed buyer. The chef becomes the "umpire" and real buyer by simple negation. He eliminates purveyors until he finally has the purchasing agent buying from the companies that he wants to do business with. This chef is the final judge from whom there is no appeal, and he can be as arbitrary as he pleases.

There is the case of the chef who takes a job which is considerably below the pay level to which he is accustomed. It is agreed, however, that he should do the meat buying. By bringing in his "friendly purveyors," he can considerably increase his income, which may be tax-free. Sometimes he will mix up the friendly purveyor with two or three other firms. This makes the purchasing look competitive.

There is a case where the person responsible for the buying is completely without product knowledge, yet sets himself up as an absolute authority. He expresses himself adequately. The average salesman fears him and settles quickly for a little of his business.

There are several devices that can be employed to run off honest salespeople. There is the case where the buyer points out that the beef shipped has an excessive amount of "gristle," which is rather nebulous. There is the case where the only acceptable beef is that which has a certain inspection number.

There is the case where the salesman is told that his quality is inadequate. Yet the buyer often cannot even define the word "quality" in objective terms. Another device is to place a "get off my back" order. It is a small sample, and when the eager salesman later walks in, he might be told that the merchandise "was not too bad, but not good enough for the high quality demands of this particular institution."

Kickbacks in some instances are made to one of the partners, or to a top level officer of a corporation, thus providing that person with an unequal distribution of the profits.

THE ACCESSORY

There are some dishonest meat purveyors, many more with great integrity. Some secure their business on both sides of the fence partly by kickbacks, partly by hard selling. Company size has nothing to do with ethics. Some very small firms engage in commercial bribery. Some of the largest do too. Belonging to the national association with its code of ethics is no assurance of company policy. Not belonging bears no relationship to lack of integrity. The conduct of any company cannot be stated in terms of a generalization.

There are very few laws governing kickbacks that are enforced. The Internal Revenue Service could be a great source of enforcement. Unfortunately, it looks more to the formality of the deduction than the morality of it. For example, the Internal Revenue Service may approve of payroll checks made to employees of purchasing units from the point of view of income tax, without any question of morality or propriety of the payment.

The future may bring about considerable change. The Department of Agriculture, under the "Stockyard and Packers Act" may become involved. The Internal Revenue Service at some point may disallow deductions for "entertainment" or for non-working employees. Sooner or later, state laws may be passed and enforced. At present there is a broad federal investigation of large-scale international kickbacks.

THE KICKBACK HURTS

Payoffs may be rationalized by management, but the fact is that where they exist, they must ultimately be reflected in the cost of doing business. In a recent national survey, it was revealed by the meat jobbing industry that meat prices are frequently arrived at by the concept of what traffic will bear. It is obvious that promiscuous kickbacks could easily lead to higher prices. If there is a kickback, and the buyer gets something for nothing, is not the purveyor entitled to make a little extra profit, too?

Overcharging

The most common situation when a kickback is in effect is to overcharge. These overcharges are usually manifested in the tonnage items such as steak meat and prime ribs.

A case of overcharging with no apparent reflection on the percentage is where the buyer accepts certain frozen beef cuts in substitution for fresh at current fresh market prices. Actually, the buyer using frozen product, could contract for these cuts in January, February, and March at the lower prices, and achieve a considerable saving in the high market summer months. In this case, prices appear in line, the purveyor enjoys an extra markup, and the buyer serves both masters.

Quality Substitution

A second device frequently employed by the supplier in a kickback situation is to invoice the purchasing unit for one quality and to ship another. The receiver must be in collusion or lack product knowledge. In one case, it was observed that a jobber invoiced for Midwest quality and shipped a lower quality. Another way of achieving the quality substitutions is to falsify the inspection number of the piece of meat. This is illegal, but it is being done.

Probably the most dramatic quality substitution is to use very fancy commercial cows, which have white cover fat, and show considerable marbling, and by using a stolen roller, roll this beef USDA Choice, thus substituting USDA Commercial beef for USDA Choice beef.

A fourth substitution for quality is to bill for one thing and ship another. For example, an actual case was observed where a restaurant was billed for 100 lbs of strip loins at $2.25 a pound. A substitution was made—bones. So for $225.00 the restaurant received five dollars worth of bones. The bones were neatly wrapped in packages, labeled "strip loins," and the collusive receiver was muted by the payoff.

Packer inspection numbers are listed in a Department of Agriculture publication, "Meat and Poultry Inspection Directory," which can be purchased through the Superintendent of Documents, Government Printing Office, Washington, D.C. 20402. The price is $3.80.

Short Weight and Improper Trim

Meat is measured on a scale for weight, and measured with a ruler for trim. If either one of these areas is overlooked, it is easy for the purveyor to overcharge.

Short Weight.—There are many devices for overcharging by short weight. In the first instance, it was observed that a purveyor uses a worksheet in his cutting room. After the order is filled, the worksheet is transcribed in his office to an invoice. At this point, there are systematic overrides of 5 to 15 lbs per item on the invoice.

A second case reveals that the overstatement for weight is achieved at the purveyor's scale. The purveyor himself acts as scaler and fills in the weights on every invoice. It is a simple matter to add weight without any employee being involved.

Short weights may be the result of a collusive arrangement between the delivery man and the person responsible for doing the receiving. Some of the merchandise is left on the truck, yet signed for by the receiver. The driver takes the merchandise elsewhere to conclude a cash sale, and makes his split with the receiver. This practice frequently occurs in retail meat operations.

There is an even more complicated arrangement to avoid being caught. The purchasing party deliberately overbuys. After it is weighed in, he decides to return a portion of it. The driver, ignorant of the arrangement, issues a credit and returns the merchandise to the plant. If no spot check happened to be made, both the purveyor and the buyer destroy the copies of the credit memo and the operator stands the overcharge.

A fifth construction, perfectly conceived to appear as a great price concession, is to sell the user a 7×10 rib at a very low price, to fix the rib oven-prepared, and to ship only the rib, crediting the user for the short ribs at perhaps ten cents per pound. Here the operator gets the advantage of the "low 7×10 price," and puts no short ribs into the freezer.

To illustrate:

Ten ribs weigh 300 lbs. Short ribs yield 20 per cent, weigh 60 lbs. Credit for short ribs is issued arbitrarily at 120 lbs. There is no way of checking on this.

The billing:	240 lbs oven-prepared ribs		
	120 lbs short ribs		
	360 lbs	@ $1.10 =	$396.00
The facts:	300 lbs 7×10 ribs	@ $1.10 =	$330.00
		Loss on 10 ribs	$ 66.00

When 7×10 ribs are bought on the "short rib credit plan," the receiving scale and specifications are valueless.

Improper Trim.—It is simple, with collusion, to make a trim substitution. For example, in the case of a bone-in strip loin, an extra inch is cut on the flank. That inch ranges from 20 to 40¢ a pound. On a single strip loin weighing 20 lbs, at 20¢ a pound, the overcharge is $4.00. There is a case where tenderloins are shaped with a wood paddle instead of a knife. This excess fat might bring the purveyor a premium of $3.00 or $4.00 per tenderloin. It is a general practice to cut the chine off of strip loins and oven prepared ribs. There are cases where the chine is left on. This is commensurate with an overcharge from 1½ to 3 lbs per primal cut.

Good Suppliers Are Eliminated

The kickback eliminates some good suppliers. The suppliers who refuse to fatten the kitty are eliminated without regard for their quality, without regard for their prices. It becomes a very simple matter for the buyer to achieve this elimination by telling management that these suppliers fail in some manner. The operator suffers by losing good sources, good products, and competitive prices.

Patrons Are Victimized

In the final analysis, the kickback affects the patron as much as anyone else. Either a substitution of quality will be made, or portions will be coming out smaller than they should be, or the selling prices must reflect the higher cost of doing business. Menu prices must be increased two or three times as much as the overcharge.

Employees' Morale Ebbs

Dishonesty must rub off on the other employees. A little bit of authority is destroyed, a little bit of loyalty is lost, and a little bit of interest in the total success of the operation is gone. The loyal and honest employees stand out in a ridiculous light socially, particularly at home, and they ultimately must lose interest, lose face, and lose heart. For the lower echelon in the organization, it is virtually a license to steal, right down the line.

RESPONSIBILITY OF MANAGEMENT

A more responsible, more intelligent management attitude would inhibit, reduce, and possibly eliminate kickbacks within any given organization. *Management that sets forth no policy or makes little effort to control kickbacks by silent assent is as guilty as the other parties.*

It is most unfortunate that there is in general an irresponsible attitude with respect to what occurs in the back of the house. It is more interesting, the course of least resistance, and much more consistent with the social stature of management, to concern itself with the merchandising and sales portion of the business.

It is not unusual today to hire a chef or a steward or a purchasing agent and to turn over the complete responsibility of purchasing to this person, which involves spending between 25 and 50 per cent of the total restaurant sales income. It is not uncommon for management to become subservient to the demands of this person, to take an indifferent attitude to what this person does, to make no attempt to supervise, manage, or control this person, and to make no investigation as to how the money is being spent.

MANAGEMENT FAILS TO MANAGE

Management rationalizes its lack of knowledge by pointing out that there are few complaints. Therefore, it must be concluded that the operation is satisfactory. An honest analysis of what is occurring involves considerably more than a negative evaluation in terms of the lack of complaints. Most patrons who are unhappy do not complain. They simply walk out of the door, and never return.

"A license to steal" is issued by indifferent management. It is almost axiomatic that the larger the organization, the more indifferent and the greater removed management is with respect to knowledge of what is occurring. For example, in a hotel, the food service operation is often treated as a necessary evil, a mere appendage of the hotel business, in spite of the fact that it represents a substantial annual dollar volume. The food service can be very profitable. It is certainly a key asset or a tremendous liability from the point of view of rooms. If the hotel has management with sufficient stature to control, operate, and manage the food service, the hotel will be commensurately better off.

Management may be guilty on several counts: there is the situation where management knows or suspects that stealing is going on, yet takes an attitude of indifference, doing nothing to arrest the situation. It is not atypical to find a relatively successful operation in which management feels that kickbacks are not uncommon, yet rationalizes in this manner: "We have a successful operation; the people responsible are doing a good job by our standards. We feel that there is some misuse of purchasing, but we are so pleased, or so bound by the staff, that we will do nothing about it." Fear is the real motivation.

Kitchen purchasing procedures are frequently indifferently set up, while purchasing procedures for the rest of the operation may be consistent with the best accounting practices. In some kitchens, no purchase orders, no written confirmations, no formal price gathering, and no systematic approach to receiving are attempted.

MANAGEMENT CONTROLS

Management can achieve considerable control by first manifesting and communicating its attitude toward kickbacks; and secondly, by setting up a system of internal control to minimize or eliminate them. Management must voice its policy that the kickback is not standard practice, and that as far as their particular organization is concerned, it will not be tolerated. Explain what is likely to happen in the event that a kickback is uncovered. It can be pointed out that the employee will be subject to dismissal and prosecution. Management should provide and announce its program. Good management might band together, and conceivably institute a bribery-prevention league to improve business morality much like that which they have in Great Britain.

PROGRAM FOR EMPLOYEES

Pay Fair Wages

A most obvious failure of management is its occasional indifference to fair wages. If a person is going to be responsible for spending money, it behooves management at least to remove the possibility of this person "requiring" a kickback in order to make ends meet. Good base wages are far more important than a bonus plan. The bonus is a fine incentive for an honest executive, but no incentive for honesty. A bonus is not too attractive side by side with 5 per cent of purchases, tax free, paid on a current basis.

Pre-hiring

Definite checks should be made with respect to pre-hiring. The job application should examine antecedents of the applicant. If large funds are to be entrusted to an employee, there should be as much information as possible, not only to help evaluate the employee's capability, but his integrity as well. One of the most important steps is to check the former employers. It is probably best that this be done through correspondence. It is important to find out why the intended employee left his former employment.

The employee may be bonded. The device is calculated to protect for embezzlement, but functions to uncover or discourage dishonest people at the outset. Put the person on notice that he is bonded. Whatever occurs in the particular employment will be a matter of record.

Establish Good Working Conditions

The working conditions, hours, vacations, and holidays should be such that the person involved is treated on an equal basis with the standards in the area. This will help create respect and loyalty in the organization. Other fringe benefits such as pensions and deferred income plans should be considered. These should be carefully examined, and evaluated in terms of what management is attempting to achieve.

If the morale of the organization is good, it will attract high type personnel, and eliminate those people from the organization who do not fit. If this is the case, it will not take long to eliminate anyone who is not working for the organization, or anyone who is involved in a kickback situation. The over-all morale of the organization is a good check for the person responsible for doing the buying.

Communicate

A kickback should be defined, and management's attitude toward it should be spelled out. It is generally true that the organization that makes it very clear what its policy is with respect to gifts and kickbacks, seldom has a problem. This is achieved in part by announcing company policy and attracting the type of personnel who are prepared to comply.

Employees' Responsibilities

Saving through prudent and wise actions are just a part of the job. The savings are the property of the company. In the event that there is a general employees' bonus plan, savings are in part the property of all the employees.

It is the responsibility of management to point out to the buyer that he cannot carry water on two shoulders; that if he is involved in taking something from the purveyors from whom he is buying, he must necessarily become muted, and ignore any of the failures of the purveyor. When he does this, he fails in his responsibility.

Management should indicate to the buyer that they intend to make a better buyer of him by training, by supervision, by providing certain formalized material.

Most of all, the employee should be told that his basic loyalty belongs to his employer. His loyalty to management will manifest itself throughout the organization, and in the long run, everyone will benefit by his good work.

Purchasing from Vendors

Kickbacks may take the form of what appears to be a personal purchasing arrangement between the employee and the vendor. Management should develop a policy to control this. Where purchasing is permitted by the employee for his own use, terms of purchase, point of delivery, and the price arrangement should be spelled out. Management may permit purchases at the company price where personal consumption is contemplated. Vendors and employees should be notified of the policy. There is always the possibility that product might be ordered, billed, and shipped to the employee with no intention or attempt made for collection. This kickback may be a regular periodic arrangement to provide a constant supply of meat for the employee's table.

PURVEYORS' PROGRAM

Approved Purveyors

Prudent management will permit business to be conducted only with management-approved purveyors. By doing this, management can control promiscuous purchasing, demonstrate to the purveyors that the buyer is only carrying out the policy of management, and that a kickback is not necessary. Management makes the purveyor responsible to management. It would seem wise that the purveyors be checked out with the Better Business Bureau, with other people in the food service industry, and with other purveyors. This interest by management can only lead to a generally healthier situation.

With respect to new purveyors, it is important that at least four things be determined: that the purveyor company is not related to the buyer; that the buyer has no financial interest in the purveyor company; that the reputation of the purveyor be beyond reproach; and that the purveyor be financially sound.

Disapprove the Kickback

Management should make it quite clear to their purveyors that they do not approve of the kickback in any form. There is no room as far as management is concerned, for the purveyors to get their business by buying it. As far as management is concerned, their business is always open. It is unethical and dishonest if any inducement is offered to the buyer. It should be pointed out that no employees, relatives, or assigns are to be on the payroll of the purveyor company, there are to be no cash gifts, no product gifts, no employee loans, and all discounts and advertising allowances shall be indicated on the invoices to the company, and the amounts remitted or credited to the company.

Gifts

The program, with respect to gifts, or what is normally classed in the business world as "wine and dine," is not quite as simple as the kickback in general. There are, for example, widely accepted practices of wining and dining customers to establish a better rapport, and induce them to do business.

Reasonable wining and dining is quite different from regular payments made secretly. Gifts are frequently made for special occasions, as marriages, births, anniversaries, and include along with this, theater tickets, and sporting event tickets, as well as taking customers out to restaurants to entertain them. These gifts may be of a reasonable nature and not surreptitious. For example, a pair of baseball tickets as compared to box seats for the season; a set of steak knives as compared to furnishing a living room.

In some cases, management has attempted to define specific acceptable limits for gifts. There has been an expanding use of the rule of thumb that the gift should not exceed that which the receiver can consume on a single day. This rule clearly makes the distinction, for example, between a quart of whiskey and a case of whiskey, a ham for Christmas and a side of beef.

Disclosure to management by the employees and vendors of gifts may be established as a company policy. In this case, employees should be notified that a report is required. Thus the employees will have the opportunity to refuse gifts that could be interpreted as an attempt to influence buying decisions, instead of a token acknowledgment that the business was appreciated.

It is not unreasonable to request that the vendors disclose gifts to employees. This would eliminate any excuse from the vendors, and create the opportunity for management to request the return of any gifts interpreted as excessive or objectionable.

An alternate course is to take the policy that no gifts shall be made, and there are many organizations that follow this. Where it is the prerogative of any organization to spell out a policy of absolutely no gifts, if in the general business world such a policy were instituted, a large volume of restaurant business would be destroyed. It is inconsistent to encourage "wine and dine" for the restaurant patrons and to outlaw it for the restaurant employees.

Non-compliance

It should be spelled out to the purveyor that unless he complies with the policy of management, two steps will be taken: first, he will be blacklisted as far as the company is concerned, and will not be permitted at any future date to do business with the company. Secondly, if sufficient evidence can be obtained that either a kickback or that false billing took place, the matter will be presented to the courts to be pursued at both the civil and criminal levels, and with the Internal Revenue Service.

Formalize Program

In order to communicate the desires and policy of management to the purveyors, it would be wise to formalize. Scandia Restaurant in Los Angeles formalized its program with a contract for the purveyors, setting the policy for both kickbacks and gifts (Fig. 10.2).

Open Door Policy

Management should admit and listen to the salesmen of the companies that have failed to sell the organization. Salesmen may have a message for management that should be heard. In general, there are many good reasons for not doing business. Management, leaving the door open, will listen to considerable criticism, mostly unjustified. These will come under the heading of inadequacies or just gripes on the part of the purveyor. On the other hand, one may learn about situations that are not consistent with policy. The purveyor may have something constructive to offer, and the buyer may be turning a deaf ear. At any rate, the buyer will be cautioned by the privilege the salesmen have of talking to management.

Spot Checks

There are many ways to spot check the buyer. Put an unidentified executive of the company in the buyer's role to check on the purveyors for an extended period of time. Use a security employee for a few months to make an evaluation of the purveyors and their habits.

Spot checks should be conducted by the management team, on quality, trim, and price. These spot checks, done periodically, will have a very good effect, not only with respect to the integrity of the buyer, but they will get back to the purveyors and keep them on their toes. If there is any doubt, a grader from the U.S. Department of Agriculture may be brought in to examine the products for quality, or the local

Scandia
RESTAURANT
9040 SUNSET BOULEVARD, HOLLYWOOD 46, CALIFORNIA

In order to establish the buying policy of Scandia, and to facilitate the approval of your firm as a purveyor to Scandia, will you please sign, date, and witness the statement below:

As consideration for approval to do business with Scandia, the undersigned purveyor agrees as follows:

1. To make no payment of any kind, either in cash, payroll payment, commission, merchandise, personal loan, or any other form of inducement or gratuity to any of the employees of Scandia, their relatives or assigns.

2. Gifts and business entertainment expenses for the employees of Scandia, their relatives or assigns, shall be limited to a reasonable amount. This paragraph refers to business expense deduction items such as occasion gifts, theatre tickets, sporting event tickets, and other similar gratuities customarily employed in sales promotion.

The purveyor hereby agrees that in the event that these provisions are violated an amount equal to ten (10) times that provided to an employee will be assessed against the purveyor as liquidated damages. The purveyor further agrees that he will pay reasonable attorneys fees and court costs in the event of a law suit brought against him for enforcement of the terms of this agreement.

Kenneth Hansen
SCANDIA

PURVEYOR ________________

BY ________________

TITLE ________________

WITNESS ________________

DATE ________________

Courtesy of Scandia Restaurant, Los Angeles

FIG. 10.2. PURVEYOR CONTRACT

Department of Weights and Measures may be called to spot check the weights.

Receiving Department

It is sometimes difficult to achieve this as a separate function in some of the smaller operations. Yet it is imperative that an attempt be made to keep receiving and purchasing completely separate, creating another obstacle for collusion.

MANAGEMENT PROGRAM

Systematic Doubts

Management will achieve a degree of success if it starts a program of systematic doubts, taking the attitude that something irregular could exist, trying to prevent and reduce its existence, and attempting to ferret it out.

(1) There should be some doubt in management's mind about the integrity of the person responsible for doing the buying. Payoff may or may not exist, but *the possibility is always there.* Doubt the buyer's motives. Is he motivated by personal reasons?

(2) Management should look suspiciously when all meat is purchased from a single firm. There are also some cases where the purchasing may be done from two purveyors, and a collusive arrangement has been made between them with respect to price. Identical prices do not happen under average competitive conditions.

(3) There should be some doubt in management's mind when certain purveyors get the bulk of the business. An attempt should be made to evaluate the amount of business each purveyor gets percentage-wise. Perhaps another aggressive competitor is needed.

(4) There should be some doubt when one purveyor is getting the profitable portion of the business, and others are getting the less profitable part.

(5) There should be some doubt when the same purveyors get the same items each time.

(6) There should be some doubt that the buyer is getting the best quality, the best price, and exact specifications. Periodic tests should be made.

(7) There should be some doubt when the buyer lives beyond his means, indicated by such things as an expensive car, expensive trips, expensive jewelry, taking extended vacations.

(8) There should be some doubt if the buyer is addicted to gambling, horses, women, or some other form of extravagance.

(9) There should be some doubt when formal prices have not been submitted. If the buyer takes all of the prices verbally, there should be some doubt that the buyer is recording the best possible price.

Back of the House Knowledge

Of all the things that management can be indicted for, the most serious is the lack of knowledge of what happens in the back of the house. Management will finally matriculate when "food cost" or "gross profit" ceases to be the keystone, becoming just another statistical guide.

The restaurant industry should take a good look at itself, and conceive of the restaurant first as a manufacturing plant, the kitchen; and secondly, a general sales area, the dining room in the front. The word "restaurant" is too freely associated with the dining area. When the operator realizes that he is a manufacturer and starts planning his restaurant from the kitchen out, the operation will become more efficient, the cost will be lower, and the control will be greater.

Make the Kickback Public

Bring the whole problem into the open. Get publicity in the trade magazines. Stop excusing and protecting friends who are involved in kickbacks, and who rationalize that they must conduct their business in this fashion in order to exist in a competitive society. These people hurt everyone in the long run. These people are issuing a license to steal, and undermining the whole buying system. The kickback, in any area, should not be tolerated, and should receive as much publicity and unfavorable advertising as possible.

The kickback is antithetical to the competitive system. It can be likened to a cancer, growing and infecting the entire system, and ultimately leading to utter destruction. It is as important a problem as any other phase involved in the operation.

Bring the kickback out into the open. This will assure its early demise.

REFERENCES

AXLER, B. H. 1974. Focus on Security for Hotels, Motels, & Restaurants. ITT Educational Publishing, Indianapolis, Ind.

BROWNE, A. E., CROW, W. C., LENNARTSON, R. W., *et al.* 1954. Yearbook of Agriculture. U.S. Dept. Agr., Washington, D.C.

CURTIS, B. 1975. Food Service Security. Chain Store Age Books, New York.

JASPAN, N., and BLACK, H. 1960. Thief in the White Collar. J. B. Lippincott Co., Philadelphia, Pa.

PRATT, L. A. 1957. Embezzlement Controls. Fidelity & Deposit Co. of Maryland, Baltimore, Md.

RUDNICK, H. L., and RUDNICK, L. G. 1962. The Status of Kickbacks under Federal Law. National Association of Restaurant and Hotel Meat Purveyors. Chicago, Ill.

11

Setting Up Shop

GENERAL LAYOUT

The physical premises for the cutting area, the equipment, and some of the rules that should be followed with respect to the butcher shop operation can be generalized. The cutting room, the cooler, the freezer, and the receiving area as well as the necessary equipment should be evaluated, in terms of particular needs.

No master design can be offered. From the items proposed, some of the ideas may be applicable to any given operation.

In the foreseeable future, automation is not likely to be achieved where meat products are disassembled, yet every operator must necessarily think in terms of the most efficient use of labor. A big factor is an efficient, well-designed plant. The flow of product coupled with step-saving devices must be planned.

Classical Look

By way of contrast, many restaurants and hotels are planned with a high built-in labor cost. The receiving may be done on a remote dock. The cooler is in an area far removed from the dock and the kitchen is frequently in the basement. Product has to be transported on an elevator. This plan may require a full time storeroom man, with part time responsibility. A butcher may be working on the same basis.

Straight Line Production

This concept recognizes that the restaurant or butcher shop is a factory assembly line.

Figure 11.1 demonstrates the plan for self-service retail meat operations. Breaking down the preparation of product into eleven major steps, the plan provides a continuous efficient assembly from receiving to display of finished products. This concept can be adapted to service

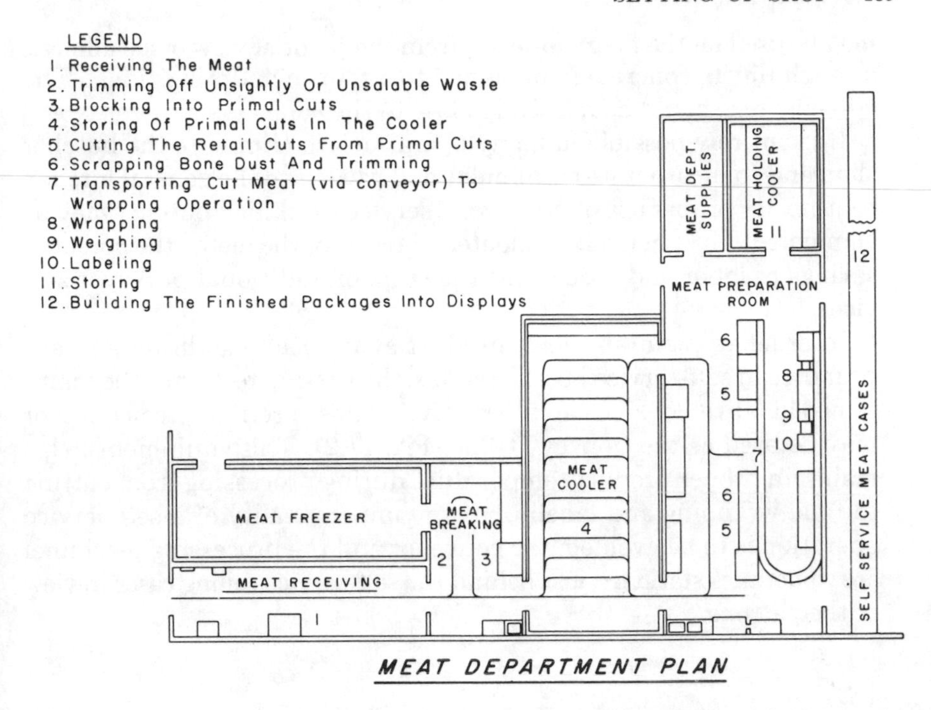

FIG. 11.1. MEAT DEPARTMENT PLAN—SELF SERVICE MARKET

meat operations and locker plants. The idea of an assembly line may be applied to a complete kitchen plan in a restaurant after an analysis of product and traffic flow.

CUTTING ROOM

In the Food Service operation the cutting room may be an area completely remote from the kitchen or a portion of the kitchen set aside for such a purpose. From the point of view of control, the cutting room, if possible, should be in the kitchen. This will improve the flow of product, reduce the labor cost, bring the cutting under closer supervision, reduce pilferage and other losses, increase the efficiency of the butcher, and permit other members of the kitchen staff to help in the butcher shop during slack periods.

The area set aside for the cutting of meat should be one in which there is little traffic to reduce the chance of bumping or other inadvertent traffic hazards. The space should be one in which there is a minimum sanitation problem, with floor drains, and adequate lighting. Sawdust may be used on the floor; however, from the point of view of cleanliness, a rough finish, concrete floor sloped to a four inch drain is ideal. The cutting area should be relatively close to the cooler.

It is entirely possible in many operations to eliminate the butcher shop and to incorporate equipment and work into the general kitchen routine. By proper use of purveyors' services, a skilled butcher may be eliminated. This should be evaluated in terms of the menu, the potential savings of labor and space, and the costs of additional purveyor services.

In order to maintain meat product at its peak condition and appearance in self-service butcher shops, the cutting room may be maintained at 50 to 60°F. As an alternative to this, a refrigerated conveyor may be used as the Convey-O-Pac (Fig. 11.2). This equipment helps maintain proper product temperature during processing from cutting to final wrapping and labeling. If the impersonality of a self-service operation is to be avoided, the conveyor and the processing personnel are sometimes stationed just behind the self-service selling cases in view of the patrons.

Courtesy of Weber Showcase and Fixture Co., Inc., Los Angeles

FIG. 11.2. CONVEY-O-PAK

COOLER STORAGE

The cooler is probably one of the most critical parts of the installation. It should have a relatively high ceiling for proper aeration; it should be adequately insulated to minimize the heat loss; and should be provided with a well-constructed refrigerator door, preferably clad in stainless steel or galvanized sheet metal, with all joints soldered for proper sanitation. There are many new types of doors and hardware available. Some of the latest innovations are relatively light, transparent doors made of plastic, doors hung on a rail that slide along the wall, and door handles provided with simple magnetic catches.

To prevent the door from riding the floor when it is opened after a few months of sagging, the cooler threshold should be raised a few inches above the floor (Fig. 11.3). As the door opens, it actually rides above the floor of the cutting room. This will also reduce wear on the gasket on the bottom of the door, as well as provide a door easy to open and close. Gaskets should be provided on all edges of the door, and should be maintained in good condition.

When a very heavy flow of traffic is anticipated, it may be well to consider a set of batten doors inside the main door which meet in the middle and swing open in either direction. The main cooler door then may be left open during heavy traffic periods. The batten doors offer this advantage: anyone passing through with full hands can elbow his way through in either direction. The doors will close gravitationally, and the cold loss is minimized.

As an alternate to the batten doors, some kind of an automatic closing device may be installed on the cooler door; an air curtain door device may be installed as an alternative. This is a special type of fan that literally sets up a narrow curtain of air permitting a minimum of heat transfer (Fig. 11.4).

Insulation

There are many acceptable types, such as Styrofoam, rock cork, sheet cork, and spun glass. The ideal insulation is one that has a low "K" factor, or heat transmission factor, is impervious to various types of infesting bugs, and moisture proof. Insulation dipped in hot asphalt emulsion makes a good installation. The cost of the insulation does not necessarily reflect its worth. It is well to investigate carefully. The thickness of the insulation should be carefully engineered. The more effective the insulation, the lower the operating, maintenance, and power costs.

Ceiling insulation should be evaluated relative to the type of roof structure. This is the area where the greatest amount of heat transfer

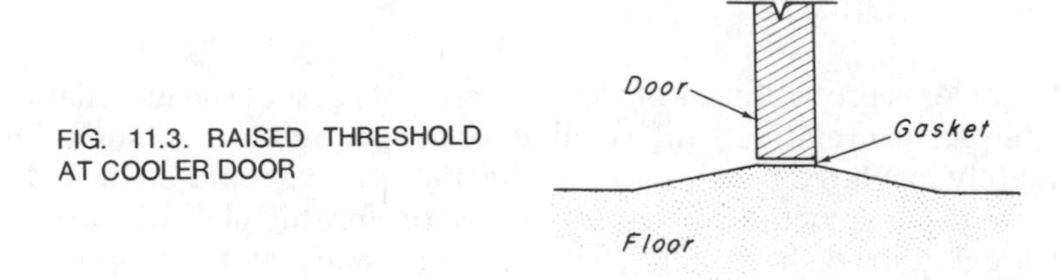

FIG. 11.3. RAISED THRESHOLD AT COOLER DOOR

will occur. Many operations have been installed without insulation in the sub-floor, when the floor is a concrete slab on the ground. When the building is above the ground, it is imperative to insulate the floor. Grey cement plaster, scored in 36-in. rectangles, is recommended for the wall finish. Ceramic tile is the very best finish but is costly. The ceiling may be finished the same as the walls. Less expensive installations are done with sheets of aluminum flexboard, or Asbestocite. The most desirable type of flooring is vertical fibre quarry tile, now available, which makes for a less slippery surface. For sanitation purposes, a pitch of $1/4$ in. per foot to a floor drain is ideal. Concrete is a less expensive, but nevertheless a satisfactory floor finish.

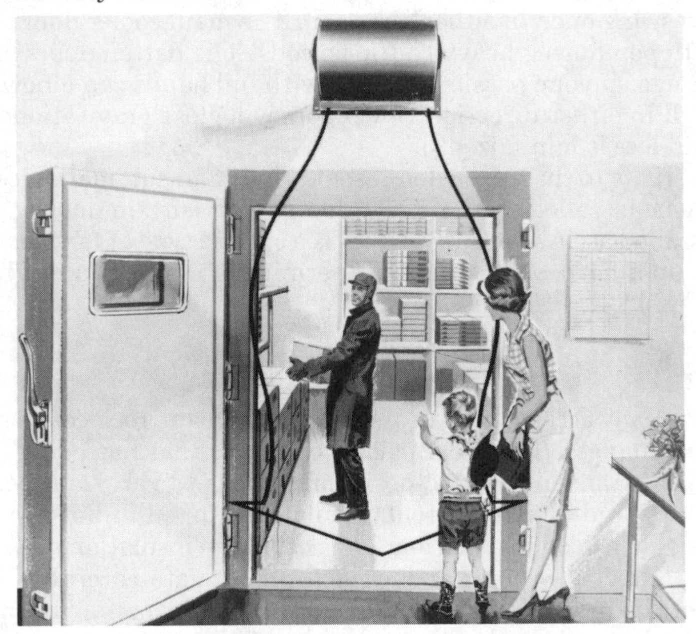

Courtesy of Ilg Industries Inc., Chicago

FIG. 11.4. AIR CURTAIN—BARRIER TO HEAT TRANSFER AND AIR CONTAMINANTS

Refrigeration

There is much fine equipment on the market. Unfortunately, in an attempt to be the low and successful bidder, equipment having inadequate capacity is often offered and purchased. This is not an unusual occurrence in restaurant installations.

The temperature in the cooler should be maintained from approximately 30 to about 38°F. In trying to narrow the range in food service operations, it is perhaps better to consider the upper end, from 35 to 37°F. This range will produce a slightly higher shrink, but will more successfully retard the growth of molds, yeasts, and bacteria than a higher temperature range. In the restaurant, where product is held for short periods, this is the best compromise. If the cooler is used for extended aging, as in a retail shop or locker plant, then a higher temperature range of perhaps 36 to 38°F should be considered to reduce shrink, and maintain higher humidity. If most of the product is protected in Cryovac or a similar wrap, the 36 to 38°F range should be considered. The higher temperature will accelerate the aging, with no adverse economic effects.

The condensing unit should be large enough so that during the heat of the summer, it will not be overworked. This can be achieved in one of two fashions: either by having a larger capacity unit or by increasing the speed at which it operates. By increasing the speed, more refrigeration can be produced, but with a considerably increased power bill, because the unit will be working with a higher head pressure. Likewise, the repair expenses will go up directly with the increase in speed. A higher initial cost installation of a low-speed, relatively high capacity condensing unit will mean a lower power bill and lower repairs in the long run. Along with the compressors, consideration of the type of condenser is very important. There are on the market at least three good types, one that is air cooled, and two that are water cooled. Depending on the size of the installation and the area in which it is going, both types should be given consideration. In relatively dry areas, where water supply is inexpensive and the water is not too hard, an evaporative condenser is very efficient and economical to operate. A sealed shell and tube treated water condenser is also good.

It is most important that the condensing unit be located in an area where the heat it creates can be easily and effectively exhausted. It is not uncommon to see these units installed in closed-in areas, where the motor generates heat, and the hot air created by the motor is supposed to cool the hot gases. The compressor fights itself. One of the most functional installations is one where the condensing units are placed out-doors in a screened-in area, protected from the elements by a roof which keeps off the sun and rainfall. In cold climates, winter protection must be provided. In very hot summer days, water running over the top of the shed will effectively reduce the temperature inside.

The cooler coils are important. The capacity should be large enough so that they will not frost up because of overwork. The T.D. (temperature difference) should not exceed 8 to 10°F. T.D. is the difference between temperature of expanded refrigerant in the coils and room air temperature. A low T.D. will help maintain a high relative humidity. A relatively new blower type coil is available. Instead of exhausting the air through the fins at the sides or the bottom of the coil, the air exhausts against the ceiling. This coil flows air along the ceiling. As the air loses velocity, it precipitates over the meat. A relatively uniform air flow is set up throughout the cooler. This is an ideal type blower for the storage of meat in that it achieves circulation without blasting the product directly with a stream of air. The ceiling height should be 8 to 12 ft.

If competitive bids are accepted on an installation, specifications should be set for all bidders on the compressor, the RPM, the condenser, the coils, and total tonnage of the system. This will help avoid a cheap, poor installation by the most competitive bidder. There are other variables which should be taken into consideration when selecting the contractor.

MONORAIL

A system should be provided where there is a heavy traffic of product, especially if there are carcasses, quarters, and primal cuts. The retail butcher and locker plant operator usually cannot operate efficiently without a monorail. Receiving scales should be incorporated in the track. In most restaurants, meat product is handled in tubs or boxes, on flats or dollies and the monorail is of little value.

Hanging product from hooks on wall bars or stationary aging racks makes for an inefficient use of space as well as a generally unsanitary condition. Removable metal shelving not only surmounts these problems, but makes for easier rotation and better aeration of product (Fig. 11.5). Shelves should be spaced away from wall for maximum air circulation. With a careful analysis of the volume of product, and by using cubes of space, the operator will discover that with shelving it takes less cooler area than anticipated to handle the load. This means saving in terms of space, installation, and operating cost.

With the help of local manufacturers portable racks that travel on the monorail system and on dollies may be devised (Fig. 11.6 and 11.7).

In addition to trolleys, there is a wide variety of equipment available

Courtesy of the Tail O' the Cock, Los Angeles

FIG. 11.5. REMOVABLE COOLER SHELVES

Courtesy of Elgee Meats, Los Angeles

FIG. 11.6. ROLLING STORAGE AND AGING RACKS—TRAVELING ON MONORAIL

to move and store product on the monorail. Where large volumes of beef ribs and loins are used, equipment similar to the rib tree (Fig. 11.8 and 11.9) which handles 18 ribs and a comparable tree handling 12 loins should be considered. Meat hooks may be needed to hang products.

Ultraviolet lights and ozone-producing equipment are sometimes used. These retard growth of microorganisms. If meat is aged too long they may be responsible for an undesirable, slightly rancid flavor. With

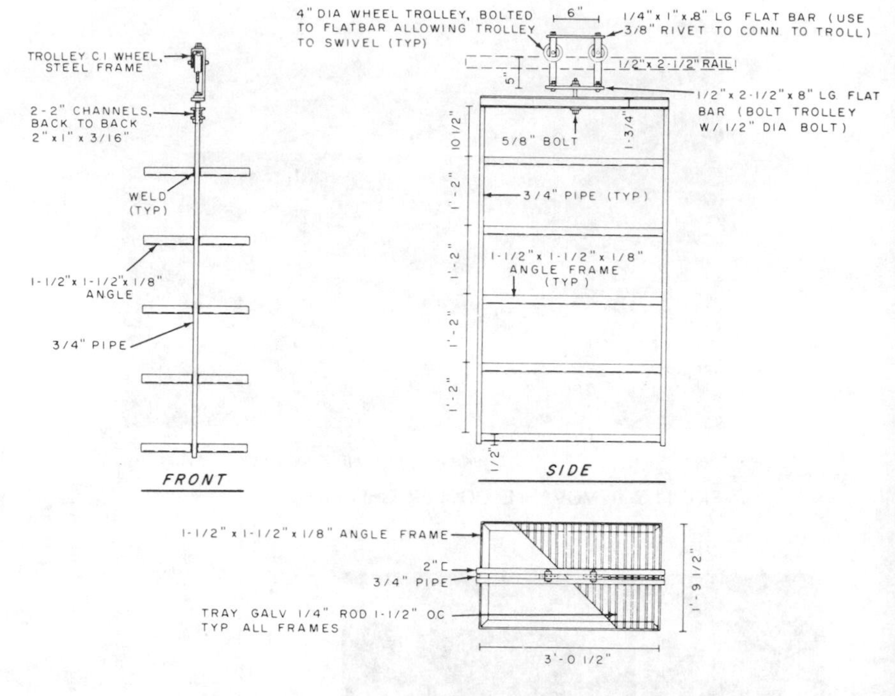

Courtesy of E. A. Stevenson Co., Los Angeles

FIG. 11.7. LAYOUT—ROLLING RACK

adequate refrigeration and 85 to 95 per cent relative humidity, no additional equipment may be needed to minimize the growth of microorganisms.

RECEIVING AREA

An adequate area should be provided to receive, inspect, and weigh products. A dramatic error from the security point of view is to put the receiving area outside the back door or on a different floor. A collusive arrangement between the person who is doing the receiving and the driver delivering the product could be arranged. The receiver could sign for the product and the product might never leave the truck. Put the receiving area some place in the kitchen, even though it appears awkward, so that the product has to be physically delivered. An ideal place for the receiving scale is next to the cooler door, with the cooler within the kitchen. Many people are at work in the kitchen and will function as censors, even without instructions from management.

FIG. 11.8. RIB TREE

LOW TEMPERATURE STORAGE

If the food service operation warrants it, a large freezer should be provided in addition to the cooler. With the increased use of frozen meats, fish, fruits, and vegetables, a large freezer is needed. It is wise to use one wall of the cooler as a common wall with the freezer. If it is practical, a door from the cooler may be cut into the freezer. If a common wall is installed between the cooler and the freezer, a relatively thin insulation will be adequate and a very light type of door can be custom ordered. A reach in freezer may be adequate.

For exterior freezer doors, an electric heater strip is available, which prevents the door from frosting shut. The insulation of the freezer should be determined by the temperature that is to be achieved. Adequate freezer insulations range from 8 to 12 in. of cork or equivalent.

The coils may either be plate or blower type. A very satisfactory coil system may be installed to be defrosted by either cold water or gas. Automatic defrosting increases the effectiveness of the system and reduces maintenance and power costs. Good freezer temperatures range from 0 to −25°F. The −10°F is ideal, with a handy margin of safety, although not necessary for the short handling and storage of meat product.

The freezer should be at least wide enough for storage on both sides

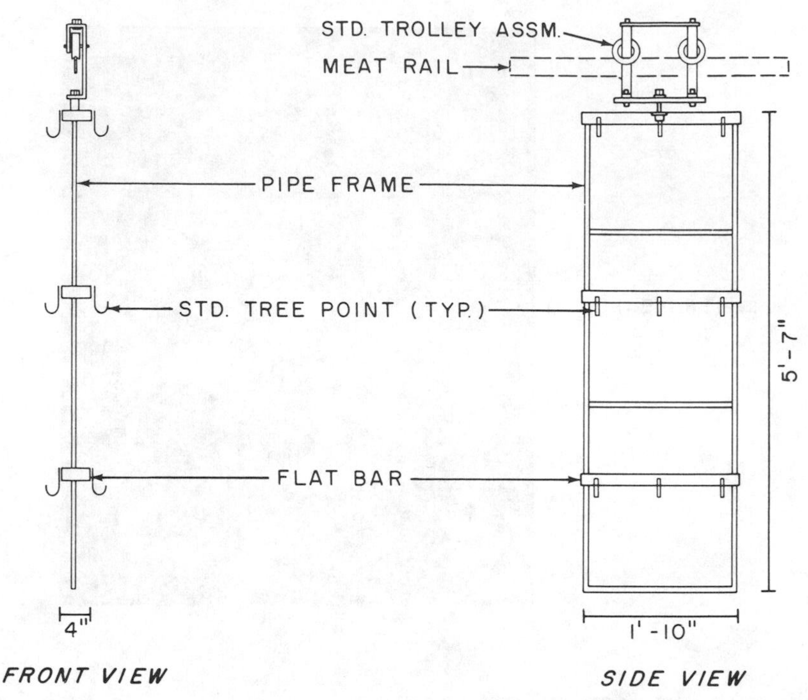

Courtesy of E. A. Stevenson Co., Los Angeles

FIG. 11.9 LAYOUT—RIB TREE

of an aisle. Thought should be given to metal shelves around all walls, which can be removed for cleaning.

The locker plant should have an area apart from the locker storage to freeze products before they are stored in the lockers. Historically, this has been some type of sharp freezer, maintained below the locker room temperature. Since recent research indicates that very quick freezing of meats contributes little to the palatability, the freezing area may be simply a portion of the locker room partitioned off behind a row of lockers with adequate air circulation supplied. There should be a door from the cutting room to move the product in, and a chain link door to the locker room to move the frozen product out.

Baskets that are made to nest and stack on a dolly may be employed to receive the product as it is wrapped. The baskets should be shallow to insure a relatively short freezing period. The product could be assembled as a work lot for each patron, rolled into the freezer, and later to the locker. This or similar equipment minimizes handling labor.

A fan should be provided in the freezer area; it should be turned on during peak product periods to accelerate freezing.

SHOP EQUIPMENT AND FACILITIES

Functional equipment should be integrated in the butcher shop. Some of it may or may not be necessary in any particular operation. Important considerations are initial cost, maintenance, and sanitation. For marginal economic situations, good used equipment is available on the market.

The raw materials, the amount of sub-assembly, and the finished products dictate the assembly line equipment for any manufacturer. Likewise, *the menu plan and meat purchase specifications must be planned prior to an intelligent decision about equipment* for any food service operation.

Scales

Heavy service type receiving scales are needed for any operation. It is imperative for receiving security that all products be weighed. Scales are now available which give a printed record of each weighing. Where a monorail is used, a track type scale should be included.

Restaurants selling portions should have some type of portion scale. An inexpensive model, about $21.00, is adequate in most cases.

Selling scales for the retail meat dealer and locker plant depend on the type of operation. Platform scales are needed for service operations. For efficiency in large volume self-service operations, automatic weighing, computing, and label printing systems should be considered. Although very expensive, these scales can be justified for their accuracy, patron confidence building, and labor saving features.

The Grinder

This important piece of equipment is frequently overlooked. The products that are processed through the grinder, the various types of ground meats and the many non-meat items ground are frequently the extended profit items and upgraded by-products. A poor tool will not cut properly. If the grinder is small relative to the amount of product that is being processed, it becomes a bottleneck with an unnecessarily high labor cost. With a small capital investment, considerable savings can be achieved by a grinder properly rated for the job.

The Patty Maker

A device to mold patties is a complementary tool to the grinder. There are many different types on the market, including a hand-operated oval cut in a piece of wood, a small stainless steel metal ring, and an automatic machine which stamps out the patty and also interleaves it with paper. It is important that some kind of device be used to make uniform patties and hamburger steaks.

The Cuber

Like the grinder, the cuber is a by-product upgrading device. Many small pieces of meat can be put through the cuber with good results. Swiss steaks, cube steaks, and chicken fried steaks, can be made. The cuber is relatively inexpensive, and readily obtained. The machine should have two features: (1) a safety device that makes it virtually impossible for the operator to put his fingers into the blades; and (2) a simple means of disassembly for cleaning and sanitation.

The Mechanical Tenderizer

This power device creates hundreds of slits with its multiple blade arrangement. By breaking down the tissue in this fashion, a meat cut can be made almost as tender as the operator desires (Fig. 11.10).

FIG. 11.10. JACCARD ELECTRIC-SUPER-TENDERMATIC

Saws

Both electric and hand type saws have a function. An electric saw should be large enough to do the job for which it is intended. The smaller sizes are generally adequate in restaurants. Saws may be equipped with two kinds of blades, one that cuts through bone, and another that slices meat. A hand saw is usually necessary. Blades are generally rented. The electric saw should be placed next to a table for receiving cut product. Meat specifications in many restaurants may preclude the need of an electric saw.

Slicers

Slicers may be needed for bacon, cheese, and cold cuts. These range from hand operated to automatic equipment with stacking devices. Bread may be sliced to a desirable thinness on the cold meat slicer or with the meat blade on the band saw.

Hand Tools and Accessory Utensils

These are usually provided by the butcher or the chef. Good knives that will hold their edges are a good investment, as is a steel to keep them sharp. A mechanical sharpening device may be rented. A larding needle for ribboning or larding can be purchased from a butcher supply house. A wide variety of functional small tools is shown in Fig. 11.11. Platters. pans, and lugs are also needed.

Lighting

This is frequently overlooked. Butchers are working with sharp, dangerous tools. Adequate lights should be provided. An even light, especially that from fluorescent tubes of approximately 50 ft-candles of light at table height is good. Special tubes can be purchased which will not distort the color of the product. The best lights over self-service retail cases are incandescent.

Cutting Tables

Tables are coming into general use in place of meat blocks. There is usually little or no reason for chopping with a cleaver. If chopping is necessary, a small, laminated wooden block may be provided and placed on a cutting table. Hard rubber cutting blocks are now available. These are easy to keep clean and last longer than wood.

The present trend is toward metal tables, on which the product can

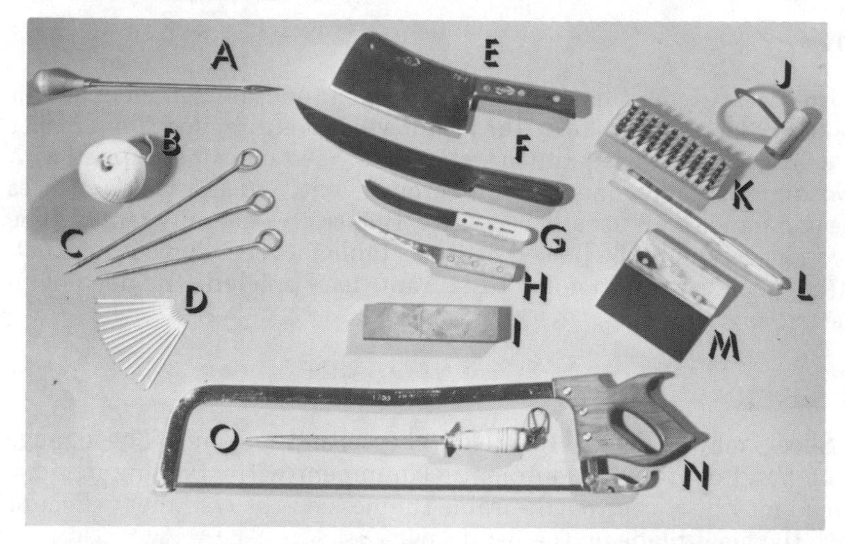

Courtesy of Cornell University

FIG. 11.11. HAND TOOLS—NEEDLE, A; STRING, B; METAL SKEWERS, C; WOODEN SKEWERS, D; CLEAVER, E; STEAK KNIFE, F; BUTCHER KNIFE, G; BONING KNIFE, H; SHARPENING STONE, I; HOOK, J; BLOCK BRUSH, K; HYDROMETER, L; BLOCK SCRAPER, M; SAW, N; STEEL, O

be moved easily. These tables are usually provided with a cutting surface 12 to 18 in. wide at the front. Metal tables will help eliminate freak accidents that sometimes occur when moving heavy products on wooden tables.

Tying Machine

A tying machine is needed if a large number of roasts are to be tied (Fig. 11.12). This equipment does a good job and is much faster than hand tying. This investment has to be weighed with the labor hours that it will save.

Plastic Bags

If the meat is to be packaged in shrink type vacuum bags, the necessary equipment must be provided. This consists of the following: (1) a loading chute to hold the product while the bag is slipped over it; (2) a vacuum pump to remove as much air as possible from the bag, drawing it tightly on the product; (3) a clip fastening device to tightly clip and seal the open end of the bag, equipped with a cutting device to trim off the excess bag; a processing machine is available to effect steps two and three (Fig. 11.13); (4) a dip tank in which the bagged product is im-

FIG. 11.12. TYING MACHINE

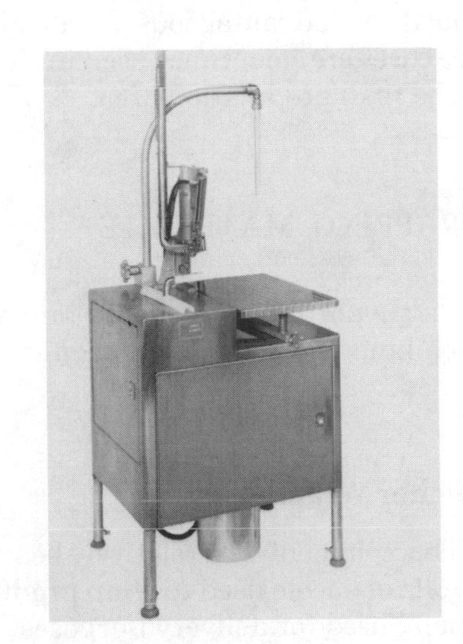

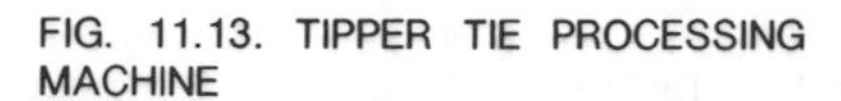

FIG. 11.13. TIPPER TIE PROCESSING MACHINE

mersed to cause the bag to shrink tight on the product (in volume operations a shrink tunnel is used, moving the product through it on an electric conveyor); and (5) a receiving area to handle the product as it comes from the tunnel or tank.

Although the process is expensive both from the point of view of the cost of the bags, as well as the labor involved, there are many operations that use it profitably. For example, some chain restaurants operate a central commissary, making portion-cut steaks which they prefer packed in poly vinyl chloride bags. Memory-type plastic bags are available at a lower cost, but are less effective.

Fly Fans

Provisions should be made at exterior doors to prevent the entry of insects. Four-bladed slowly revolving fans are acceptable. Use of an air curtain, however, is more effective, eliminating the need of any exterior screen door, and effectively repelling all air contaminants, including dust.

Refrigerated Cases

There are two general classes of display cases, service and self-service. Service cases are made with a single glass for maximum display, and multiple glasses for maximum holding. Restaurants sometimes use a meat case advantageously to display product in the dining area. Service cases are sometimes used in the kitchen to store short order items in the final preparation area.

WRAPPING MATERIALS

Only a portion of the broad range of paper goods will be considered. Each kind of paper has a specific function, and is best utilized as intended.

Butcher Wrap

This commonly is a relatively heavy, unbleached, white, or pink paper in rolls or sheets used to wrap product for sanitary reasons only and is usually used for delivery purposes.

Wrapping Wax

This is usually a light weight, brown colored wax kraft paper used as an inner wrap with the butcher wrap, especially for ground beef or wet items as corned meats. It is not an effective long run freezer wrap.

Freezer Wrap

There are a wide variety of effective freezer wraps. Cryovac, Pliofilm, and aluminum foil are all very effective, though sometimes expensive. A Kraft base paper laminated with paraffin wax blend, or with foil, or with resins such as polyethylene, or in combination with each other, are generally effective and economical freezer papers. Waxed buckets or other moisture proof paper containers are sometimes used.

The undesirable palatability changes that occur due to dehydration and oxidation of the fats, can be minimized with a good quality, tight, moisture-proof, oxygen excluding wrapping material.

Bacteriostatic Wraps

Where it is necessary to cut and store fresh steaks, the problem of loss of bloom and surface bacterial action is manifest. In the average restaurant cooler it is not uncommon to find the use of unsanitary wet cloths or a paper which is totally ineffective. There are at least two kinds of paper products on the market, treated chemically to have a bacteriostatic effect. Restaurants and service butcher shops will find peach platter wrap in rolls or sheets easy to use and effective. Steaks should be placed on and inter-leaved with this paper. Other cuts may be wrapped. The self-service shop should use a specially treated cellophane, such as MSAD 80 which combines good display while maintaining the fresh cut characteristics of the product. Some restaurant operators may select this cellophane, although it requires more labor to wrap each steak individually than to use the platter wrap. These materials are effective for but a few days. Cryovac may be used for extended periods.

Patty Paper

Special papers are available for machines making ground beef patties. There are different kinds and qualities for special purposes. For example, a double sheet of laminated paper may be used for patties to be frozen, facilitating separation while still frozen by splitting the two sheets of paper.

Miscellaneous

Some of the other necessary wrapping supplies that may be required, depending upon the type of operation involved, include twine, gummed tape, and for self-service display, tubs, boats, boards, cellophane wrap, and heat sealing equipment.

SANITATION

Most equipment is difficult to maintain in a sanitary condition. Proper sanitary practices are vital to a good operation, and most critical where highly perishable meat products are concerned. A few basic rules should be followed: (1) demand personal cleanliness, clean fingernails, clean garments; (2) provide a wash basin in the shop with hot and cold water, liquid soap, and a tool sterilizer, if possible; (3) provide steam or hot water; (4) clean all utensils daily—air dry; (5) hose down and clean all equipment daily, including grinder head and cuber knives; (6) use germicides containing chlorine to destroy bacteria; equipment should be prewashed before chlorine sanitization, as organic material will interfere with chlorine action; water temperature should be 130°F or higher; and (7) laminated blocks cannot be soaked, but should be scraped and wirebrushed daily; the surface should be bleached and chlorinated periodically.

The U.S. Department of Agriculture "Guide to Construction and Equipment" should be consulted where federal inspection is contemplated. The local office should be contacted during the planning stage. As a construction guide for a sanitary plant, even where no inspection is contemplated, the information is most valuable.

SAFETY

The processing of meat involves some very dangerous equipment, with a high contingent insurance cost. Rules of safety should be formalized, posted and enforced. A few basic rules follow:

(1) Knives are at their best when sharp—and very dangerous too
Keep handles clean
Hold them firmly
Provide a storage place just for knives
Store them all with handles in the same direction
Don't grab for a knife when it drops
Don't carelessly leave knives on table or block
Don't put a piece of meat on top of a knife
Don't rinse a knife in a sink full of soapy water

Don't carry a knife while moving something else
Walk about holding knife handle, blade down

(2) Permit no "horse play" in butcher shop

(3) Provide adequate work space and aisles around power tools and work tables

(4) Every power tool is dangerous if abused
Don't force the electric saw
Don't reach into the grinder head—use a stomper—provide a safety device
An electric patty machine is responsible for many serious accidents

(5) Keep the floors clean to prevent slipping

REFERENCES

ANDERSON, D. L., and SHAFFER, P. F. 1954. Principles of layout for self-service meat departments. Marketing Research Rept. *77,* U.S. Dept. Agr., Washington, D.C.

ANON. 1945. Meat Handbook of the U.S. Navy. U.S. Navy Dept., Washington, D.C.

ANON. 1961. U.S. inspected meat processing plants. M.I.D.-A.R.S. U.S. Dept. Agr., Washington, D.C.

ANON. 1962. What is happening to refrigeration. Institutions Magazine, *51,* No. 6, 99–100.

BRODNER, J., MASCHAL, H. T., and CARLSON, H. M. 1962. Profitable Food and Beverage Operation. 4th Edition. Ahrens Publishing Co., New York.

DUKAS, P., and LUNDBERG, D. E. 1960. How to Operate a Restaurant. Ahrens Publishing Co., New York.

EDINGER, A. T., MEWIS, B. H., MUMFORD, H. D., *et al.* 1949. Retailing pre-packaged meats. A circular. U.S. Production and Marketing Admin., Washington, D.C.

HARWELL, E. M., ANDERSON, D. L., SHAFFER, P. F., and KNOWLES, R. 1953. Packaging and displaying meats in self-service meat markets. Marketing Research Rept. *44,* U.S. Dept. Agr., Washington, D.C.

LASCHOBER, J. A. 1960. A Short Course in Kitchen Design. Institutions Magazine (printed in book form) Chicago, Ill.

ROMANS, J. R. and ZIEGLER, P. T. 1977. The Meat We Eat, 11th Edition. Interstate Printers and Publishers, Danville, Illinois.

STOKES, J. W. 1960. Food Service in Industry and Institutions. Wm. C. Brown Co., Dubuque, Iowa.

TRESSLER, D. K. and EVERS, C. F. 1957. The Freezing Preservation of Foods, Vol. 1. Fresh Foods. *2:* Cooked and Prepared Foods. Avi Publishing Co., Westport, Conn.

TRESSLER, D. K., EVERS, C. F., and EVERS, B. H. 1953. Into the Freezer—and Out. Avi Publishing Co., Westport, Conn.

12

Beef Hindquarter

The following chapters illustrating meat cutting are intended only as a guide, not as a comprehensive course. For the beginner, they may help provide additional material. For the experienced cutter, they may provide some variations of technique. The procedures in the text sometimes approximate the U.S. Department of Agriculture's acceptance specifications; at other times they differ. There are many regional variations and the text is no exception. Percentage yields are based on (1) USDA Choice carcasses, (2) the particular specifications of the text, and (3) the slight trim and yield variations beyond definition.

In some instances yields are intentionally omitted to avoid confusion; at best they are only intended as approximations. Knife cuts are generally made perpendicular to the surfaces cut. Refer frequently to the bone structure charts. These will help to clarify and expand the photographs and the cutting descriptions. Bone references to anterior is the point closest to, and posterior is the point farthest from, the head of the carcass.

CUTTING AND TYING

Definitions

Cubing.—This generally refers to passing the product through a mechanical device which partly cuts the tissues and knits pieces together. Cubing is done by machines that have two sets of knives which counter-rotate. As product is fed between the knives, the counter rotation advances the product, and the series of small knives make short, shallow cuts. The product is usually fed from the top and received at the bottom.

Scoring.—This is done with the sharp edge of a knife. The primary function is to shorten long (and usually tough) muscle fibers. Frequently, scoring will be done in a diamond (diagonal) pattern, which adds to the attractiveness of the piece of meat. A typical piece of meat that may be scored would be a flank steak. The knife cuts are just deep enough to break the surface.

Pounding.—This is usually a manual operation, with some kind of a device resembling a mallet. It can be done with the flat side of a butcher cleaver, creating a flat surface. In general, when it is done in this manner, the piece of meat is covered on the top and on the bottom with a piece of wax paper. There are special tools for pounding which have a waffle pattern or a set of small knives. In any case, the pounding is generally done to shape and to actually break down the tissues.

Dicing.—Dicing refers to cutting the meat into uniform cubes as, for example, with beef stew and braised sirloin tips.

Grinding.—Grinding is done with a mechanical device, which literally chops or cuts the meat into small pieces. Frequently grinding is done two or more times to adequately mix the product. Sometimes it is done first through a coarse plate, which is described as "chili grind," and then through a fine plate. For sausages or hors d'euvres the meat is generally ground several times to make it very fine and to cut any connective tissue. There are a variety of grinder plates and sizes that may be used and a variety of opinions on their selection. The grinder should cut quickly and not be allowed to become hot. Prechilling meat improves the grind.

Tying.—Tying is usually done to make a roasting item conform to a given shape, as for example, a boneless leg of lamb. Sometimes a string is run around the product several times and tied at the end. More professionally, it is achieved by tying a series of single loops around the product. The single strings are fashioned in such a manner that they can be drawn up and tied tightly.

The "butcher's knot" is used widely because it is easy, quick, and cinches well. It is a combination of two single knots, a slip knot, over which is placed a half hitch. The execution looks complicated, but by studying the following directions along with the how-to illustrations (Fig. 12.1A through 12.1K), the butcher's knot is easily mastered.

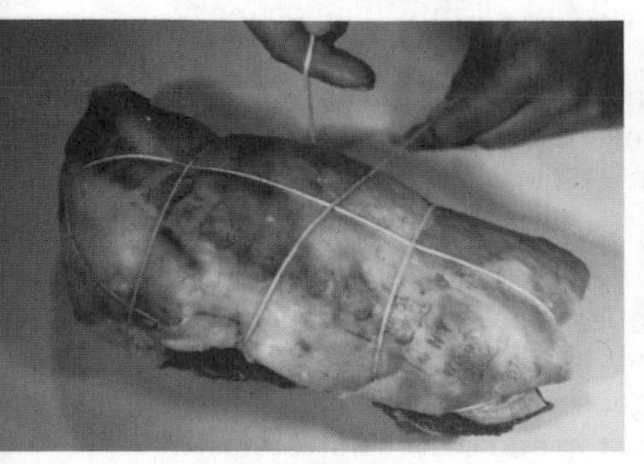

FIGURE 12.1A

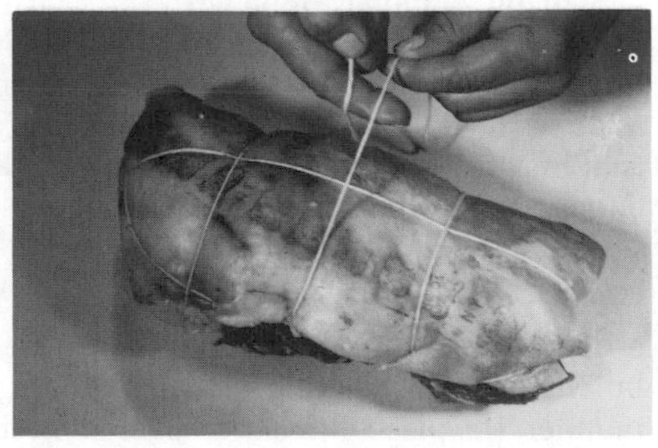

FIGURE 12.1B

(1) Run the end of the string under, around, and over the roast. Hold the end of the string in the left hand and place the right index finger under the standing part (Fig. 12.1A). (2) Bring the end of the string over

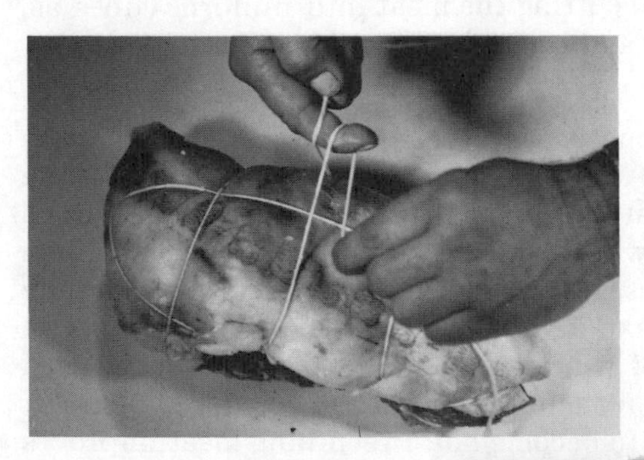

FIGURE 12.1C

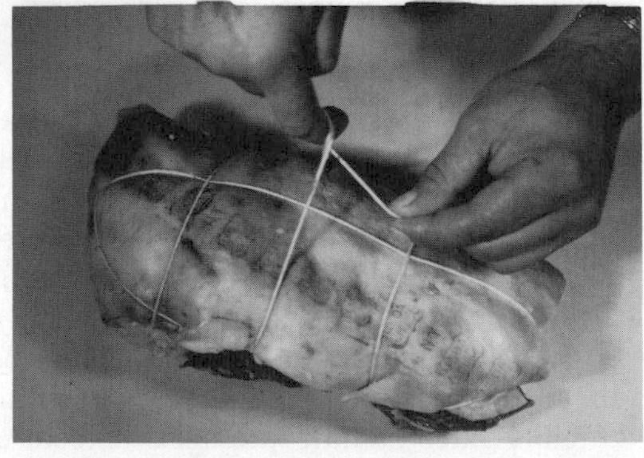

FIGURE 12.1D

the index finger parallel with the standing part (Fig. 12.1B), (3) around the index finger and back in front of it (Fig. 12.1C). (4) Twist the right hand, forming a loop (Fig. 12.1D).

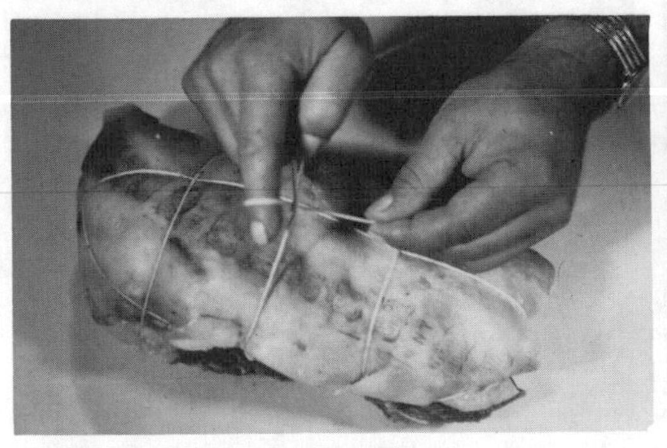

FIGURE 12.1E

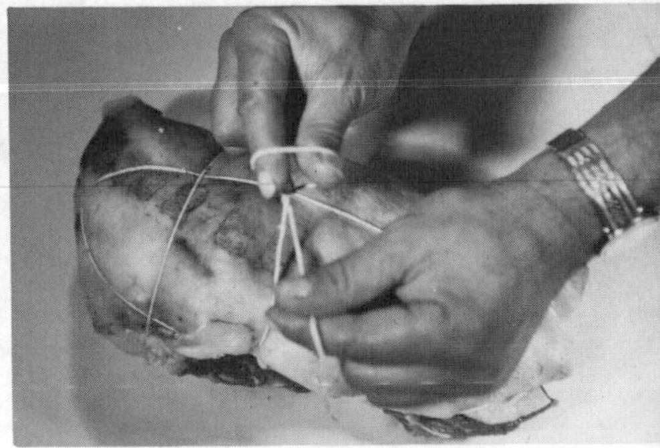

FIGURE 12.1F

(5) Continue to twist until the back of the hand is up, the standing part slides along the finger and rests between the loop and the end (Fig. 12.1E). (6) Place the thumb in the loop with the index finger and reach through for the end (Fig. 12.1F).

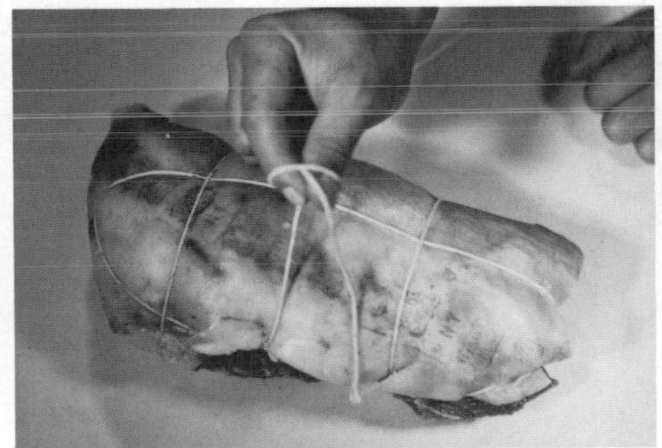

FIGURE 12.1G

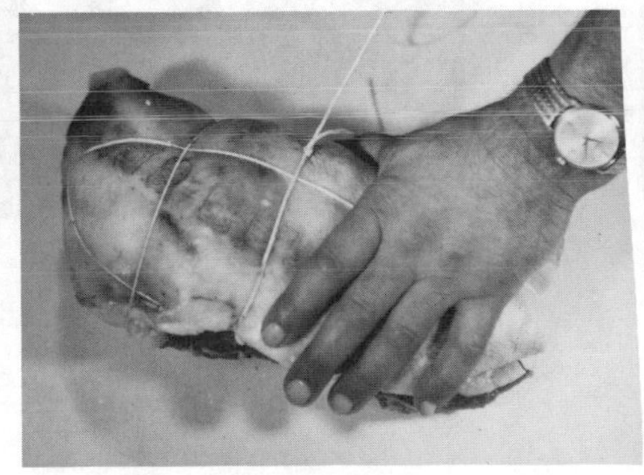

FIGURE 12.1H

(7) Draw the end through the loop forming a slip knot (Fig. 12.1G).
(8) Draw up and position the knot (Fig. 12.1H).

This completes the first step, the slip knot. Now start the half hitch.

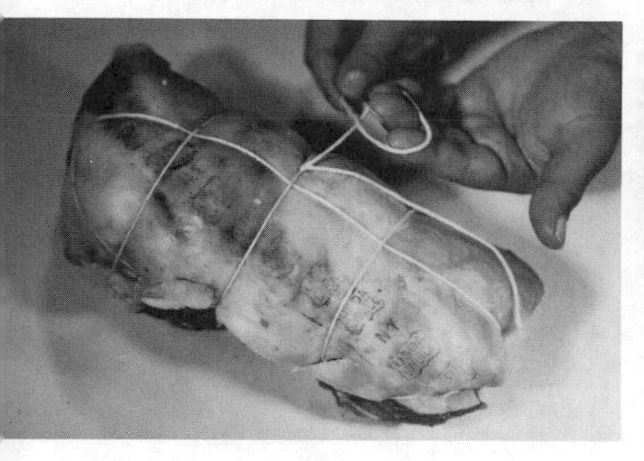

FIGURE 12.1I

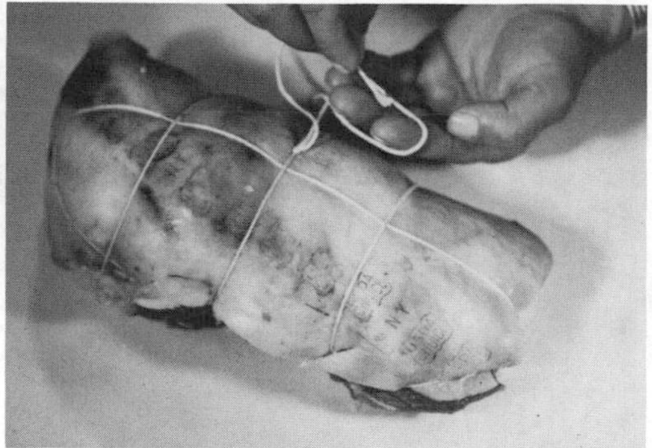

FIGURE 12.1J

(9) Cast a loop in the standing part on two fingers of the left hand (Fig. 12.1I). **(10)** Twist the loop and draw the end through (Fig. 12.1J). **(11)** Draw up the knot (Fig. 12.1K). Cut the tied ends short.

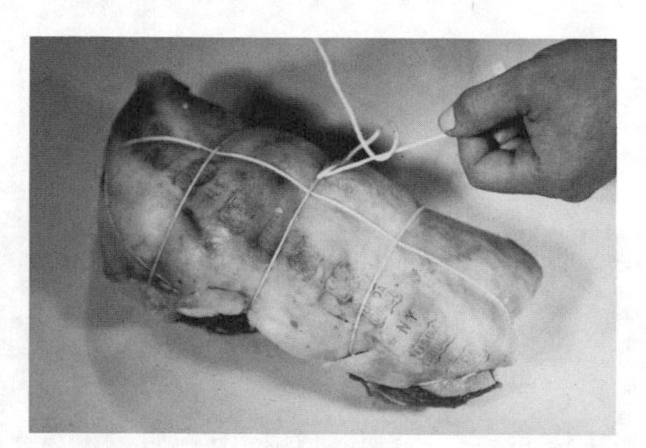

FIGURE 12.1K

USE OF BY-PRODUCT

Offal, according to Webster, is the "waste parts; the entrails of a butchered animal." Today the name is a misnomer. The meat industry has long since learned that waste and profits cannot coexist. Offal today has been merchandised and upgraded. In many instances, as in the case of veal sweetbreads, it is the most valuable product. Every industry is on the same threshold. Many operators have already learned to merchandise their "offal."

An attempt should be made to get maximum advantage from the by-product, either in the dining room, for employee's meals, or in the retail display case. Here are a few simple rules: (1) separate fat, bone, and lean; (2) separate the thick lean; (3) upgrade the more tender cuts for chicken fried steaks, cube steaks, swiss steaks, tenderettes; (4) steak by-products as tenderloin tails should be used for steak sandwiches, brochettes, grenadine of beef, stroganoff, etc.; (5) use lean cubes for stewing and braising; (6) use fat and bones for rendering and stock (round bone rings are a fine garnish for broiled steaks); and (7) use the small fat pieces of trim for grinding. A good quality ground beef should have visible fat. It should range from 15 to 30 per cent. Very lean ground meat is relatively tough, flavorless, and dry.

The average restaurant operator and retail market customer has some idea that ground meat should look red and lean. To satisfy this notion considerable cow and bull meat is used. This does not make an especially palatable hamburger, but it does have the "right appearance." In this case, taste tests should prevail over the visual examination. Many very successful coffee shop operators grind USDA Choice boneless chucks, with virtually no defatting. The ground meat looks very fat and pink, and it is most palatable.

BEEF SIDE

The beef side (Fig. 12.2 and 12.3) is separated into forequarter and hindquarter at the packing house. A point between the 12th and 13th rib is located with the tip of a knife. The knife cut is extended to the backbone. The bones are sawed through and the forequarter hangs by the flank. The forequarter is hooked, and the flank is cut through separating the quarters. As a rule, one rib bone, as described, is left in the hindquarter (Fig. 12.4–12.6) to shape it. This rib bone is the one found in the strip loin portion of the loin.

	Per cent
Forequarter	52
Hindquarter	48
	100

BEEF HINDQUARTER

Primal Cuts and Yields:

Cut	*Per cent*
Trimmed Loin	37
Round	49
Flank, fat, kidney	14
	100

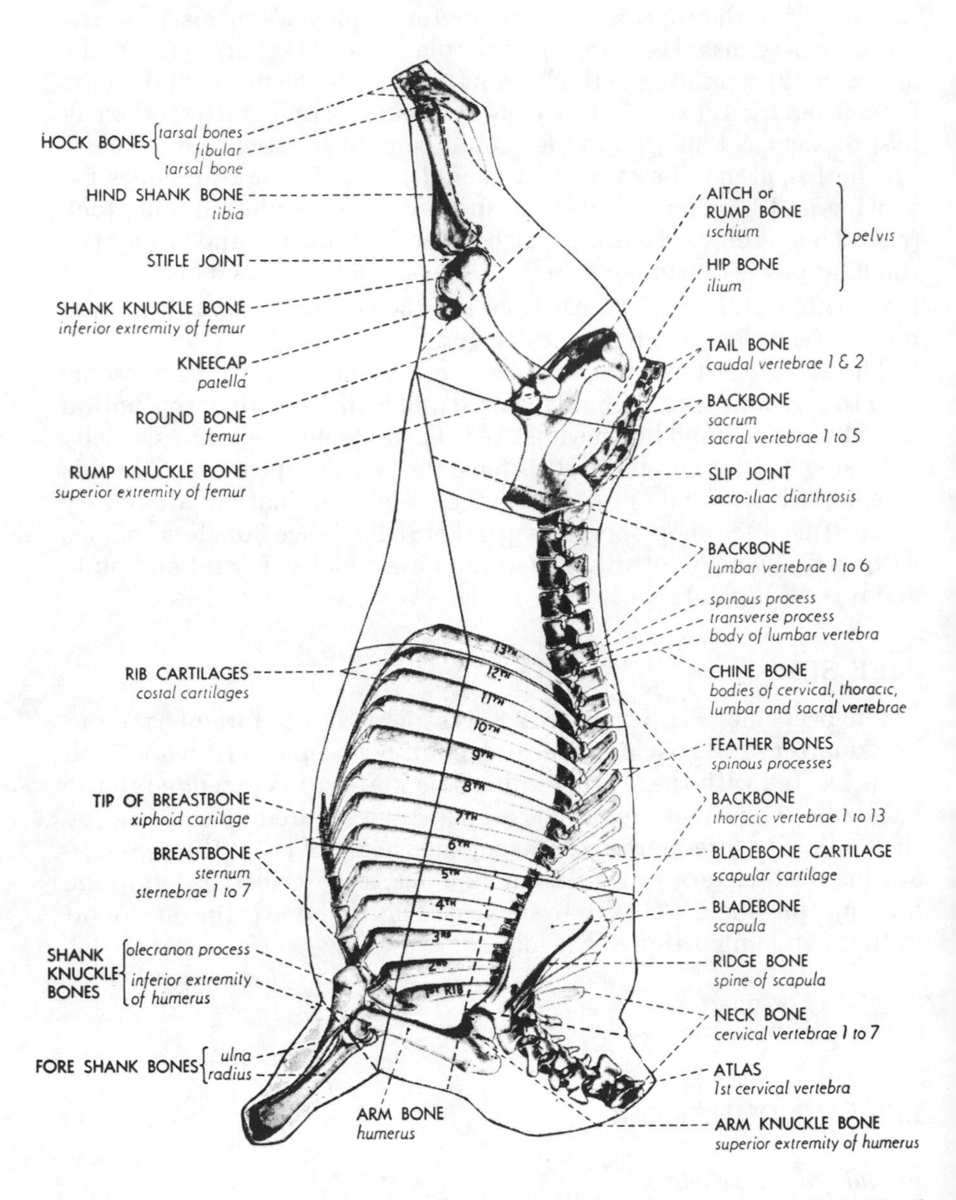

Courtesy of National Live Stock and Meat Board

FIG. 12.2. BEEF SIDE—SKELETAL STRUCTURE

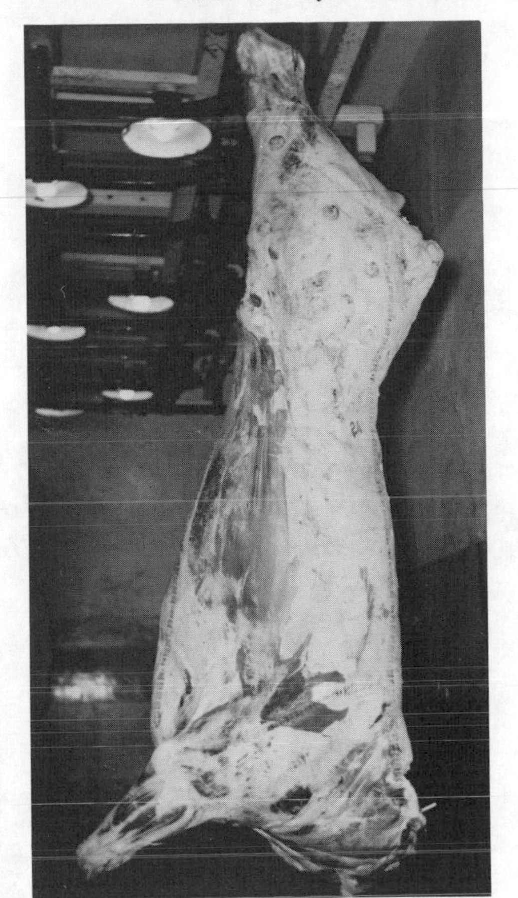

FIG. 12.3. BEEF SIDE

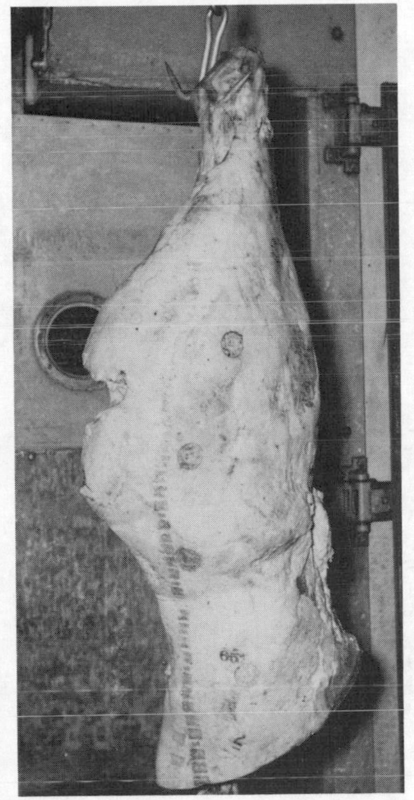

FIG. 12.4. BEEF HINDQUARTER

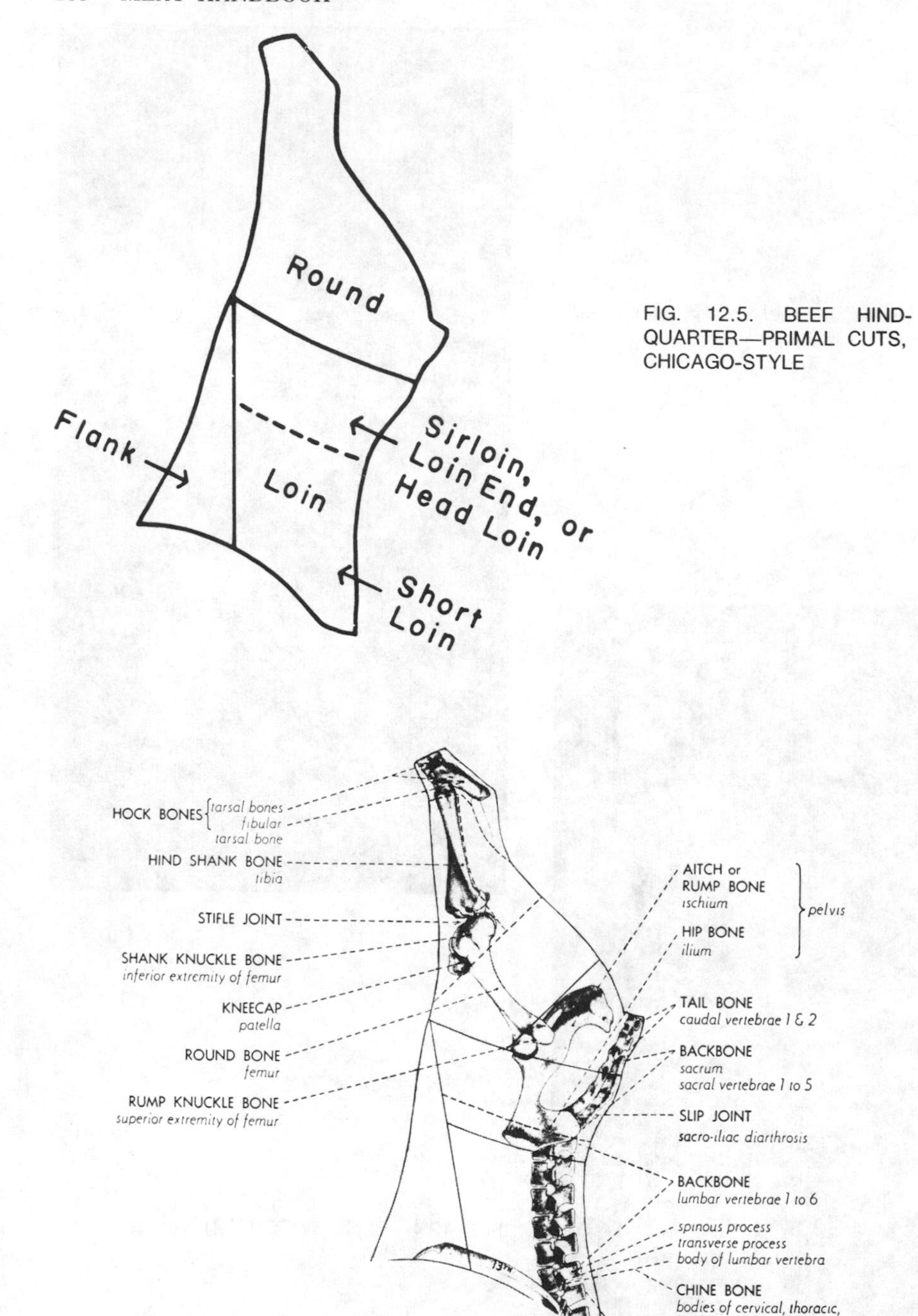

FIG. 12.5. BEEF HINDQUARTER—PRIMAL CUTS, CHICAGO-STYLE

Courtesy of National Live Stock and Meat Board

FIG. 12.6. BEEF HINDQUARTER—SKELETAL STRUCTURE

BREAKING BEEF HINDQUARTER

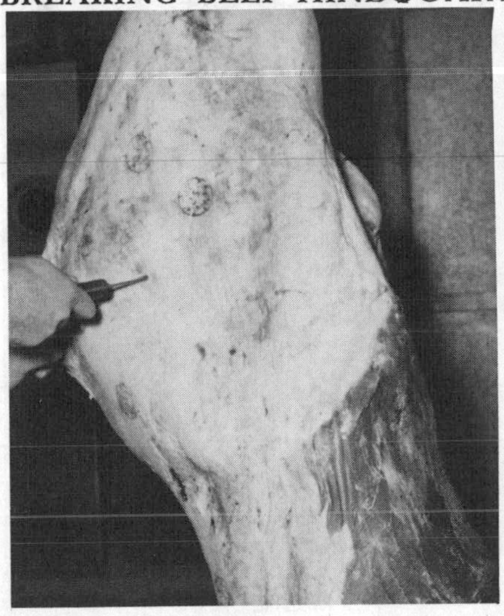

FIGURE 12.7A

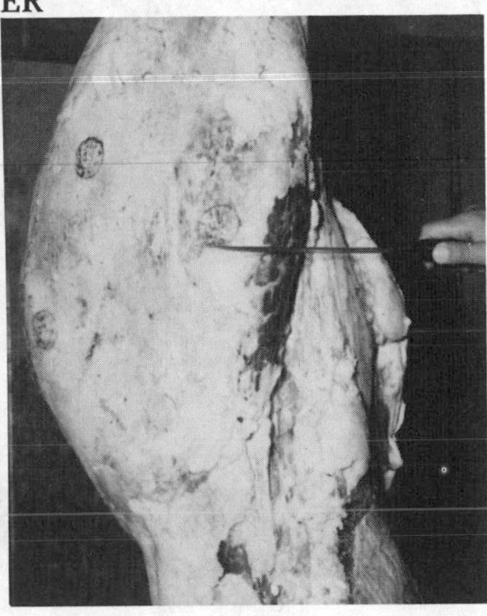

FIGURE 12.7B

(**1**) Locate protuberance of rump knuckle bone with knife tip (Fig. 12.7A). (**2**) Extend an imaginary straight line from second caudal vertebra through point in step one to opposite side of hindquarter (Fig. 12.7B).

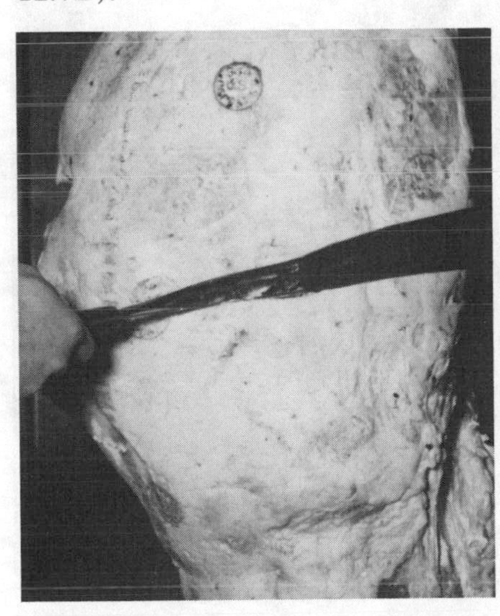

FIGURE 12.7C

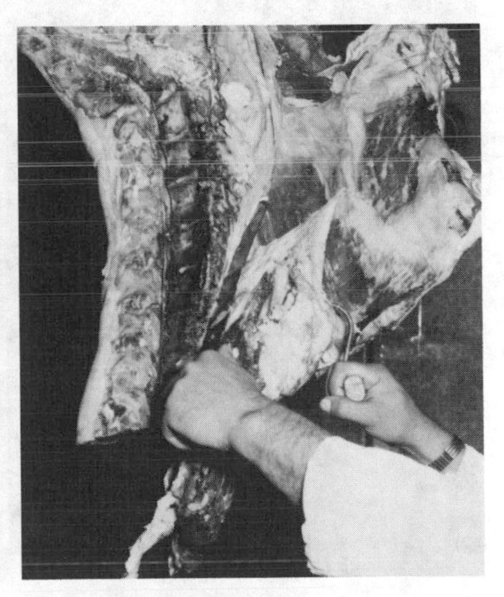

FIGURE 12.7D

(**3**) Cut along this imaginary line to bone, partially separating loin and round (Fig. 12.7C). (**4**) Remove kidney knob and "hanging tenderloin" (Fig. 12.7D).

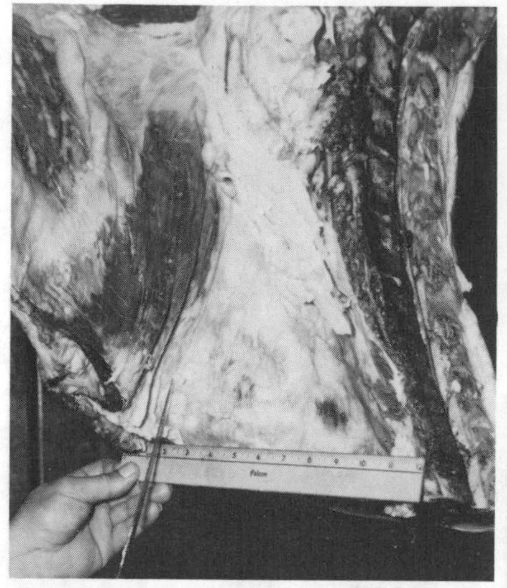

FIGURE 12.7E

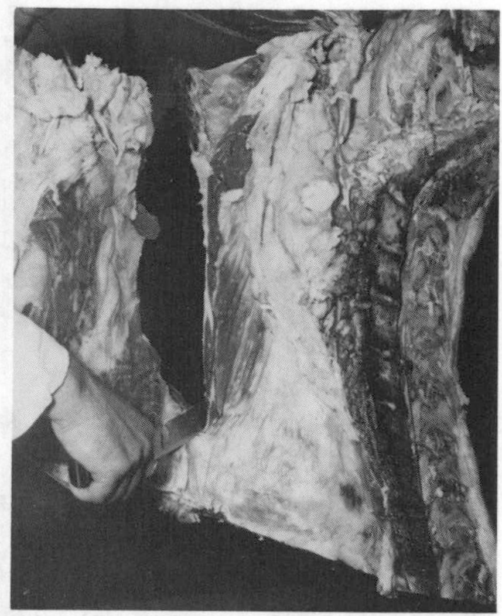

FIGURE 12.7F

(5) At the rib end of loin, locate a point 10 in. from protruding edge of chine (Fig. 12.7E). (6) From a point where the knuckle and cod fat naturally separate, make a slanting knife cut in a straight line to point marked on the rib end of loin (Fig. 12.7F).

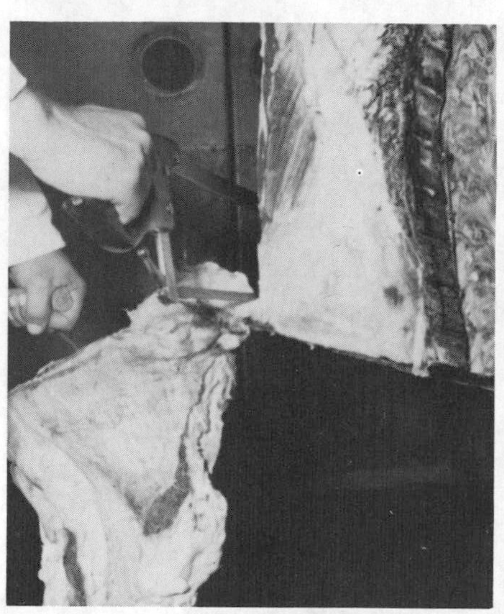

FIGURE 12.7G

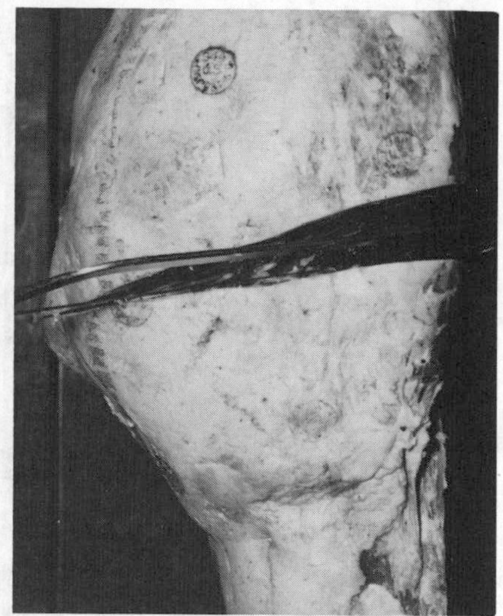

FIGURE 12.7H

(7) Hook flank and saw through rib bone to remove flank (Fig. 12.7G). (8) Continue separation of loin and round by extending the knife cut to the tail vertebrae and saw through the knuckle bone. Hook loin before separation is completed (Fig. 12.7H).

Breaking Variations

Trimmed Hindquarters.—Flank is off and kidney out. Consists of round and loin in one piece.

Long Round.—Cut is made at the top of the ilium, leaving the loin end attached to the round, and a short loin.

The kidney conformation varies with the side of the animal. They are called "closed" and "open" sides (Fig. 12.8 and 12.9).

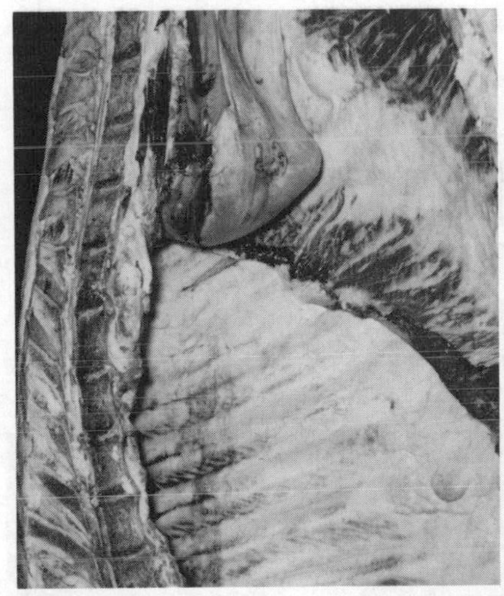

FIG. 12.8. BEEF KIDNEY, RIGHT SIDE OR CLOSED SIDE

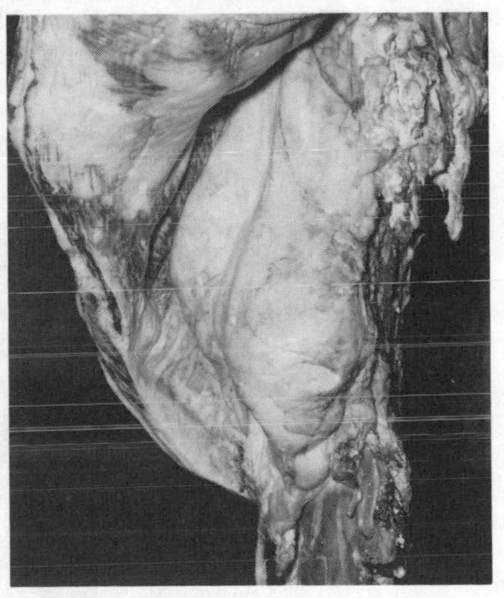

FIG. 12.9. BEEF KIDNEY, LEFT SIDE OR OPEN SIDE

LOIN

Packing House Trim

The loin, as it drops out of hindquarter, is usually trimmed in a more or less standardized manner before it is marketed. The fat protruding from the chine bone is trimmed to a plane level with chine. The loin is then placed on the bench, fat cover side down.

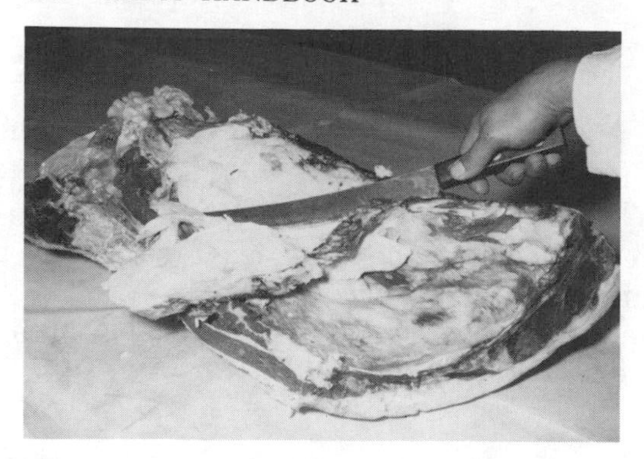

FIGURE 12.10

(1) From an imaginary point 1 in. above protruding edge of chine and in a straight line to a point 2 in. below cut on the flank side, all fat is removed extending above plane so described (Fig. 12.10).

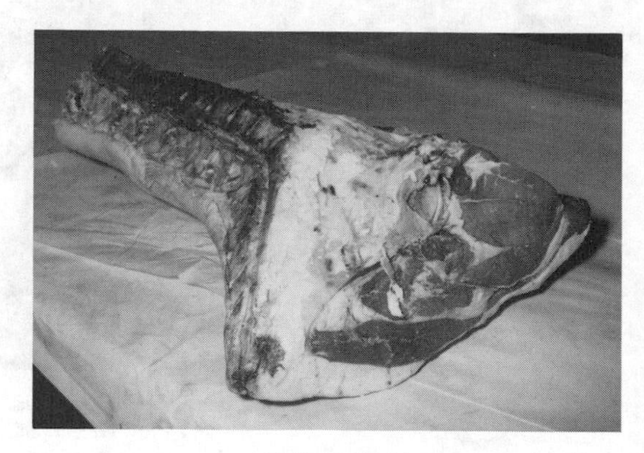

FIGURE 12.11

(2) Fat in pelvic region is trimmed to a maximum of 1 in. (Fig. 12.11).

From the buyer's point of view, some control of trim can be achieved by testing yields. Buyer should develop specifications, describing an acceptable trim.

Muscle Boning the Loin

"Stripping," or muscle boning the loin (Fig. 12.12), is most popular in the food service industry and with the "carriage trade" meat retailers. The cuts and yields are as follows:

Cut	*Yield Per cent*
Tenderloin	12
9 in. Boneless strip loin	21
Top sirloin	19
Bottom sirloin	15
Trim, fat, and bone	33
	100

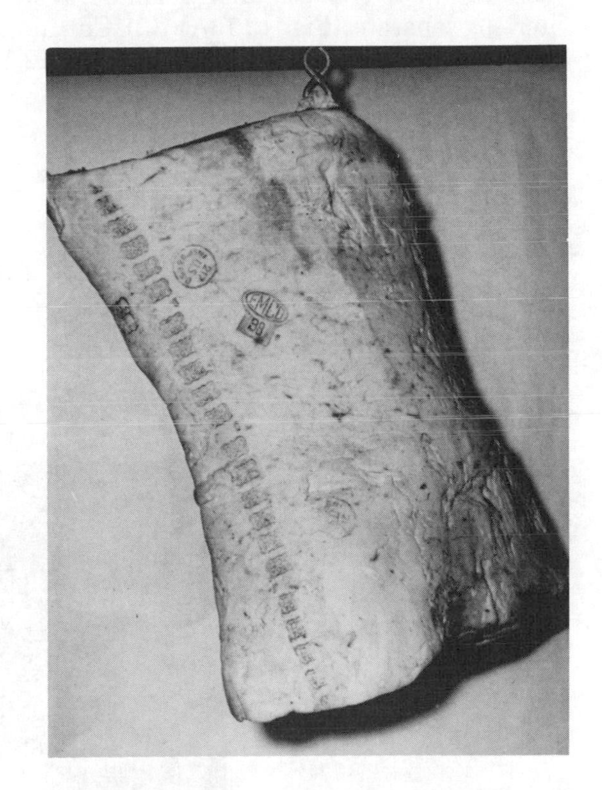

FIG. 12.12. BEEF LOIN

The usual steps are to "pull" tenderloin, sever strip loin from loin end, and finally separate top and bottom sirloin.

Pulling Tenderloin (Filet)

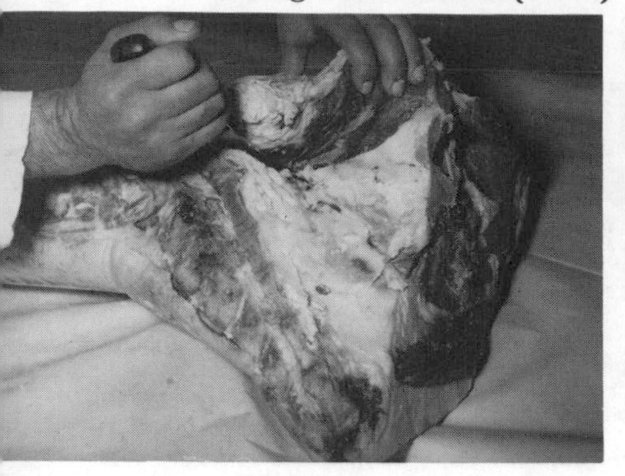

FIGURE 12.13A

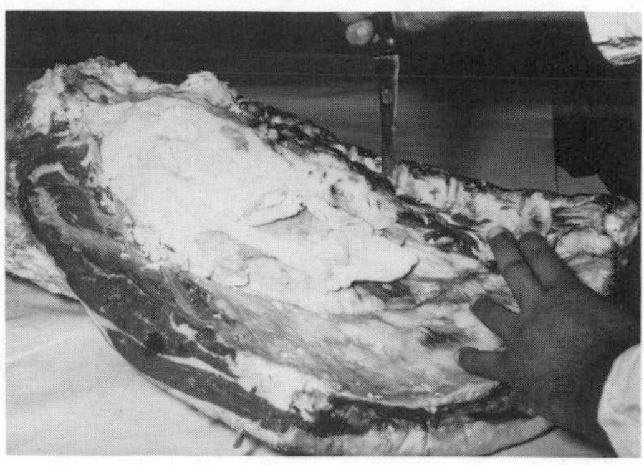

FIGURE 12.13B

(**1**) Start at loin end. Cut along seam between tenderloin and hip bone. Start to separate (Fig. 12.13A). (**2**) Cut along inside of chine bone, carefully following contour of bones (Fig. 12.13B).

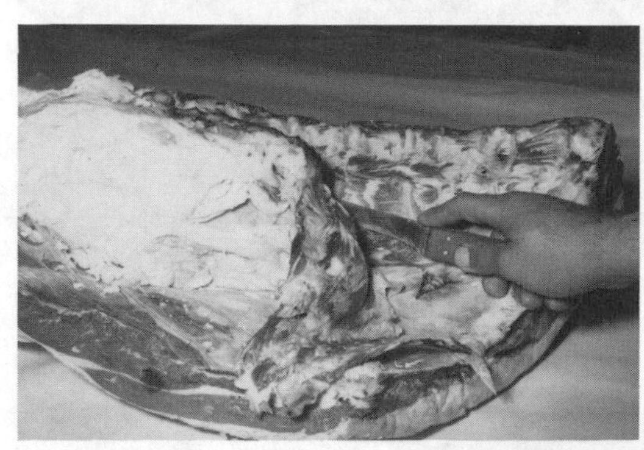

FIGURE 12.13C

FIGURE 12.13D

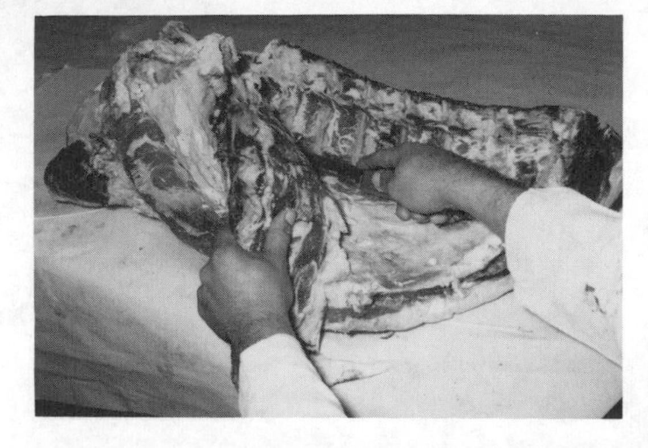

(**3**) Starting at rib end of loin, follow contour of backbone (Fig. 12.13C). (**4**) Free the tenderloin by cutting and pulling it (Fig. 12.13D).

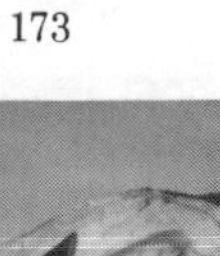

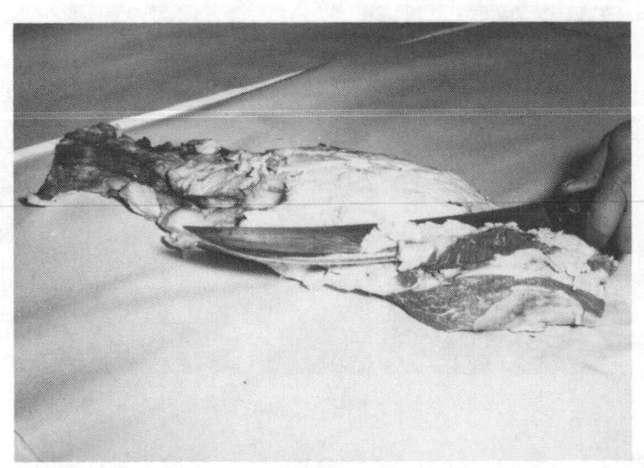

FIGURE 12.13E

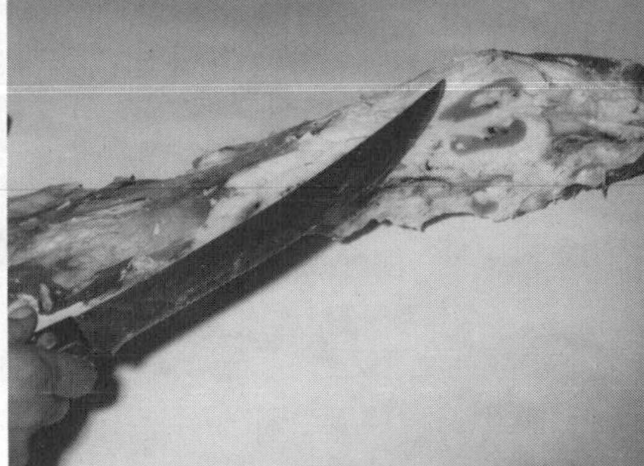

FIGURE 12.13F

(5) To approximate a jobbing house trim, (a) trim the head muscle (Fig. 12.13E); (b) expose and split lymph gland on the top of the butt end (Fig. 12.13F); and (c) trim surface fat to this plane, tapering fat to silver seam at a point not over three quarters of length of the tenderloin. Shape both sides by trimming in straight lines to conform with lean. Trim off all ragged edges.

Making Strip Loin

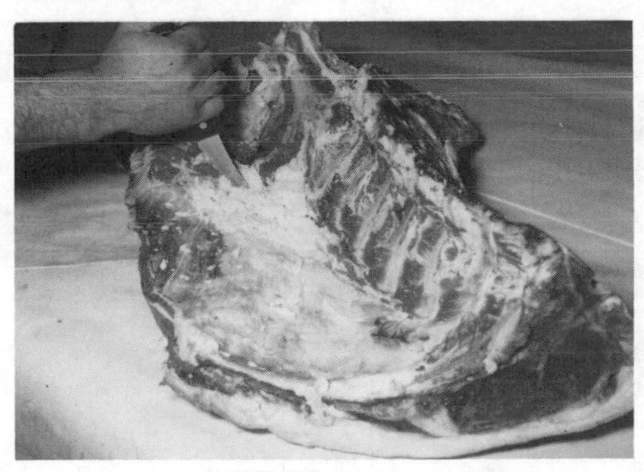

FIGURE 12.14A

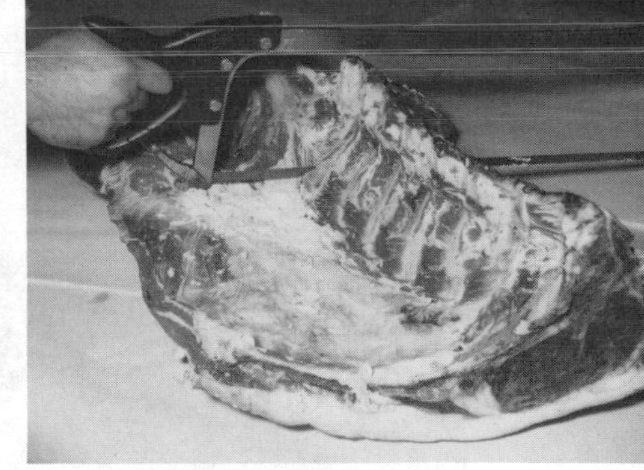

(1) With a knife tip, locate tip of hip bone. Extend cut perpendicular to chine to sixth lumbar vertebra (Fig. 12.14A). (2) Saw through backbone (Fig. 12.14B).

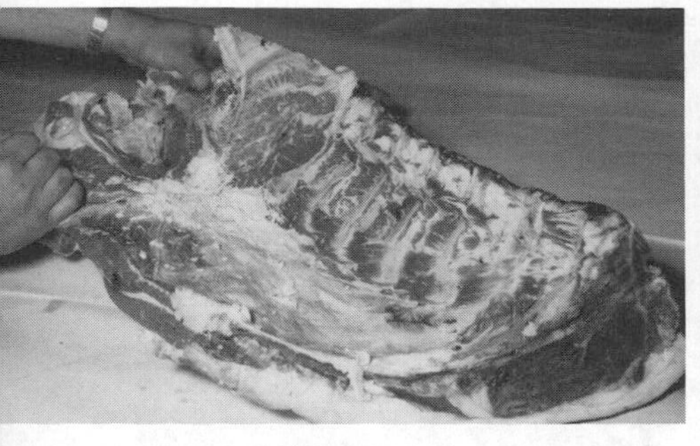

FIGURE 12.14C

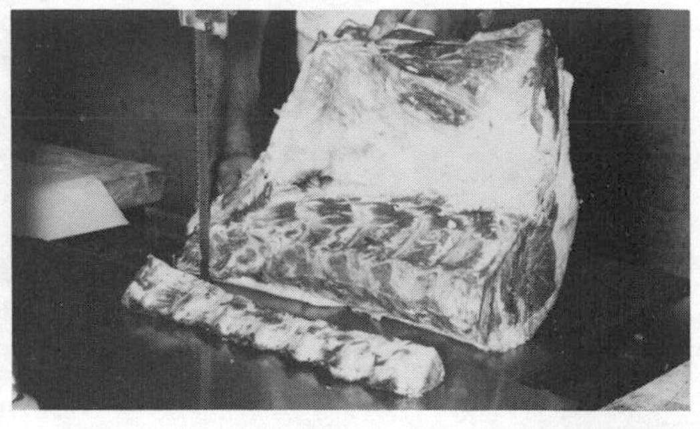

FIGURE 12.14D

(3) Sever strip loin with knife cut through flank (Fig. 12.14C). (4) Saw off protruding chine bone (Fig. 12.14D).

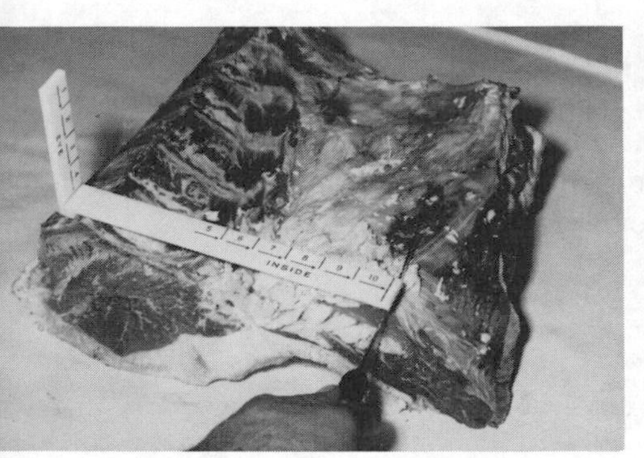

FIGURE 12.14E

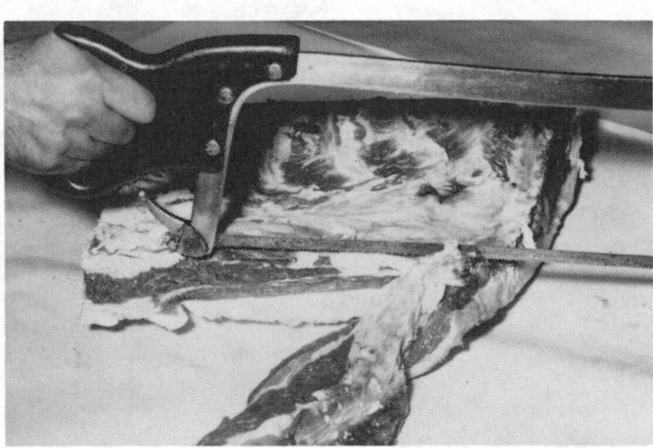

FIGURE 12.14F

(5) Measure and score 10 in. from inside edge of chine at both ends of strip loin (Fig. 12.14E). (6) Between score marks, cut off flank in a straight line. Saw through single rib bone (Fig. 12.14F).
(7) This is the average trim of a 10 in. bone-in strip loin (Fig. 12.14G).

FIGURE 12.14G

Boning Strip Loin

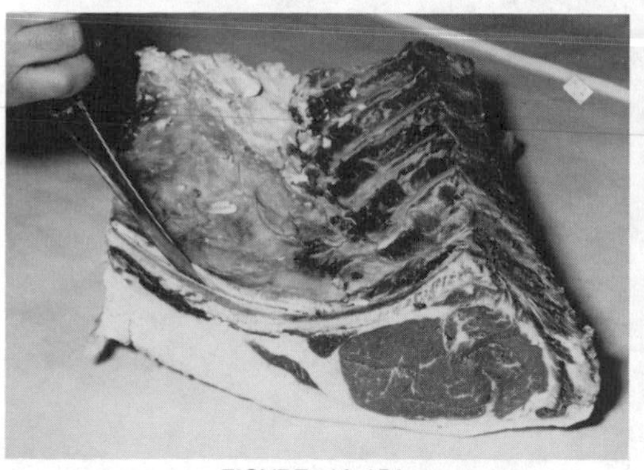

FIGURE 12.15A

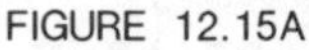

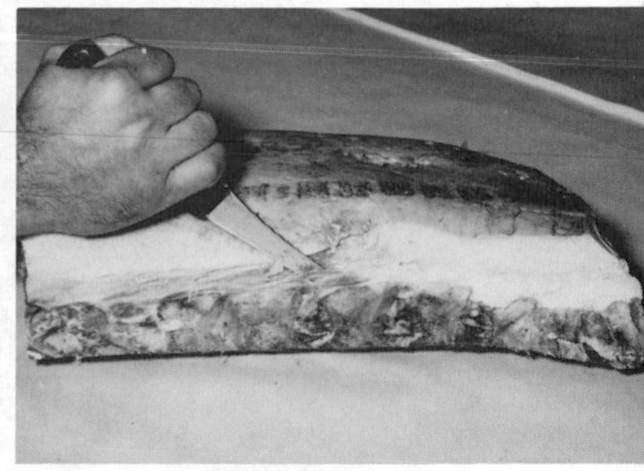

FIGURE 12.15B

(**1**) Loosen rib bone by scoring along both sides. Cut free or pull with a meat hook. Bone will twist away from back bone (Fig. 12.15A). (**2**) Place strip loin fat side up and score along inside of chine bone (Fig. 12.15B). (**3**) Using a flexible pork trimming knife, free finger bones. Rest

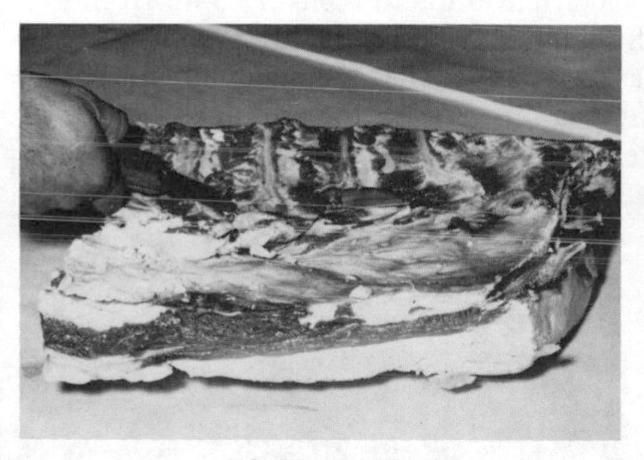

FIGURE 12.15C

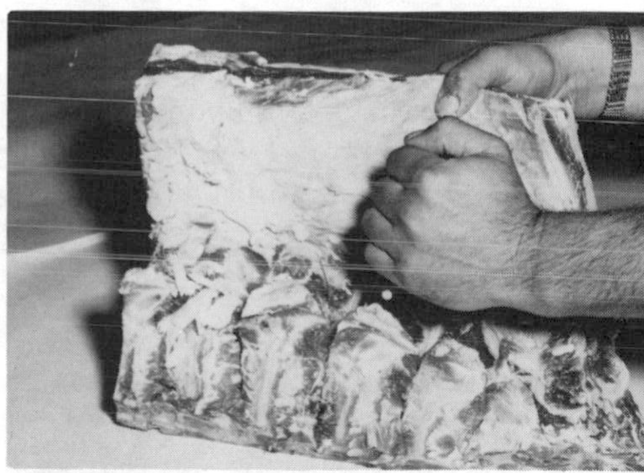

FIGURE 12.15D

tip of knife on one bone and bend to undercut finger bone next to it (Fig. 12.15C). (**4**) With lean side up, complete cuts under feather bones. Run knife along chine bone to cut made in step 2. When cut is complete, strip loin muscle will fall away from bones (Fig. 12.15D).

The final trim of boneless strip loin is to trim flank and bevel surface fat. Measuring flank trim varies widely regionally, and frequently, within a given jobbing house. There are at least two satisfactory approaches:

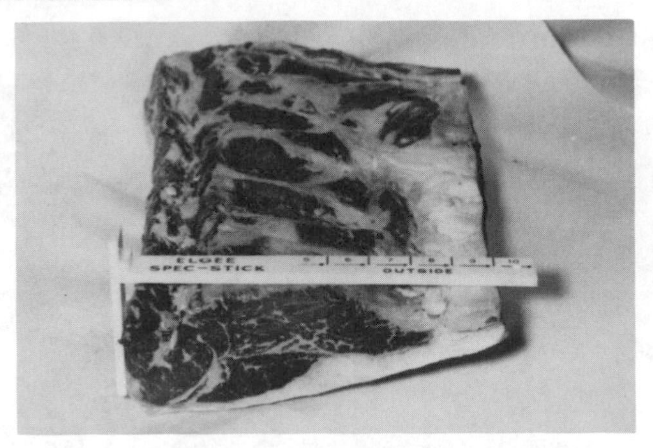

FIG. 12.16. BONELESS STRIP LOIN—9 IN. OVERALL

(1) The measurement may be overall from the back of strip loin to a point where the flank is cut off (Fig. 12.16).

(2) Measure from tip of eye on both ends to a point where flank is cut off. The acceptance service specification is to measure 3 in. from eye on rib end (Fig. 12.17), and 2 in. from eye on hip end (Fig. 12.18).

There are other ways of measuring a boneless strip loin. For example, make an 8 in. boneless strip loin by simply boning out an 8 in. bone-in strip loin. However, the index for measuring is destroyed when bone is lifted, and the user loses any element of receiving control.

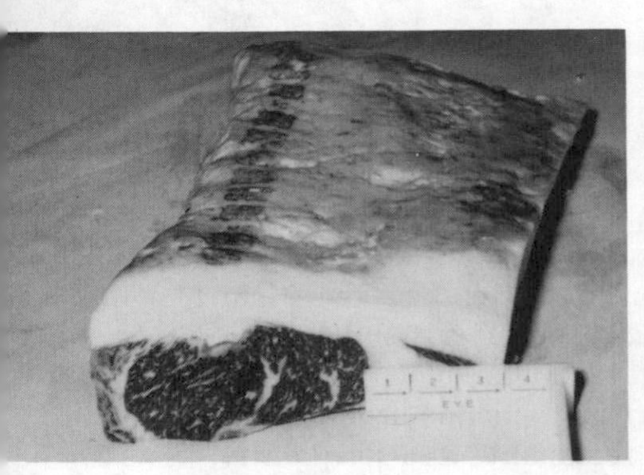

FIG. 12.17. BONELESS STRIP LOIN—3 FROM EYE (RIB END)

FIG. 12.18. BONELESS STRIP LOIN—2 IN. FROM EYE (HIP END)

The approximate yields of various boneless strip loins from a 10 in. bone-in strip loin are:

Cuts and Yields:

Cut	Per cent
Boned (no further flank trim)	80
10 in. boneless (overall)	72
9 in. boneless (overall)	65
8 in. boneless (overall)	58
IMPSA item No. 180	58

Separating Top and Bottom Sirloin

This cut is sometimes called the head coquille, head loin—tenderloin out, or sirloin butt.

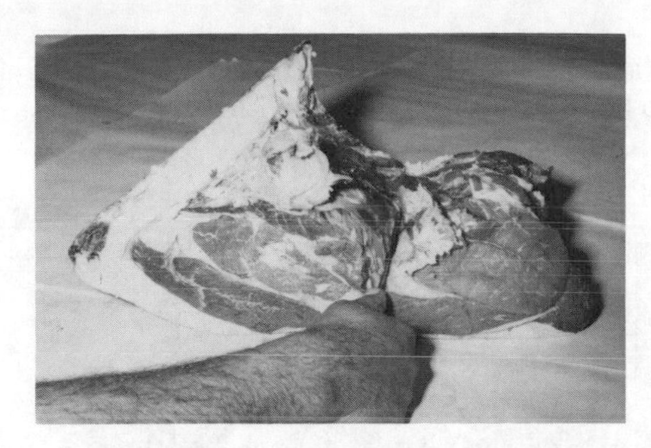

FIGURE 12.19A

(1) Starting at the rump knuckle bone joint, knife along natural seam (Fig. 12.19A). (2) Seam until top and bottom are separated (Fig. 12.19B).

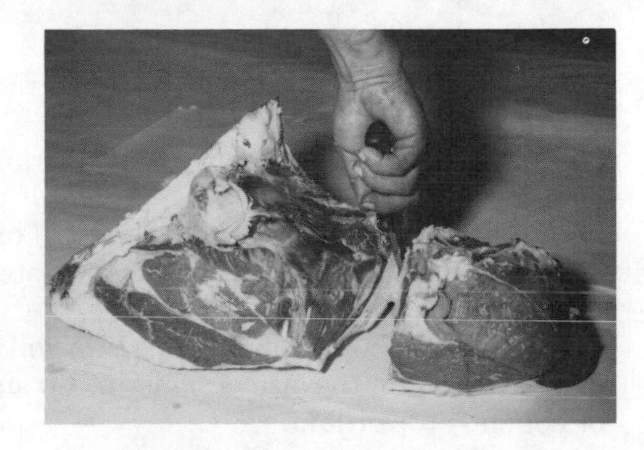

FIGURE 12.19B

Boning Top Sirloin Butt

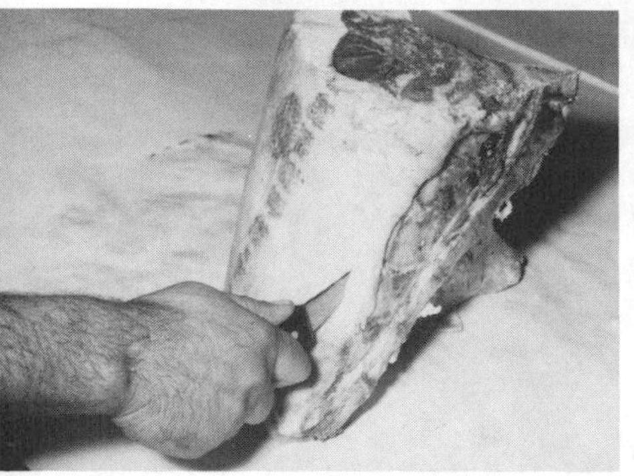

FIGURE 12.20A

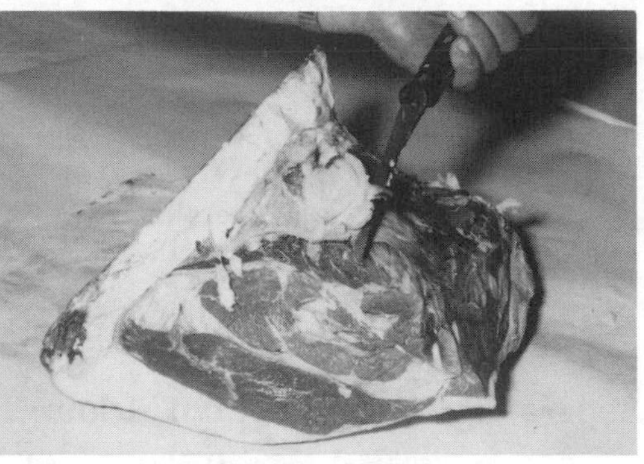

FIGURE 12.20B

(**1**) Cut inside backbone (Fig. 12.20A). (**2**) Start at rump knuckle bone joint and score along rump bone with knife tip (Fig. 12.20B).

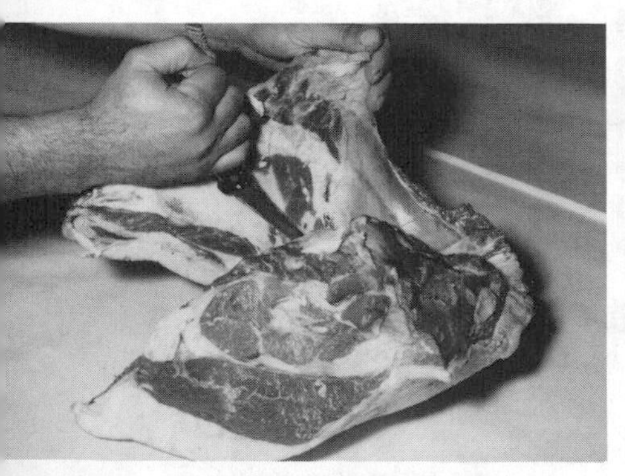

FIGURE 12.20C

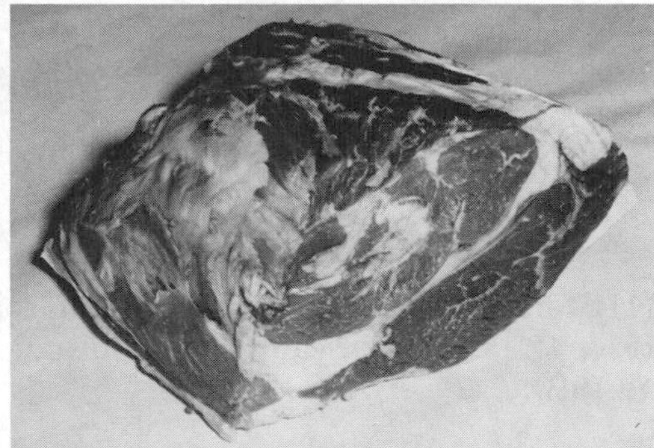

FIGURE 12.20D

(**3**) Continue to seam along bone. Top sirloin butt and bone will finally separate (Fig. 12.20C).
(**4**) Square up, block, or trim top butt. This is arbitrary and regional. The test of the trim is the final yield of steaks cut (Fig. 12.20D).

Boneless top butt yield from bone-in will range from 66 to 72 per cent, depending upon the size of bone-in top sirloin and the final blocking of boneless top sirloin.

Dividing the Bottom Sirloin

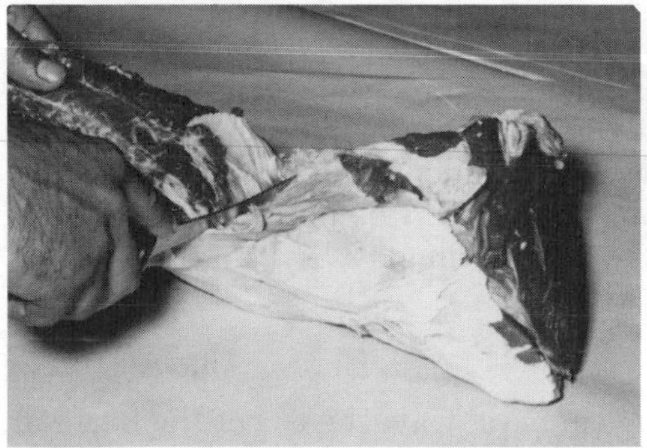

FIGURE 12.21A

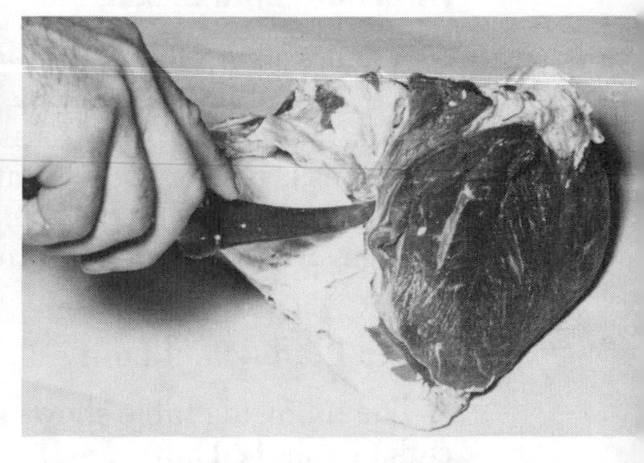

FIGURE 12.21B

(1) Seam out thin "flap" muscle. This is a continuation of flank, left on loin (Fig. 12.21A). (2) Locate natural seam of ball tip with knife tip (Fig. 12.21B).

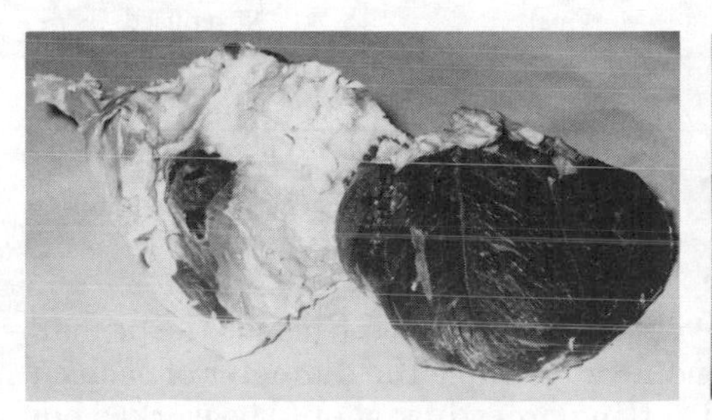

FIGURE 12.21C

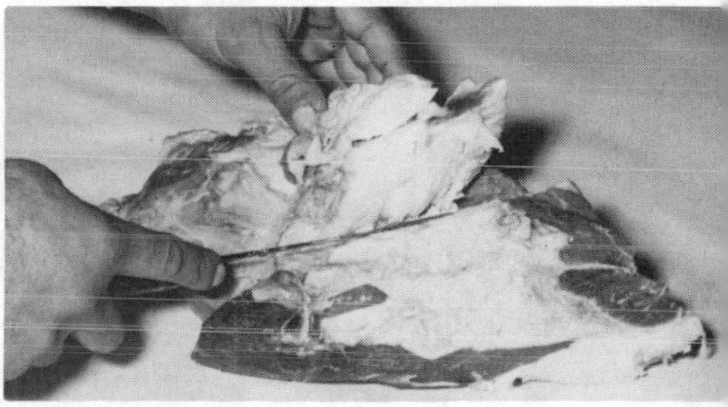

FIGURE 12.21D

(3) Seam ball tip away from fat (Fig. 12.21C). (4) Trim off deckle fat on inside of triangle tip to expose lean and bevel surface fat of triangle tip to conform to lean (Fig. 12.21D).

Cuts and Yields:

Cut	*Per cent*
Flap	15
Ball tip	25
Triangle tip	30
Fat	30
	100

Alternate Loin Break

When a short loin is required for club, T-bone, and porterhouse steaks, the style of breaking must be altered. The tenderloin is not pulled. The first step is to locate the tip of the hip bone and saw off short loin. This is steaked on the saw. The balance is referred to as the head loin or loin end. The butt tenderloin should be pulled, then top and bottom sirloin separated as in muscle boning.

Menu Plan—Beef Loin

The following table shows in a general way, the use of the beef loin cuts (Table 12.1).

TABLE 12.1. MENU PLAN—BEEF LOIN

Cut	Steak	Oven Roast
Tenderloin	Excellent	Gourmet item
Strip loin	Excellent	Gourmet item
Top sirloin	Very good	Excellent
Bottom sirloin		Good
Flap	Marginal	Never used
Triangle tip	Fair	Good
Ball tip	Fair	Good
Short loin steak	Excellent	Rarely used

Beef Tenderloin

This cut offers a variety of very acceptable uses. Steak is the primary use, and usually the best economically. Since the diameter of each cut will vary, a carefully planned cutting procedure should be worked out. There are many variations. The steak is rated very high because of tenderness, although it has little flavor (Fig. 12.22).

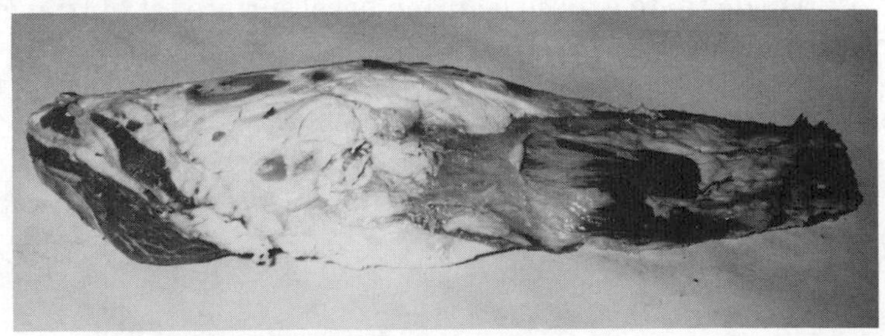

FIG. 12.22. BEEF TENDERLOIN

The tenderloin may be roasted whole or with the thin tail portion cut off. The roast may be larded by inserting thin strips of back fat with a larding needle to improve flavor and increase juiciness. The cover fat is sometimes trimmed before roasting.

Steaking Beef Tenderloin

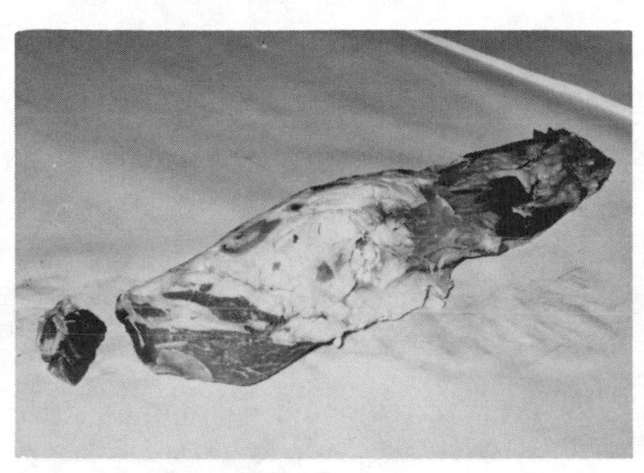

FIGURE 12.23A

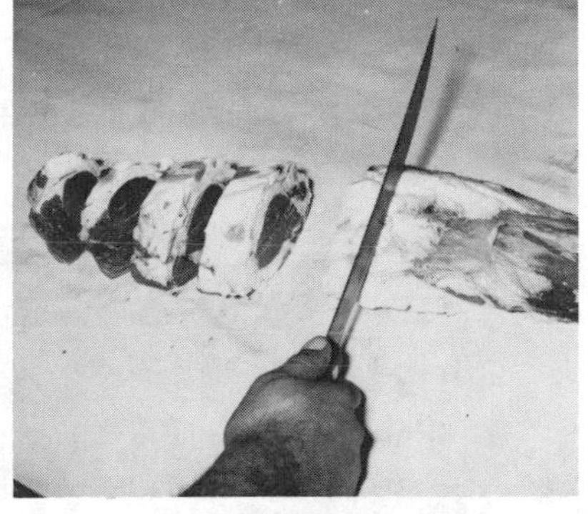

FIGURE 12.23B

(1) Square off nose, by removing a 1 in. cut (Fig. 12.23A). **(2)** Cut steaks to point on tail where the circumference is too small for intended steaks (Fig. 12.23B).

Trim the excess fat from the steaks. The amount trimmed is arbitrary. If all of the fat is removed, it may not look like a steak. On the other hand, some in-flight airline caterers have found it most acceptable to remove all fat and underlying silver skin. For an average situation, $1/4$ to $1/2$ in. fat appears most desirable.

FIGURE 12.23C

(3) Trim excess fat from tail portion and separate small side strip muscle from main muscle (Fig. 12.23C).

The nose and tail cuts may be sliced thin for grenadine of beef, cut into cubes for brochettes, butterflied or "booked" for steak sandwiches. Cuts for medallions or tournedos of beef are usually defatted. Sometimes the tournedos are bacon wrapped on the outside.

Cuts and Yields:

Cut	*Per cent*
Steaked Tenderloin	
Center cut steaks—fat on	48
Tail cut, side strip muscle off	10
Trim and nose cut	7
Fat	35
	100
Defatted tenderloin	65 to 70
Center cut tenderloin roast	about 60

Strip Loin

The strip loin (Fig. 12.24) has a variety of regional steak names, such as "sirloin steak," "New York cut," "strip steak," "top loin steak," "strip loin steak," "loin strip steak," "Kansas City steak," "individual loin steak," and sometimes when cut with the bone-in, "Delmonico," or "boned sirloin." Though not as tender as the tenderloin, nor as flavorful as the top sirloin, it is the favorite among steak eaters.

For roasting use whole boneless strip loins including the hip end. Cut flank close to eye and trim excess cover fat. This is one of the most delicious of beef roasts, usually featured as roast sirloin of beef or entrécôte de boeuf.

FIG. 12.24. BONE-IN STRIP LOIN

Steaking.—Buying specifications should be determined to provide steaks uniform in size, consistent with the intended portion cost, large enough to fill the plate, and thick enough to broil properly.

There are three important cutting premises: (1) pre-trim primal cut before steaking; (2) cut with a plan; and (3) trim individual steak into a first rate finished product.

Pre-trim the strip loin:

(1) The flank should be cut to desired length, either measured from eye, or measured overall. From the patron's point of view, if flank is too short, the strip loin loses its conformation and the patron might feel there is some substitution. If flank is too long, the patron may feel that he is getting too little lean for his money.

(2) The cover should be pre-trimmed. As steaks are cut, they will then more closely approximate the intended final weight.

(3) The connective tissue should be removed on the lean side. If this is done after the steaks are cut, it is sometimes easier, with less trim loss.

There are two cutting approaches to steaking for portion control: cutting for a given weight, or cutting for a given thickness with an average weight. The conventional method is to cut for a given weight range. This maintains a uniform portion cost. What happens in practice because of the conformation of the muscle, is that each steak will vary in appearance. The thickness will vary. Large surface steaks will be thin, others comparatively thick.

A progressive approach is to use a ruler rather than a scale. By buying a uniform weight strip loin and by cutting the same number of uniformly thick steaks from each, the portion cost can be averaged and controlled. Two objectives can be achieved: (1) the steak will appear uniform for the patron; and (2) the cost will be consistent for the operator based on an average yield (Fig. 12.25).

Cut with a plan: (1) mark off steak portions before cutting; (2) start steaking at the rib end; (3) cutting may be done with a knife or meat blade on the electric saw (bone-in steaks are generally cut with an electric saw); (4) cut uniformly thick steaks, wedging the tail portion slightly to square off the angle of the rib end; after a few steaks are cut, the cuts should be parallel with the hip end; and (5) to avoid miscuts at the hip end, redivide the last few steaks before cutting.

In the hip end, there is considerable epimysium tissue (sheath of connective tissue surrounding a muscle) clearly defined in the lean. In roasting, this will tend to partially gelatinize, but when broiled, this tissue becomes very chewy. Quality operators will down-grade the hip end, using it as a by-product rather than as a dinner steak or for "well done" steak orders. A satisfactory minute cut sirloin, or a steak sandwich

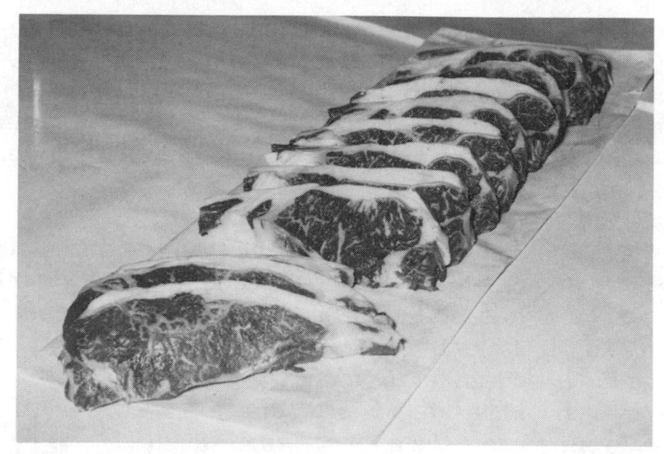

FIG. 12.25. STRIP LOIN STEAKS

of less than ¾ in. may be cut from this end. The cut can be used for many other inexpensive entrées. The epimysium will vary from 1 to 3 in. in depth into the strip loin muscle.

Trim the cut steak carefully: (1) remove all connective tissue and "age" from the lean side; (2) remove the back strap (the yellow ligament); (3) trim tail where necessary; a good plan is to show a small amount of flank lean in tail where possible; and (4) trim the cover fat and tail from the patron's point of view. The patron measures by eye the uniformity of the steaks served, the thickness, and the lean-fat ratio. When considerable fat is left on the steak, and considerable fat is consequently left on the plate, a poor impression is usually made.

The comparative photographs show the same strip loin steak, cut from the center portion or "dip part" of the strip loin. Figure 12.26 is

FIG. 12.26. STRIP LOIN STEAK—UNTRIMMED

FIG. 12.27. STRIP LOIN STEAK—TRIMMED

the untrimmed portion of a 9 in. boneless strip loin on a large dinner plate. Figure 12.27 is the same steak but carefully trimmed, on the same dinner plate. Observe how the psychological lean-fat ratio is improved by a closer trim. The steak to plate ratio is another important appearance factor. The same steak on a smaller dinner plate would appear even larger.

The fact is that 10 to 12 ozs of lean is as much or more than the average person wants or requires. What is important is that the guest visually experiences a good value.

Cuts and Yields:

Cut	*Per cent*
Strip steaks from 9 in. boneless strip loin	
½ in. maximum cover	
1½ in. tail from eye	
including all cuts	60
Where hip cuts are separated	
primary steaks	48
hip cuts	12

Top Sirloin Butt

The top sirloin butt is an excellent piece of meat for oven roasting, and very good for steak meat. Properly aged top sirloin is reasonably tender when cut from the higher grades.

As a roast, it may be cooked whole (Fig. 12.28), or split (Fig. 12.29). To achieve maximum flavor, the cover fat may be removed; salt, pepper,

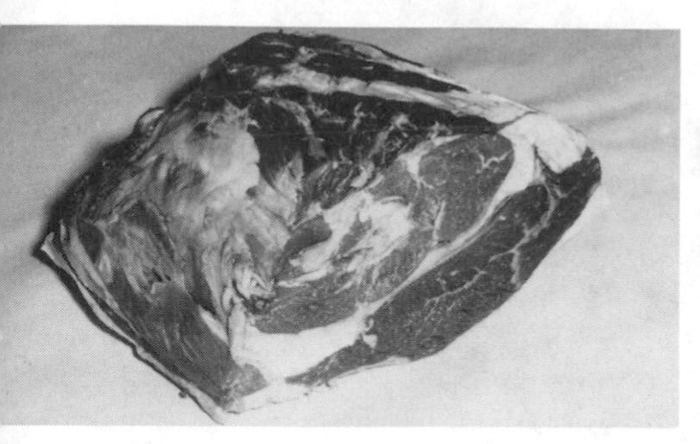

FIG. 12.28. TOP SIRLOIN BUTT

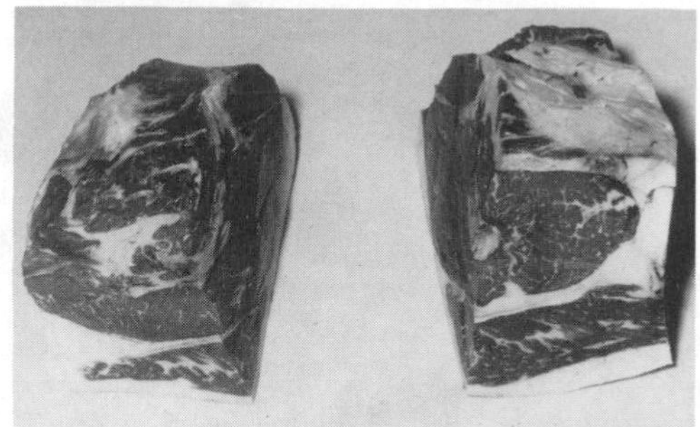

FIG. 12.29. TOP SIRLOIN BUTT SPLIT TO ROAST

and other condiments should be applied to all surfaces. The crusting that occurs during roasting will enhance the flavor. Following the two basic cutting rules, pre-trim and cut with a plan when steaking. Pre-trim top sirloin by trimming off excess cover, remove blue seam tissue, trim off interior fat, and trim off fat on chine side by undercutting. Cut with a plan. There are at least three methods:

Full Face Cut Steaking Technique

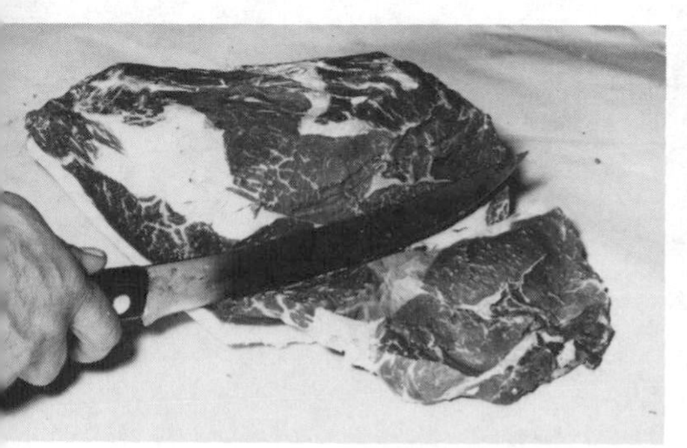

FIGURE 12.30A

FIGURE 12.30B

(1) Make a full face cut which will achieve uniformly thick steaks (Fig. 12.30A). (2) Weigh full cut and set cutting plan (Fig. 12.30B). (3) Cut

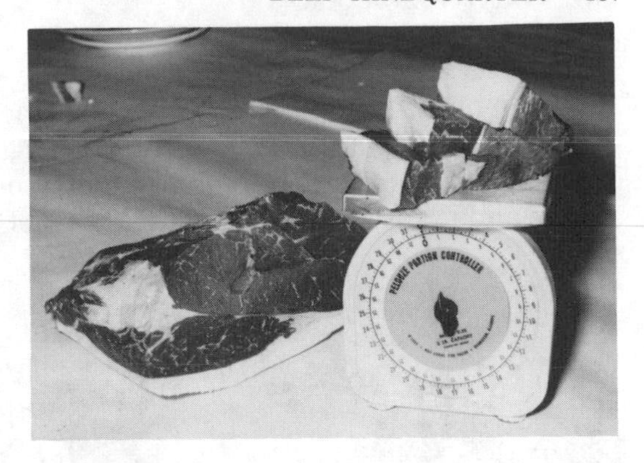

FIGURE 12.30C

the portions from full cut. This works especially well when more than one size steak is used (Fig. 12.30C).

Splitting Technique

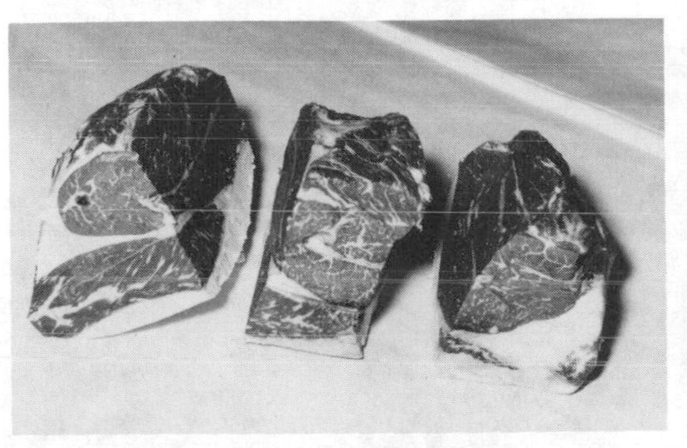

FIGURE 12.31A

FIGURE 12.31B

A second technique is to split the top sirloin into two or three pieces (Fig. 12.31A) and steak out each piece (Fig. 12.31B). Uniform thickness will be sacrificed. The yield probably will be greater and portions more uniform in weight. The small muscle on top of lean side is not tender. Frequently, this is seamed off before splitting. The face cut, because of an excessive amount of connective tissue, should never be used as a steak. The connective tissue may be cut out and the lean used for a small steak, cube steak, brochette, or other by-product. The total steak yield is about 50 to 65 per cent.

Center Cut Technique

When uniform appearing, relatively thick top sirloin steaks are desired, and the operator is tolerant of a slightly lower yield and higher portion cost, a center cut technique may be used. Usually top sirloins 10 lbs and under are desirable for this method.

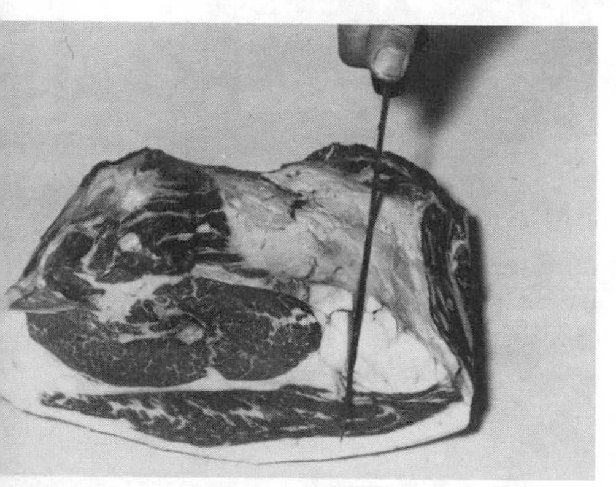

FIGURE 12.32A

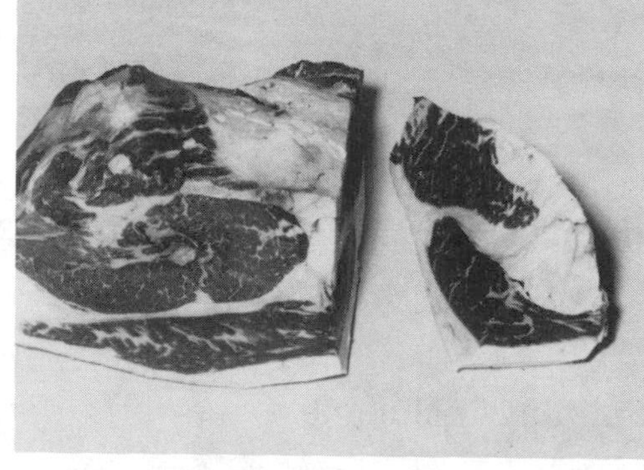

FIGURE 12.32B

(1) Locate heavy fat in face on chine side of top sirloin (Fig. 12.32A). (2) Square off (Fig. 12.32B). (3) Square off bottom sirloin side slightly

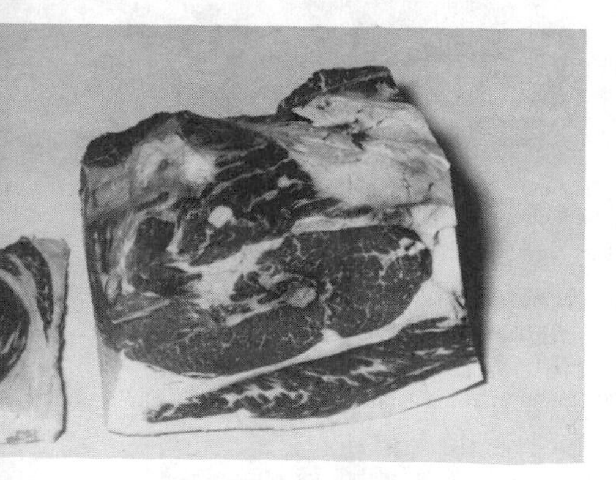

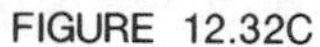

FIGURE 12.32C

FIGURE 12.32D

(Fig. 12.32C). (4) Split into two equal pieces (Fig. 12.32D). (5) Pre-trim by removing blue seam and excess cover fat. For any connective tissue on the face, make one cut deep enough to remove it. At top, undercut

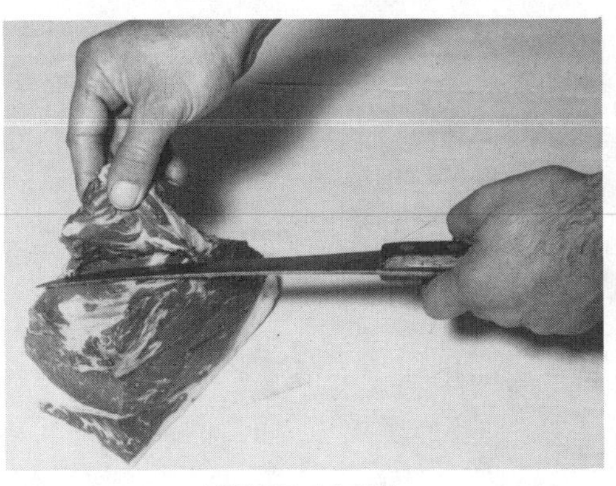

FIGURE 12.32E

FIGURE 12.32F

or seam off small muscle showing considerable connective tissue (Fig. 12.32E). (**6**) Set up cutting plan for thickness and start to slice steaks (Fig. 12.32F). (**7**) Very uniform portions are achieved (Fig. 12.32G).

FIGURE 12.32G

Bottom Sirloin

The bottom sirloin (Fig. 12.33) may be defatted and tied to oven roast whole. It is reasonably tender and flavorful. Upper grades are sometimes presented as "boneless roast beef."

The flap (left front, Fig. 12.34) may be cubed and used for Swiss steak, chicken-fry steak, or cube steak. The thin surface tissue should be removed before cubing.

The ball tip (rear, Fig. 12.34) may be roasted or steaked. The ball tip makes a very good "cheater steak," especially for inexpensive steak sandwiches, sometimes featured as a "club steak" or "country club steak." When cut thin and very quickly grilled or pan broiled it is improved in flavor and tenderness. To steak the ball tip, place it flat side

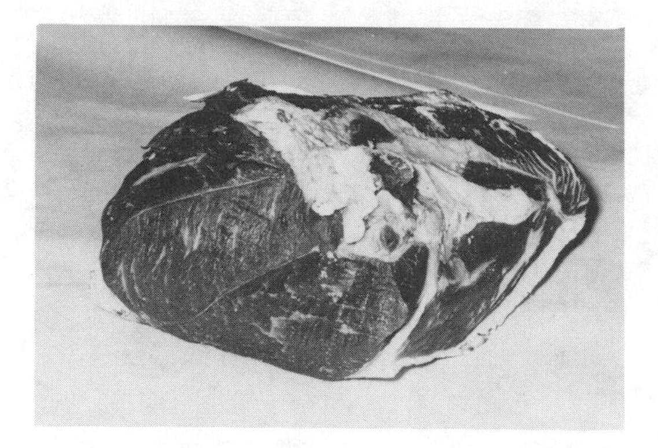

FIG. 12.33. BOTTOM SIRLOIN—WHOLE

FIG. 12.34. BOTTOM SIRLOIN—SEAMED OUT INTO COMPONENTS

down on the cutting board, making cuts perpendicular to the cutting board. Steaks may be butterflied to increase the perimeter or change the conformation.

The triangle tip (right front, Fig. 12.34) may be oven roasted, pot roasted, or marinated for sauerbraten. It is seldom steaked for broiling. Low priced steak operations often mechanically tenderize ball tips and triangle tips from USDA Choice and USDA Prime beef for "New York steaks" and "top sirloin steaks," rather than using lower grades of top sirloin and strip loins with an enzyme tenderizer. To slice triangle tips when potted or roasted, remove all cover fat, place deckle side down, and slice perpendicular to cutting board. Start slicing across any point of triangle.

Short Loin

Three types of steaks are cut from the short loin (Fig. 12.35). Each one is based on the relative amount of tenderloin. The porterhouse steak

FIG. 12.35. SHORT SIRLOIN—PORTERHOUSE END

shows a full tenderloin, the T-bone steak shows some tenderloin, and the club steak shows little, if any. Short loin steaks should be cut with a saw starting at porterhouse end (Fig. 12.36). They should be wedged slightly to conform to rib end of short loin. Tail and cover fat should be trimmed. The club end may be upgraded by removing the small piece of tenderloin and cutting Delmonicos, bone-in New York steaks, or removing bones and cutting boneless New York steaks.

FIG. 12.36. SHORT LOIN WITH ONE PORTERHOUSE STEAK CUT

BEEF ROUND

A typical packing house round (Fig. 12.37) is usually sold as dropped out of the hindquarter. It should have no more than two tail bone sections. The cod fat or udder fat may or may not be trimmed, according to regional custom. The "hanging tenderloin" (see Fig. 12.47) should be excluded by cutting it off at the juncture of the first and second lumbar vertebrae.

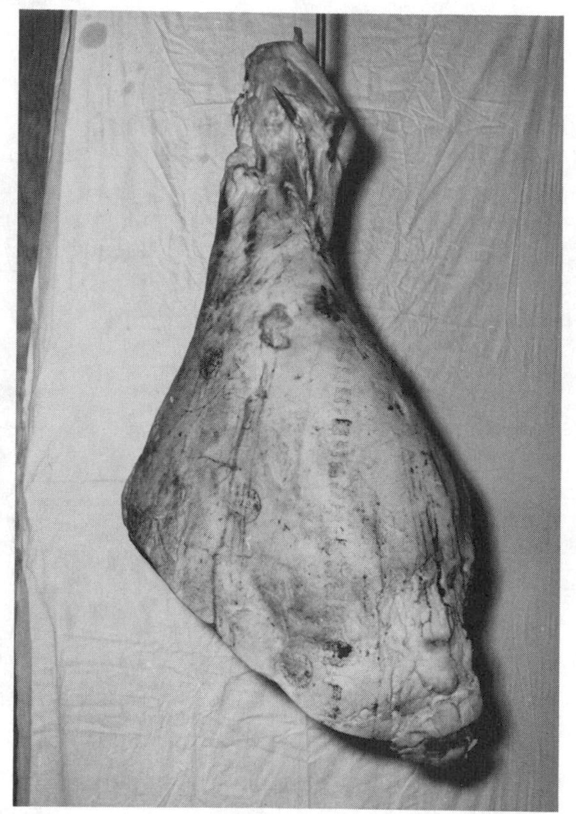

FIG. 12.37. BEEF ROUND—PACKING HOUSE STYLE

Muscle Boning the Round

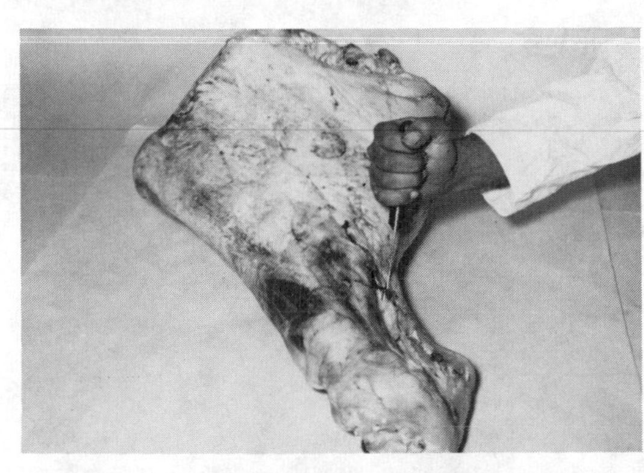

FIGURE 12.38A

(**1**) Insert knife tip between shank and tendon. Sever tendon (Fig. 12.38A). (**2**) Seam along hind shank bone and shank muscles (Fig. 12.38B). (**3**) Repeat on other side of shank bone (Fig. 12.38C).

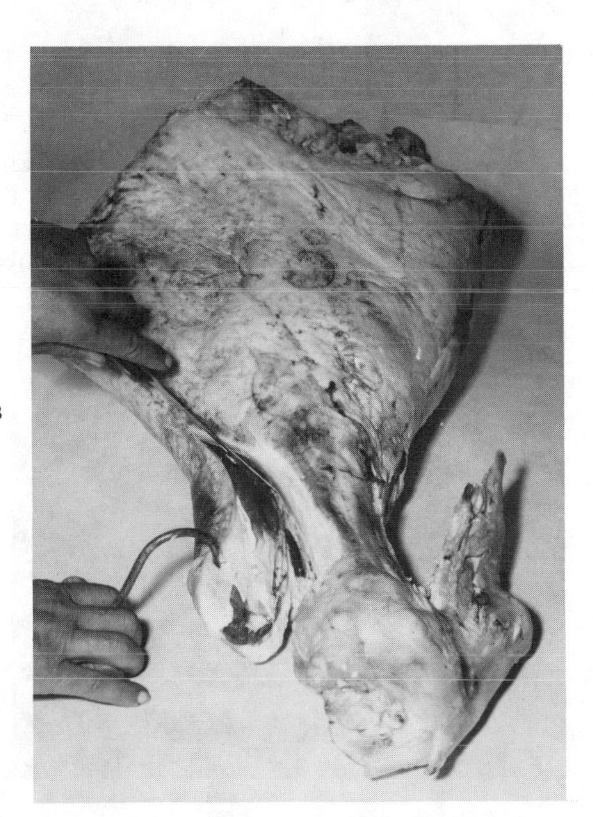

FIGURE 12.38B

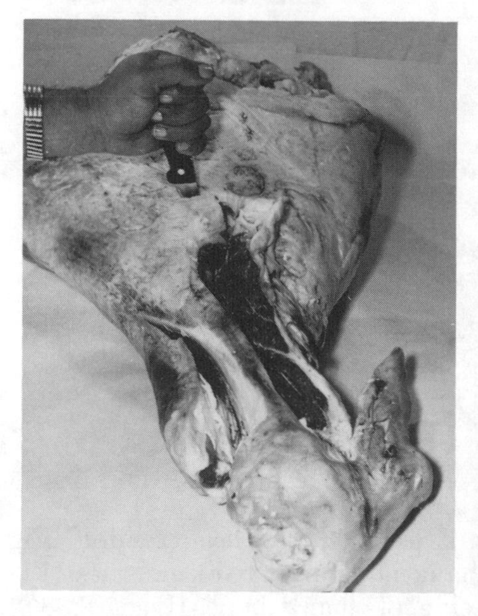

FIGURE 12.38C

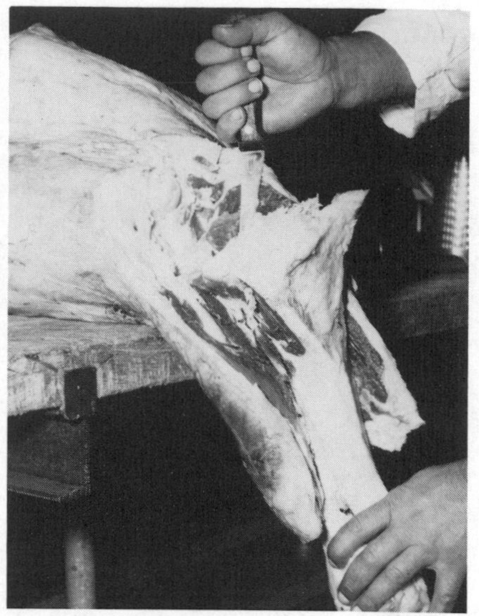

FIGURE 12.38D

(**4**) Continue freeing meat from bone. Run knife through stifle joint. Free hind-shank bone by exerting pressure on it over edge of cutting table (Fig. 12.38D). (**5**) Locate natural seam and remove lower shank muscles (Fig. 12.38E). (**6**) On the top of round, locate a point behind aitchbone

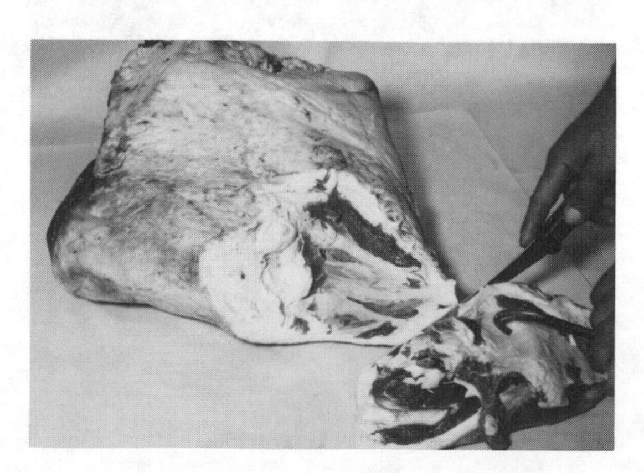

FIGURE 12.38E

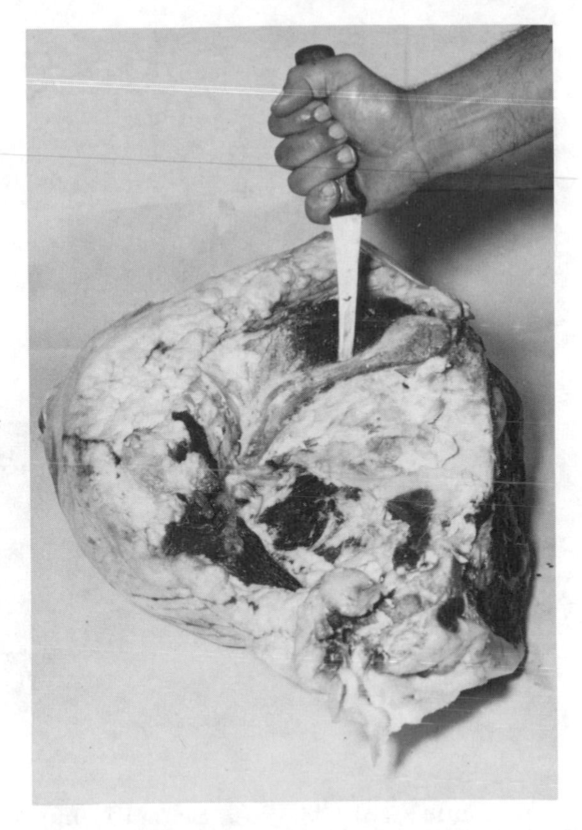

FIGURE 12.38F

(Fig. 12.38F). (7) Seam out aitchbone and with it the two tail bones (Fig. 12.38G). (8) With knife tip, locate natural seam between knuckle and

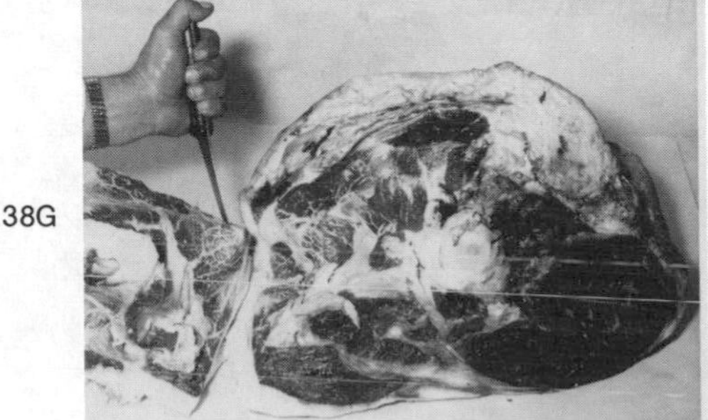

FIGURE 12.38G

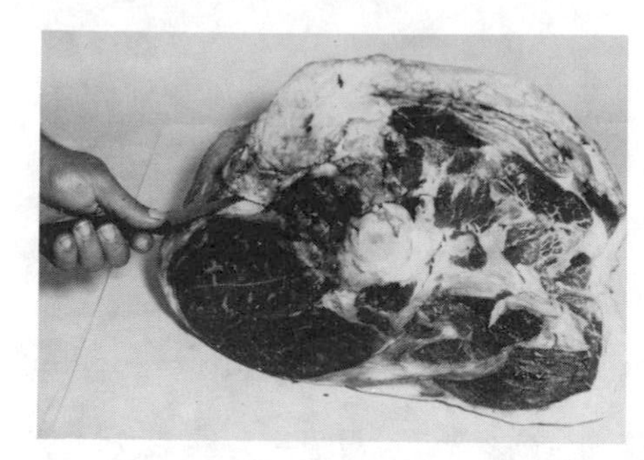

FIGURE 12.38H

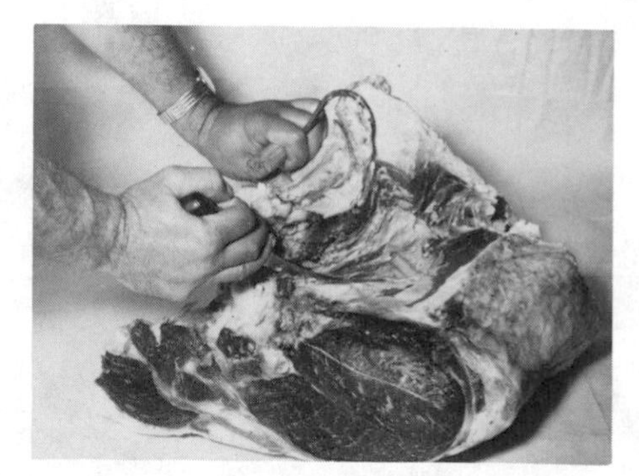

FIGURE 12.38I

top round (Fig. 12.38H). (**9**) Seam to rump knuckle bone and pull top round (inside round) (Fig. 12.38I). (**10**) The trim on top round will vary regionally. The cod or udder fat is frequently trimmed or shaped and aged surface may be removed (Fig. 12.38J). (**11**) Bone out rump knuckle bone (Fig. 12.38K). (**12**) Locate with knife tip seam between knuckle

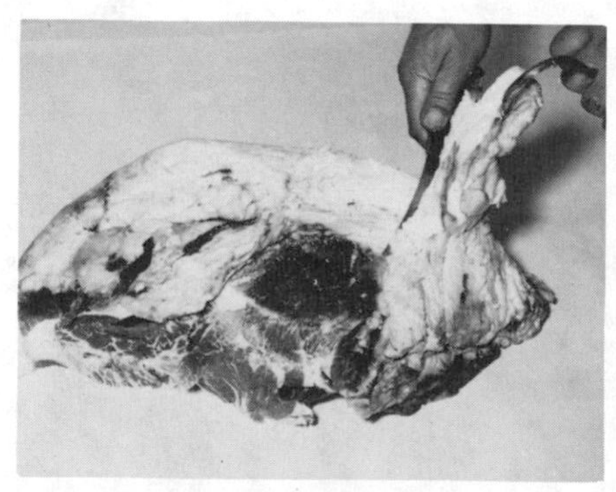

FIGURE 12.38J

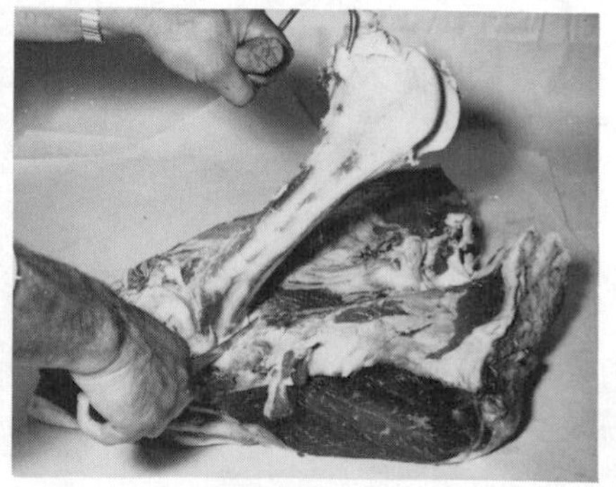

FIGURE 12.38K

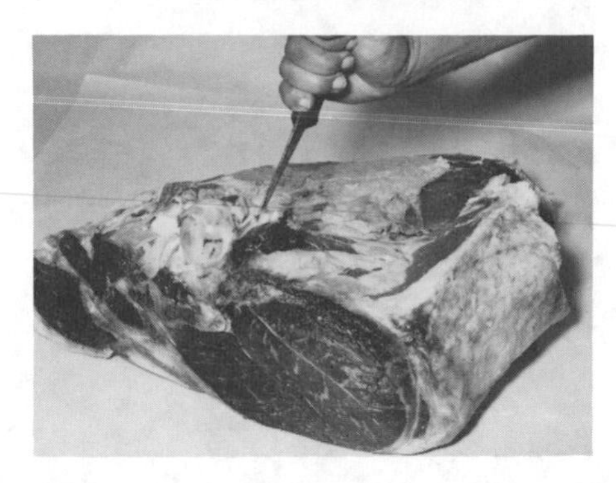

FIGURE 12.38L

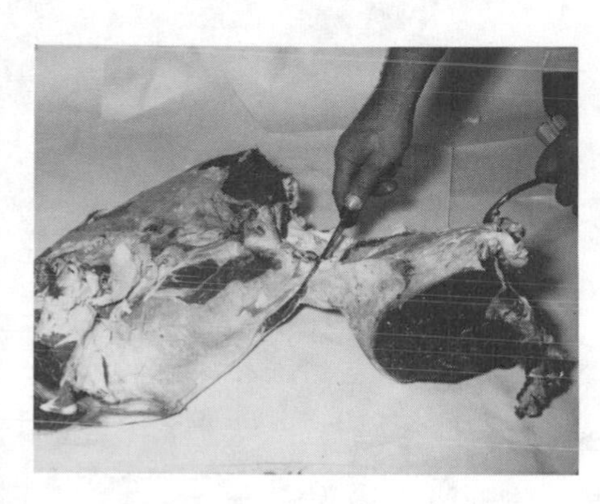

FIGURE 12.38M

and bottom round (Fig. 12.38L). (**13**) Seam and separate the two cuts (Fig. 12.38M); on left, bottom round; on right, knuckle. Trim off loose lean pieces, excess fat and age. The soft knee cartilage and heavy connective tissue should be trimmed from knuckle.

Cuts and Yields:

Cut	*Per cent*
Top round (inside)	25
Bottom round (outside)	30
Knuckle	10
Shank meat and trim	8
Fat and bone	27
	100

Cafe Round

A simple cutting approach to the cafe round, ship round, steamship roast, or buffet round is illustrated (Fig. 12.39).

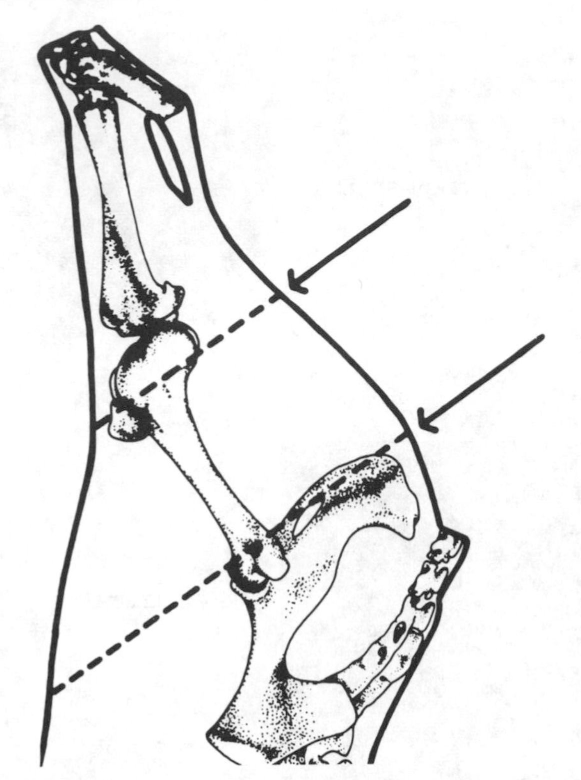

FIG. 12.39. CAFE ROUND—SKELETAL ILLUSTRATION

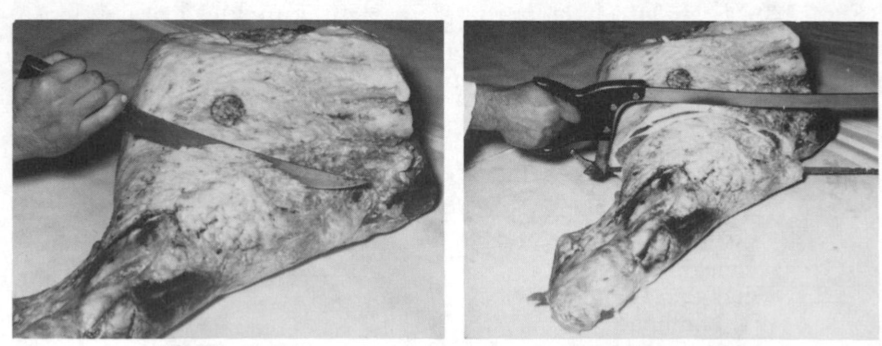

FIGURE 12.40A

FIGURE 12.40B

(**1**) Knife to round bone (Fig. 12.40A). (**2**) Saw off shank (Fig. 12.40B). (**3**) With knife describe cut inside and parallel with aitchbone (Fig.

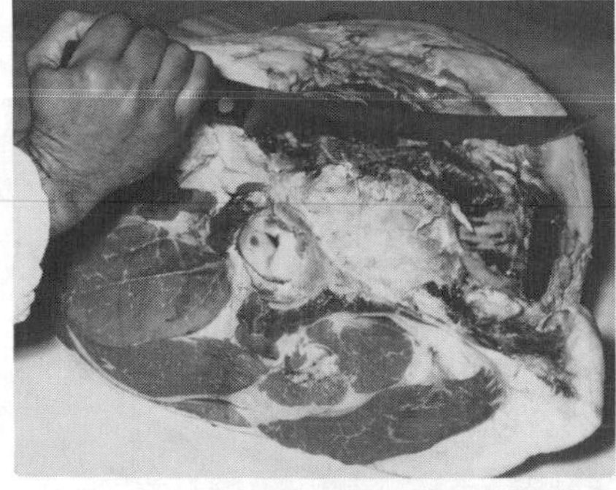

FIGURE 12.40C

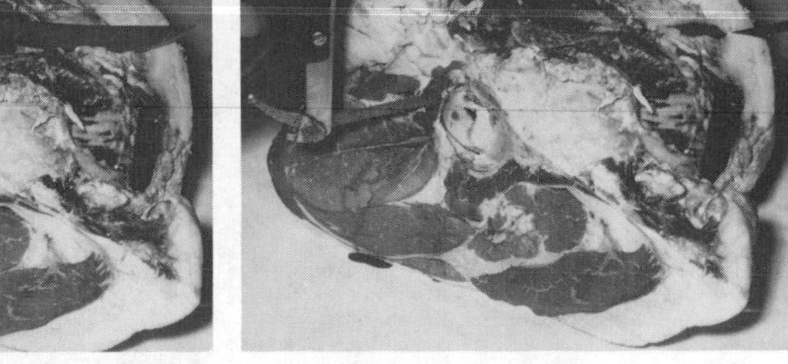

FIGURE 12.40D

12.40C). (4) Saw through bones (Fig. 12.40D). (5) Complete knife cut to separate bone-in rump from cafe round (Fig. 12.40E). The cafe round portion yields 70 per cent. An alternate approach to the cafe round might be made by following through the first seven steps of muscle

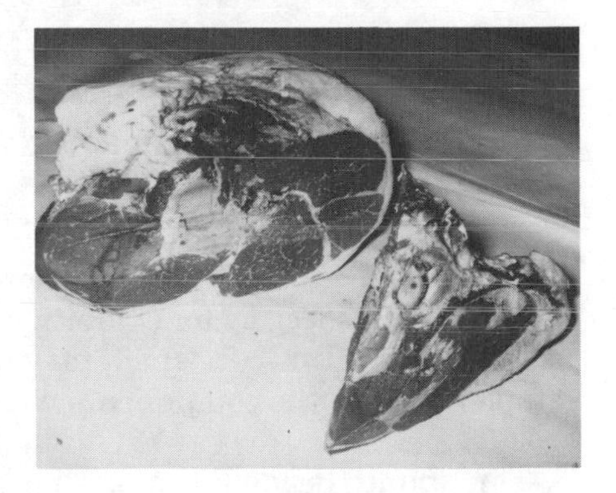

FIGURE 12.40E

boning. The large end of the round can then be squared off by knifing off a portion of the rump. The shank end may be sawed off at any arbitrary point. This yield will be higher.

Menu Plan—Beef Round

Cuts from the round are generally regarded as "rough cuts." They are marginal for broiling and oven roasting. Evaluation and upgrading will depend upon the economics of the operation.

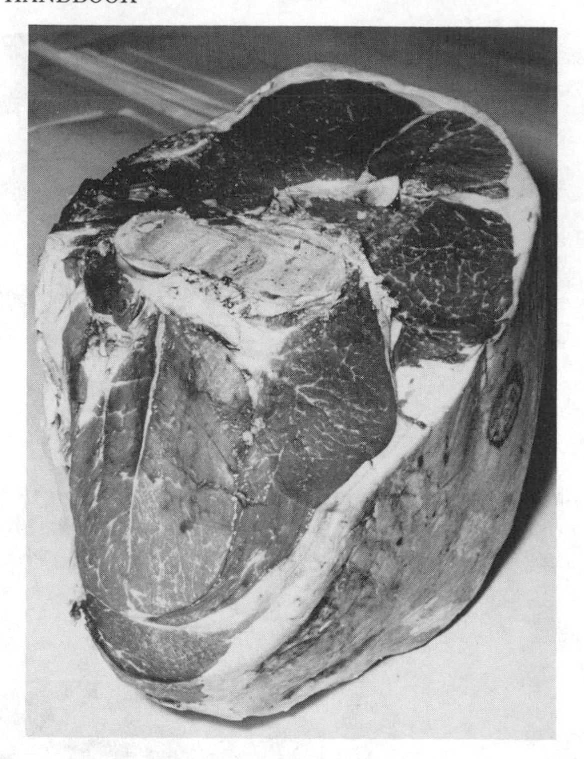

FIG. 12.41. CAFE ROUND

Cafe Round.—This is a popular presentation for medium priced, rare roasted beef, especially for exhibition slicing. Considerable roasting time is required for this cut, which may make for a more tender, more palatable cut than the components when roasted separately (Fig. 12.41).

Top Round (Inside Round).—This is a marginal oven roast. Because of the uniformity of the whole piece, and the minimum of connective tissue, it is popular with many operators. Sometimes ground, the top round (Fig. 12.42) or other round cuts are less flavorful and drier than many other rough cuts. "Ground round steak" is more of a sales device than a practical menu plan. Sometimes two top rounds are tied together for rare roasting.

Bottom Round (Outside Round).—Not as tender or clear as the top round, the bottom round (Fig. 12.43) is sometimes used for roast beef by barbecue operations. The slow moist heat of a covered barbecue contributes to the palatability of the finished product.

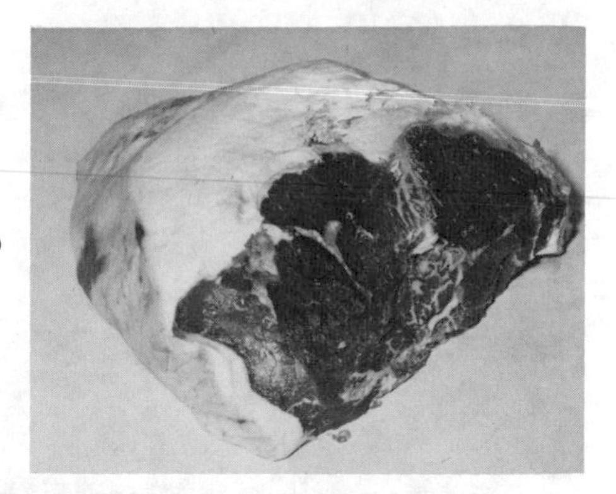

FIG. 12.42. TOP ROUND

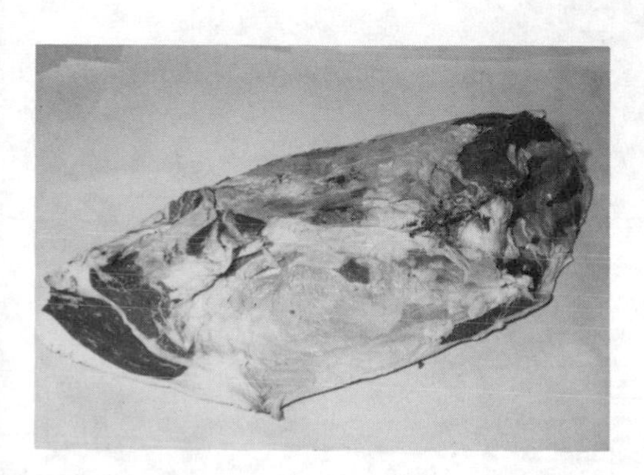

FIG. 12.43. BOTTOM ROUND

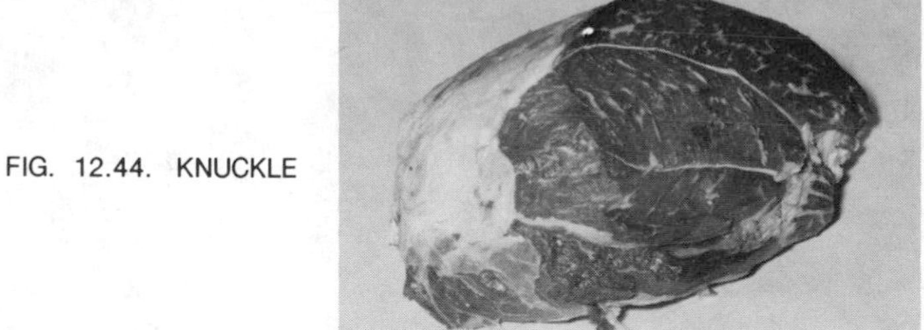

FIG. 12.44. KNUCKLE

Knuckle.—Though smaller than the other cuts, the knuckle (Fig. 12.44) makes a reasonably good oven roast, and is sometimes sliced into marginal broiling steaks. It is especially good for a wide range of by-product dishes.

FLANK AND KIDNEY

The whole flank is a by-product of breaking a hindquarter. The components are a flank steak, some beef trimmings, and considerable fat (Fig. 12.45).

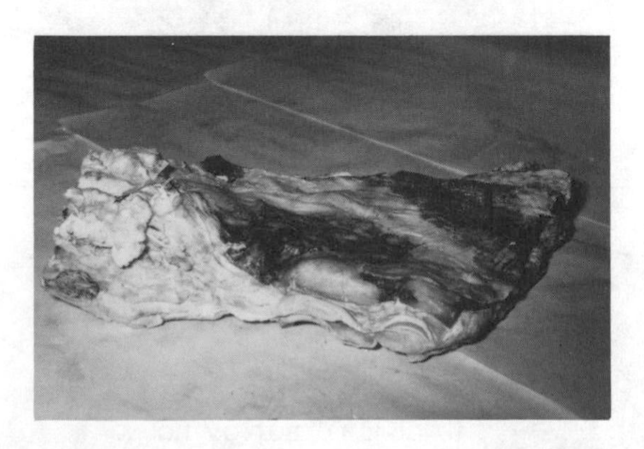

FIG. 12.45. FLANK

Pulling Flank Steak

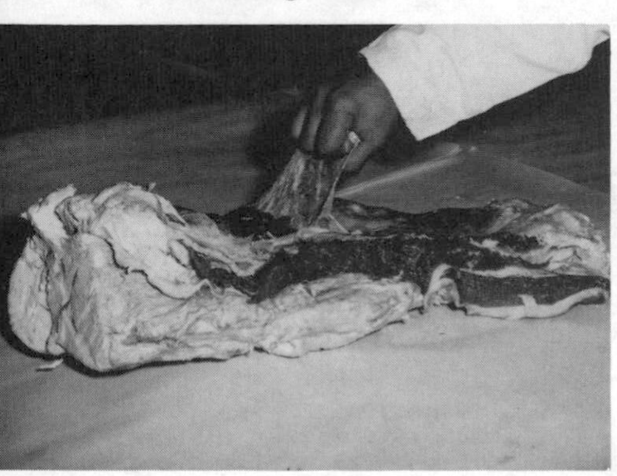

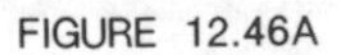

FIGURE 12.46A

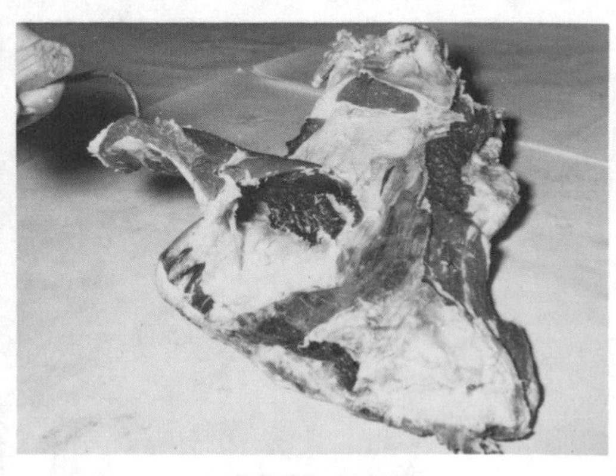

FIGURE 12.46B

(1) Locate flank steak and score around it. Pull serous membrane covering steak (Fig. 12.46A). (2) Score and pull steak from thick membrane below it (Fig. 12.46B). The flank steak is used for London broil, beef roulades, Swiss steak, and other by-products.

Kidney and Hanging Tenders

The hanging tender is a by-product item (Fig. 12.47—left).

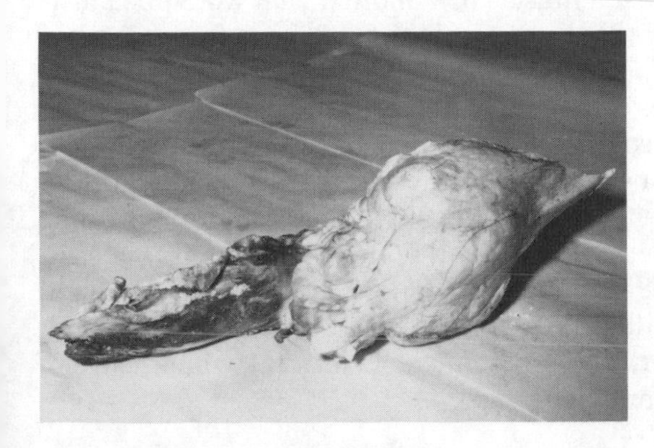

FIG. 12.47. KIDNEY AND HANGING TENDER

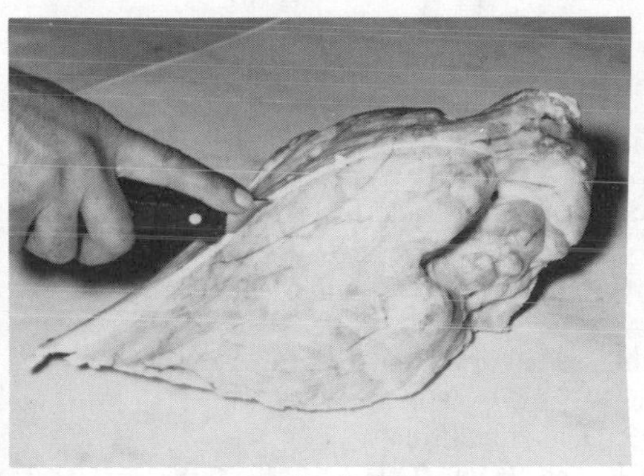

FIGURE 12.48A

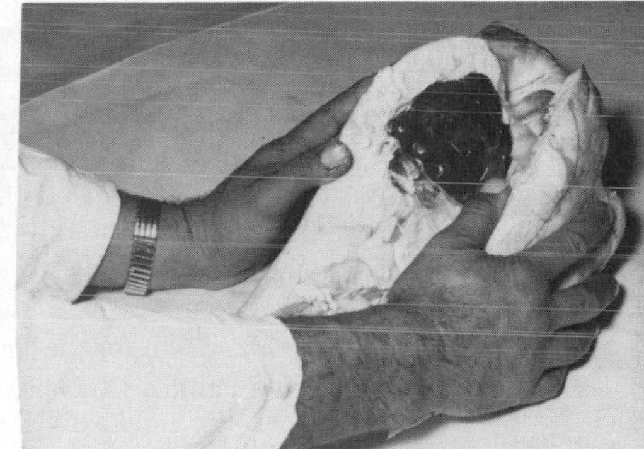

FIGURE 12.48B

(1) To remove kidney, split fat (Fig. 12.48A). (2) Pull fat away from kidney (Fig. 12.48B). Refer to Chapter 17 on variety meats for a menu plan.

MEAT HANDBOOK OF THE U.S. NAVY

The United States Navy has prepared a complete handbook on meat for training and use of commissary personnel, NAVSANDA Publication

No. 55. The text is well illustrated and contains many photographs. The Navy approach, though quite different from the civilian one, is functional and certainly worth investigating. The basic theory is one of following natural muscle seams to produce boneless cuts. Two objectives are achieved: (1) the use of each muscle in a manner that would be most valuable economically, and (2) the conservation of space for transportation and storage.

There are five basic steps involved, which should be evaluated, partially in terms of a total program or modified to improve an existing plan: (1) divide along major seams to make carving easier and result in a greater yield; (2) separate thick and thin cuts, tender and rough cuts, to the maximum advantage; (3) remove all bones to make carving easier—utilize bones for stock; (4) remove all tendons, ligaments, and tough serous membranes; this will make carving easier, and the steaks and ground meats more tender; and (5) remove excess fat which can be utilized raw for rendering and cooking.

REFERENCES

ANON. 1945. Meat Handbook of the U.S. Navy. U.S. Navy Dept., Washington, D.C.

ANON. 1960. Institutional meat purchase specification for fresh beef, Series *100*. U.S. Dept. Agr., Washington, D.C.

ANON. 1961. Meat Buyer's Guide to Standardized Meat Cuts. National Association of Hotel and Restaurant Meat Purveyors, Chicago, Ill.

ANON. 1962. Facts about beef. National Live Stock and Meat Board, Chicago, Ill.

BULL, S. 1951. Meat for the Table. McGraw-Hill Book Co., New York.

ROMANS, J. R., and ZIEGLER, P. T. 1977. The Meat We Eat, 11th Edition. Interstate Printers and Publishers, Danville, Illinois.

WANDERSTOCK, J. J., and WELLINGTON, G. H. 1961. Let's cut meat. Cornell Extension Bull. *1053*. Ithaca, N.Y.

13

Beef Forequarter

PRIMAL CUTS

The forequarter, which is 52 per cent of the beef side, consists of three primal cuts: cross cut chuck, 7 × 10 rib, and plate (Fig. 13.1). The skeletal structure of the forequarter is shown in Fig. 13.2; photograph of forequarter in Fig. 13.3; relationship to the entire beef side can be seen in Fig. 12.2. in Chap. 12.

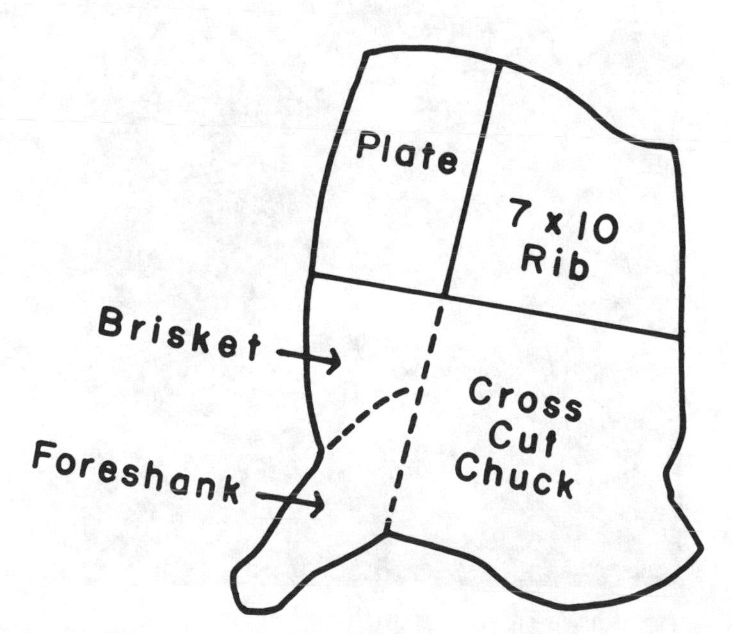

FIG. 13.1. BEEF FOREQUARTER—PRIMAL CUTS CHICAGO-STYLE

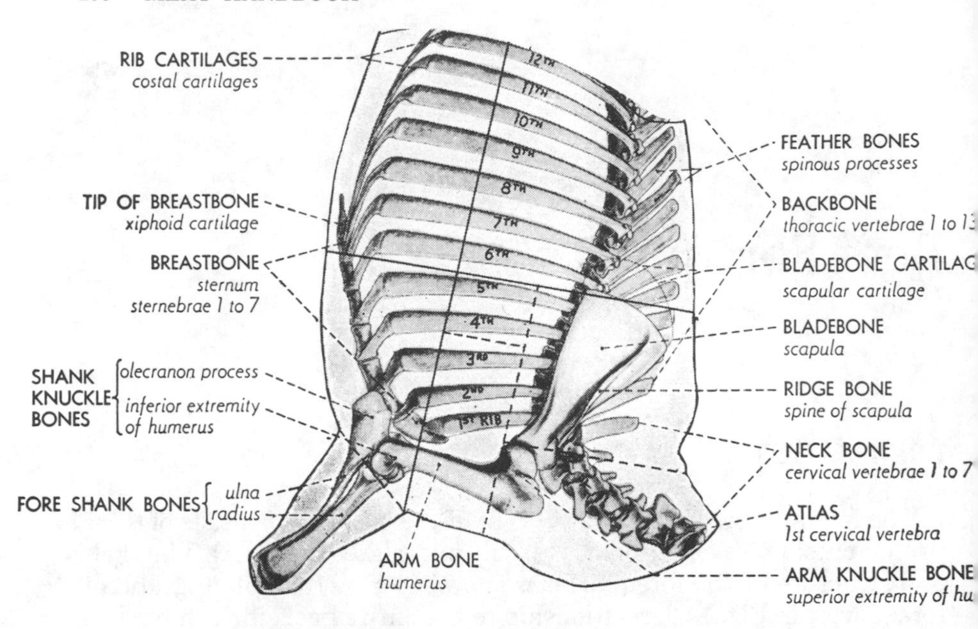

FIG. 13.2. BEEF FOREQUARTER—SKELETAL STRUCTURE

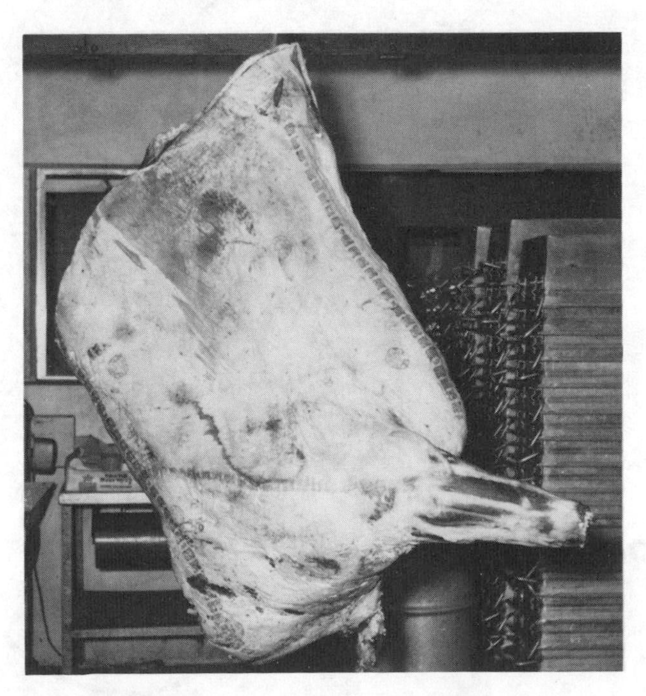

FIG. 13.3. BEEF FOREQUARTER

Cuts and Yields:

Cut	*Per cent*
Packing house style rib (7 × 10 rib)	17
Plate	15
Cross cut chuck	68
	100

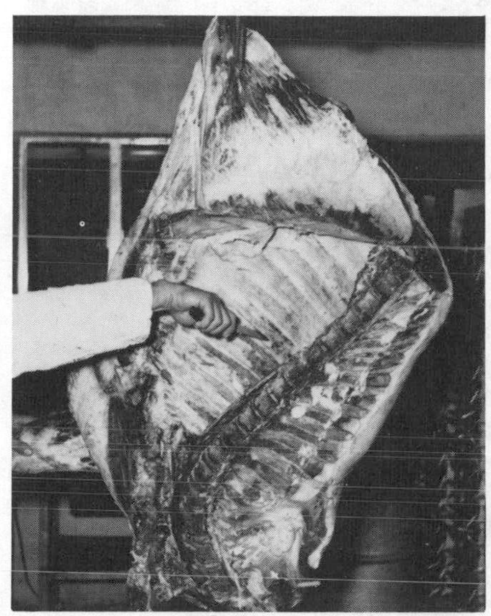

FIGURE 13.4A

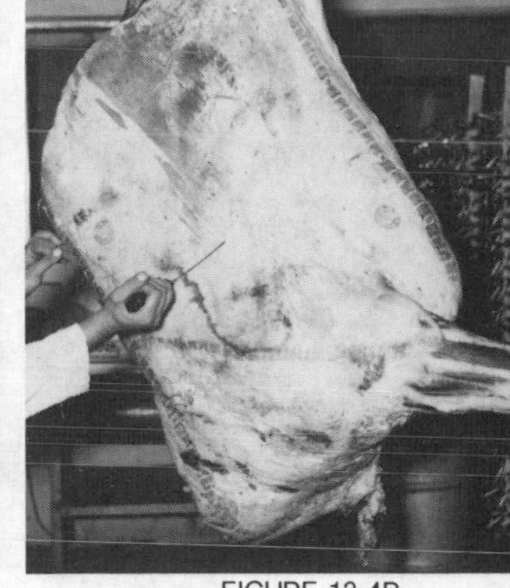

FIGURE 13.4B

(1) Insert knife between 5th and 6th ribs (Fig. 13.4A). (2) Cut in a straight line between ribs (Fig. 13.4B). (3) Cut through shoulder blade

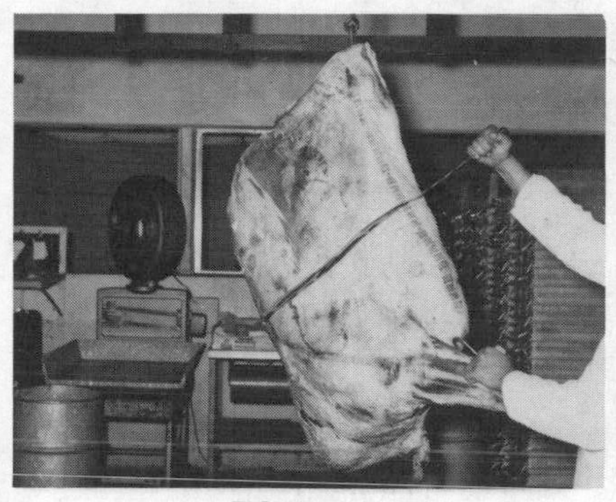

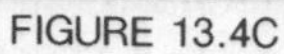

FIGURE 13.4C

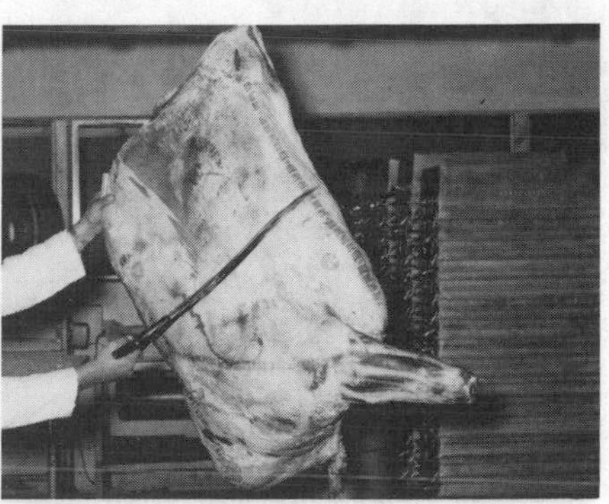

FIGURE 13.4D

to backbones, extending cut across quarter through breastbone (Fig. 13.4C). (4) Saw through backbone (Fig. 13.4D). (5) Saw through

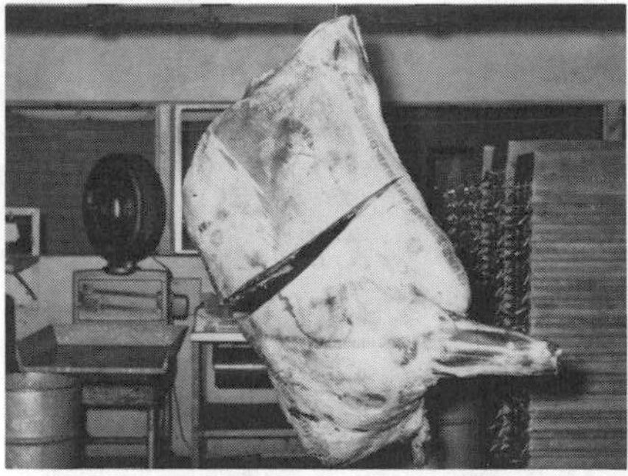

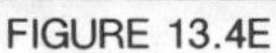

FIGURE 13.4E

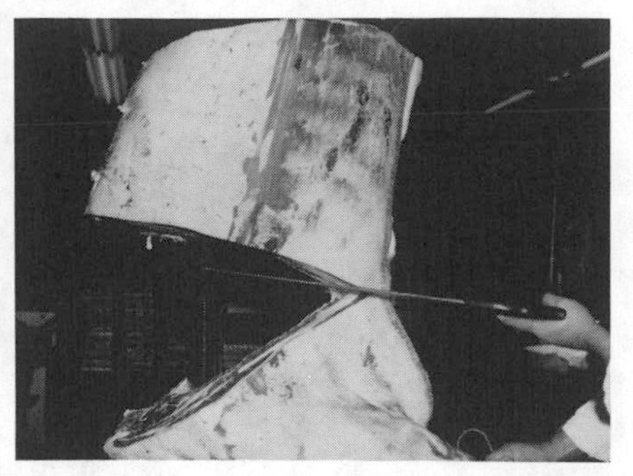

FIGURE 13.4F

breastbone (Fig. 13.4E). (**6**) Drop off cross cut chuck (Fig. 13.4F). (**7**) Remove plate and rib portion to cutting table or saw (Fig. 13.4G)

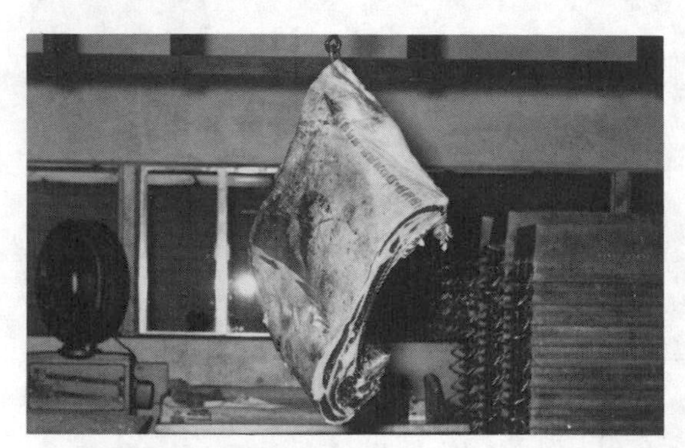

FIGURE 13.4G

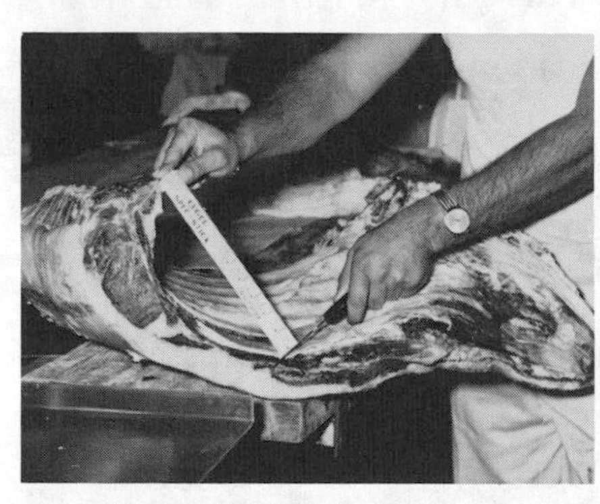

FIGURE 13.4H

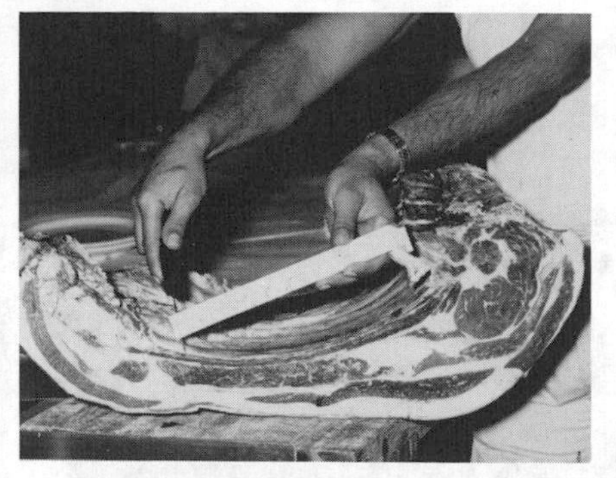

FIGURE 13.4I

(**8**) Measure 10 in. from protruding edge of chine at loin end (Fig. 13.4H). (**9**) Measure same distance on chuck end (Fig. 13.4I). (**10**) Saw in straight

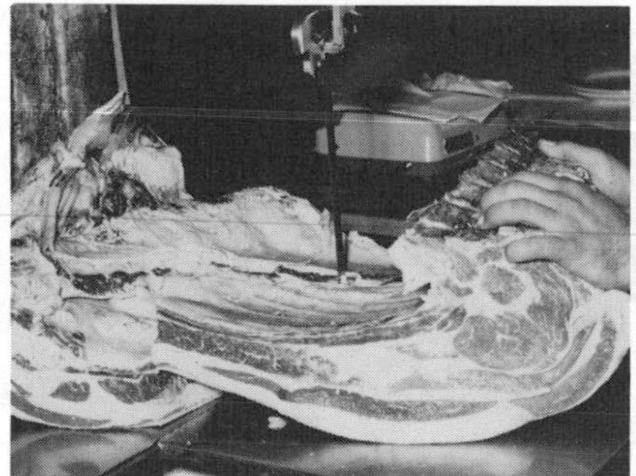
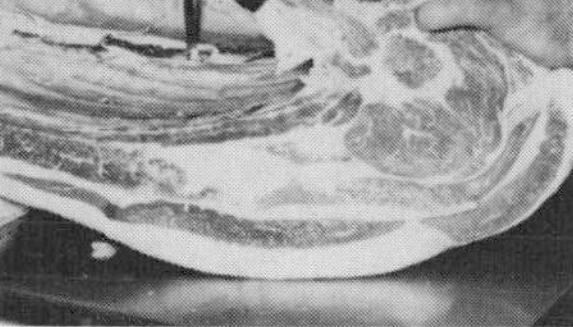

FIGURE 13.4J

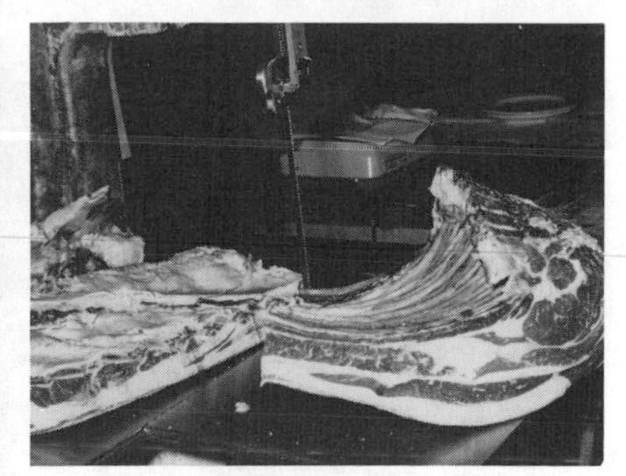

FIGURE 13.4K

line between two marks (Fig. 13.4J). (**11**) Separate 7×10 rib (right) and plate (left) (Fig. 13.4K).

Breaking Variations

Some of the regional breaking variations are:

Arm Chuck.—Cross-cut chuck with brisket off.

Square Chuck.—Cross-cut chuck with brisket off and shank sawed off in same plane with brisket cut.

Long Plate.—Brisket and plate in one piece.

Beef Back.—Square chuck and 7×10 rib in one piece.

Triangle.—Cross-cut chuck, plate on.

MUSCLE BONING THE CROSS-CUT CHUCK

Among the institutional trade, muscle boning is most generally followed.

Cuts and Yields:

Cut	*Per cent*
Boneless chuck	42
Clod	18
Boneless brisket	9
Shank meat and trim	8
Fat and bone	23
	100

The usual procedure is to remove brisket, shank, seam out clod, and bone out chuck.

Separating Brisket

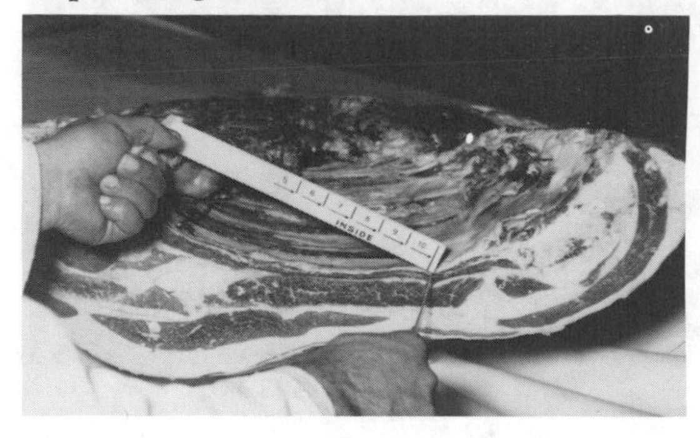

FIGURE 13.5A

(**1**) Locate point 10 in. from protruding edge of chine on rib end (Fig. 13.5A). (**2**) Locate tip of first segment of sternum. Saw through rib bones in a straight line between two points (Fig. 13.5B). (**3**) Extend saw cut

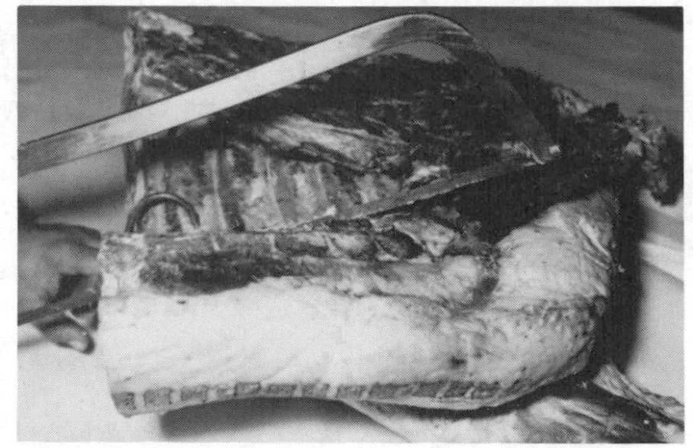

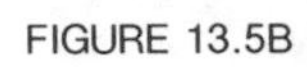

FIGURE 13.5B

with a knife to natural seam between brisket and shank and seam brisket out (Fig. 13.5C).

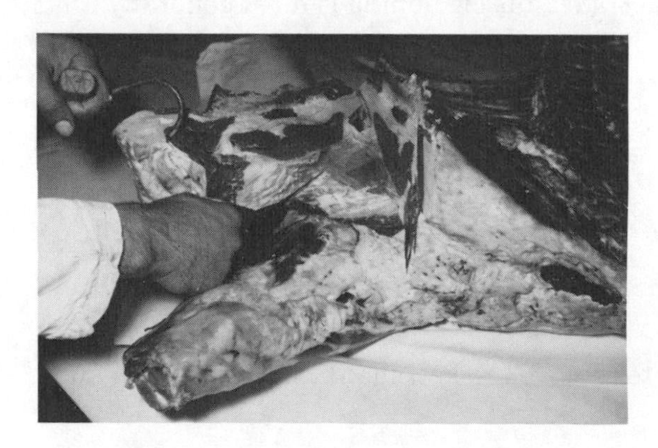

FIGURE 13.5C

Boning the Brisket

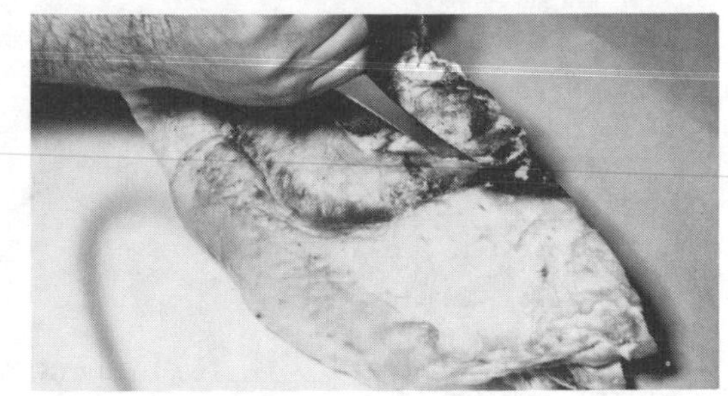

FIGURE 13.6A

(1) To shell out interior bones, start by inserting knife tip below breastbone (Fig. 13.6A). (2) Score the full length of the breastbone (Fig.

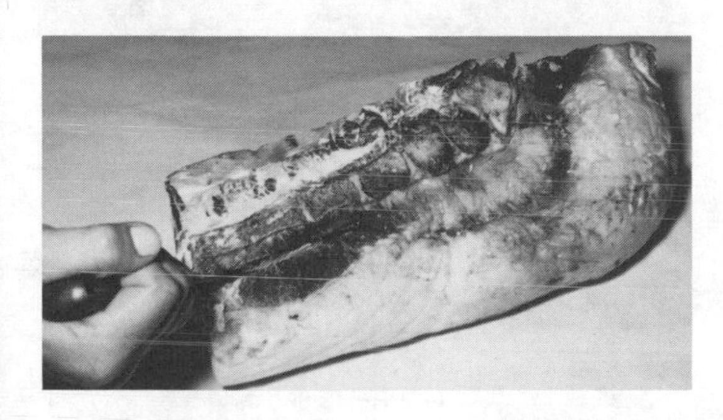

FIGURE 13.6B

13.6B). (3) Knife between the brisket and the rib bones (Fig. 13.6C). (4) Cut along breastbone and drop away bones (Fig. 13.6D). (5) Remove deckle fat on the inside of brisket by scoring just below breastbone to

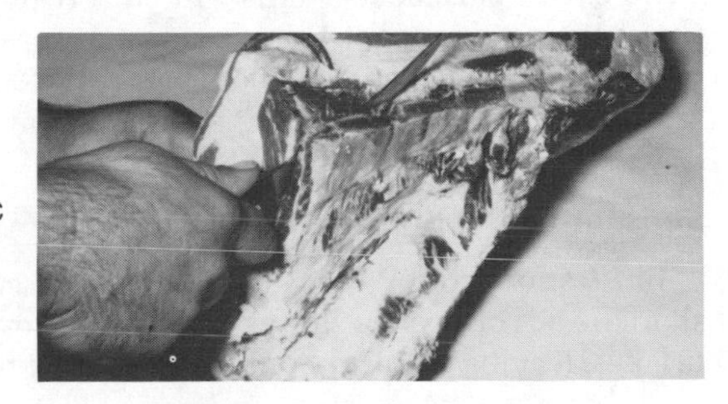

FIGURE 13.6C

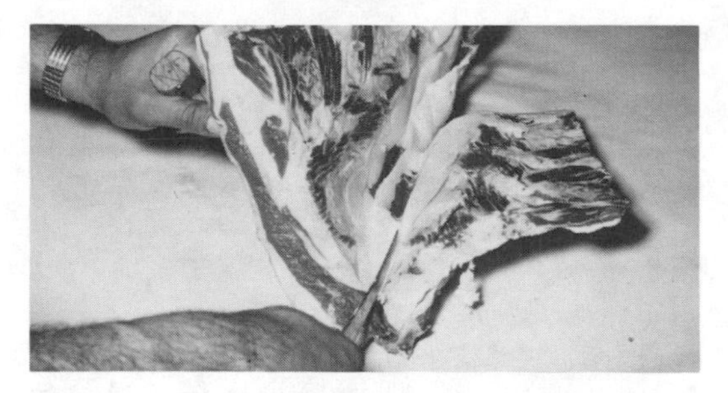

FIGURE 13.6D

the lean of brisket (Fig. 13.6E). (**6**) Pull and remove deckle along natural seam (Fig. 13.6F).

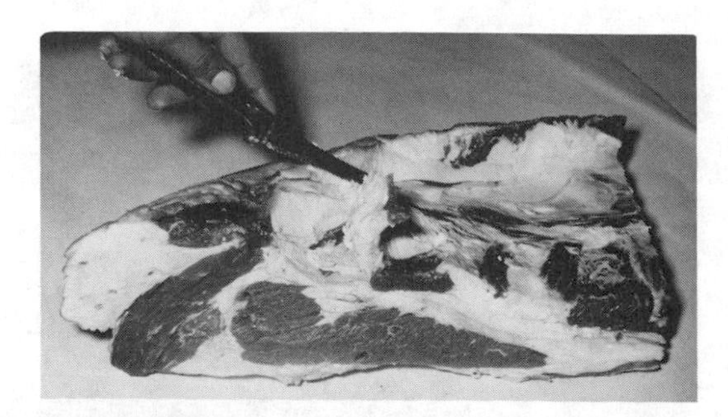

FIGURE 13.6E

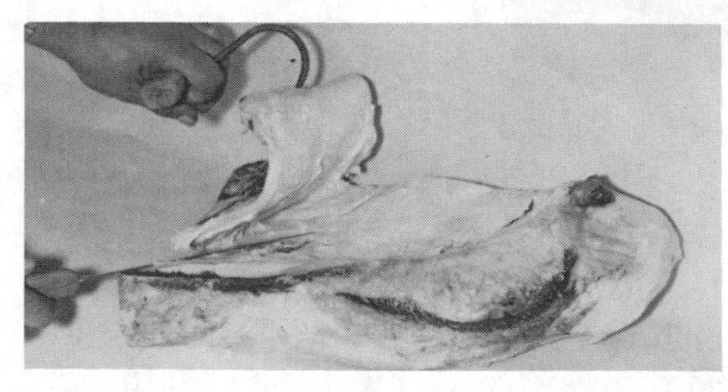

FIGURE 13.6F

Trim fat below deckle, leaving only flecks of fat, and exposing lean. Trim excess surface fat. This will vary regionally. The web muscle on the outside point of the brisket is sometimes left intact, sometimes removed when specified. The boneless brisket will yield 45 to 60 per cent relative to the trim and amount of cover fat.

Separating the Foreshank

Illustrations follow the natural bone structure. Frequently, the foreshank is cut off with a saw as a continuation of the cut, taking off the brisket, leaving the arm bone in the chuck to be boned later.

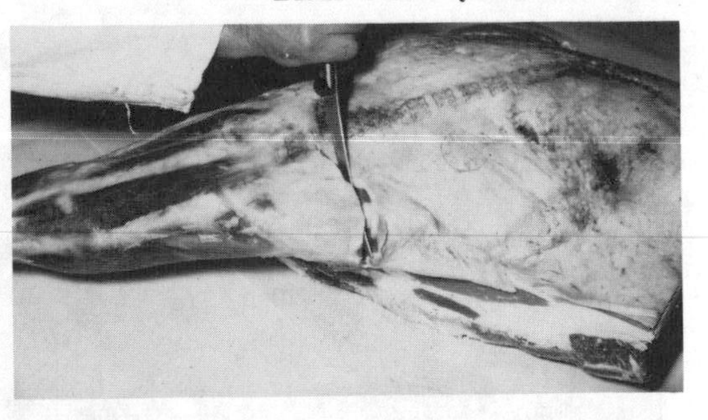

FIGURE 13.7A

(1) Locate the tip of shank knuckle bone and knife through lean following bones to arm bone (Fig. 13.7A). (2) Separate lean from arm bone

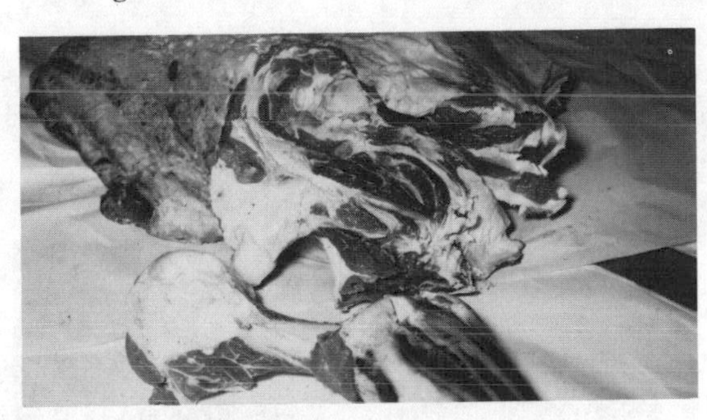

FIGURE 13.7B

by knifing around superior extremity (see skeletal chart), dropping arm bone and foreshank away from chuck (Fig. 13.7B).

Pulling Clod

(1) Locate ridge bone with knife tip where arm bone has been removed (Fig. 13.8A). (2) Score along ridge bone with knife to rib end of chuck (Fig. 13.8B). (3) Locate blade bone joint where it connects to arm bone in area where shank was removed. Find natural seam and start to sep-

FIGURE 13.8A

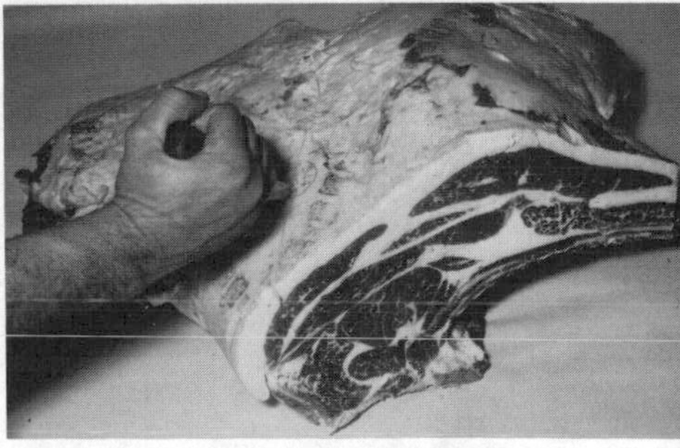

FIGURE 13.8B

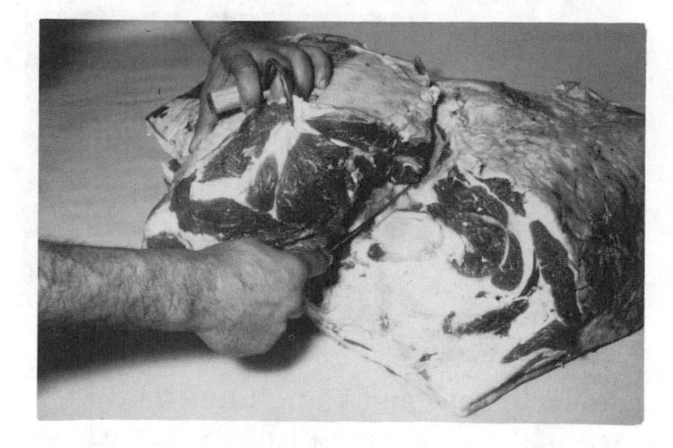

FIGURE 13.8C

arate clod (Fig. 13.8C). (4) Continue seaming between clod and rib bones (Fig. 13.8D). (5) Pull clod away from blade bone (Fig. 13.8E).

FIGURE 13.8D

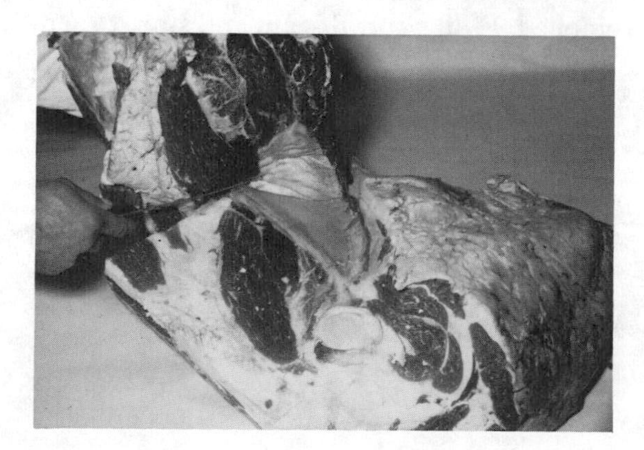

FIGURE 13.8E

Boning the Chuck

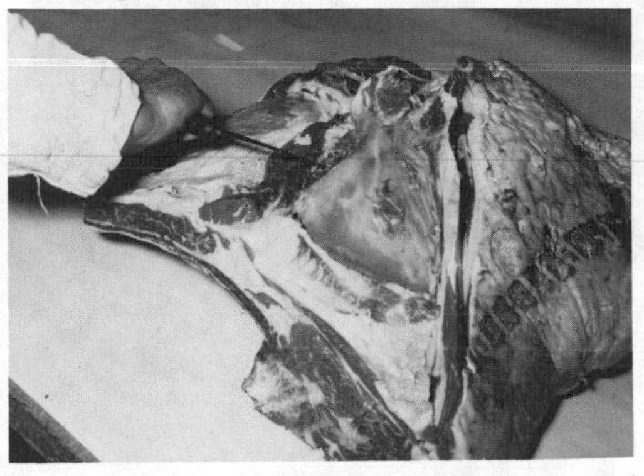

FIGURE 13.9A

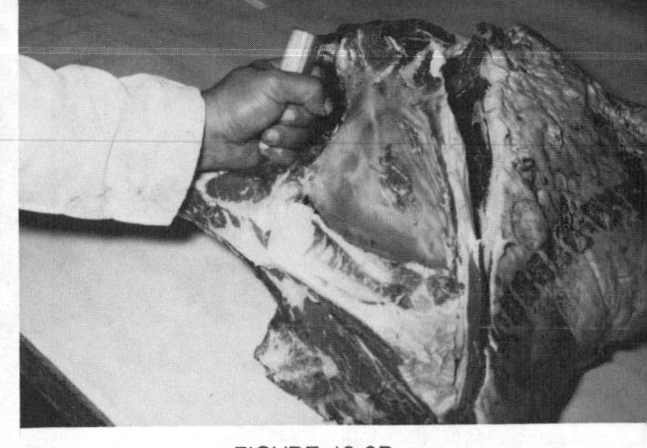

FIGURE 13.9B

(1) Score along both sides of blade bone (Fig. 13.9A). (2) Run boning hook between shoulder blade bone (scapula) on both sides to loosen. Pull with hook (Fig. 13.9B). (3) Snap and cut blade bone loose (Fig. 13.9C). (4) Trim out meat on inside of chine bone (Fig. 13.9D).

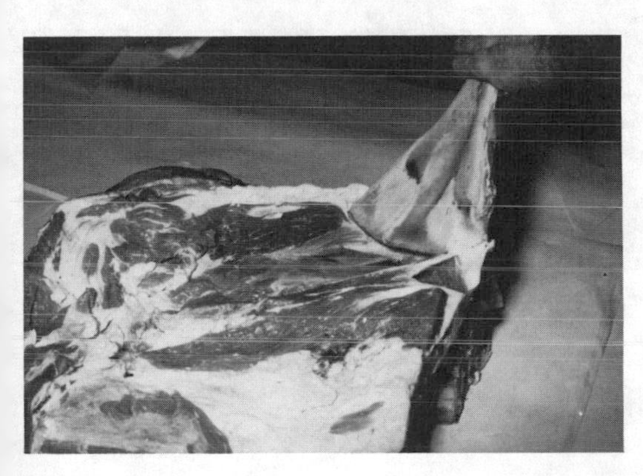

FIGURE 13.9C

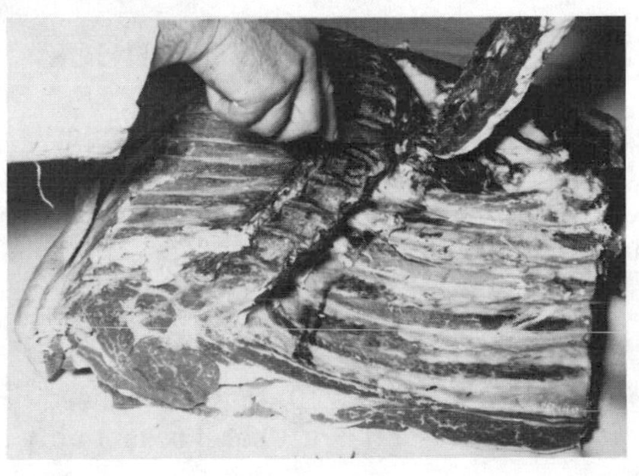

FIGURE 13.9D

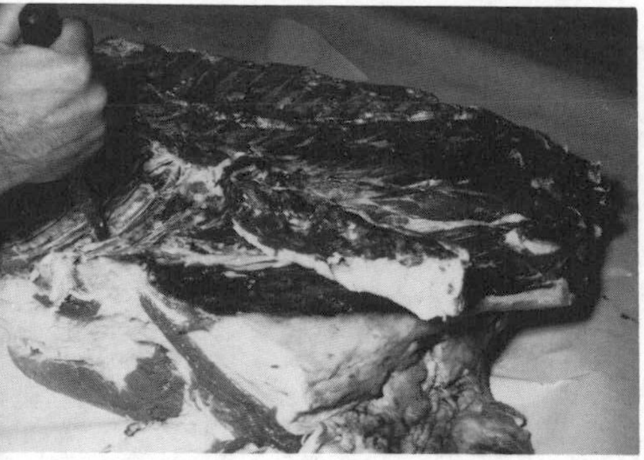

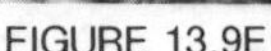
FIGURE 13.9E

FIGURE 13.9F

(5) With tip of knife score between all rib bones (Fig. 13.9E). (6) Separate vertebrae, knife through feather bones and again through vertebrae. Lift bones one at a time (Fig. 13.9F). (7) Remove back strap. Trim off excess age and fat (Fig. 13.9G).

FIGURE 13.9G

MENU PLAN—CROSS-CUT CHUCK

The cross-cut chuck and its components generally are classified as "rough cuts." The clod, or parts of it, is sometimes oven roasted. This is a marginal oven roast. Most cuts are best adapted to moist heat cooking, pot roasting, braising, stewing, and grinding.

Brisket

The brisket (Fig. 13.10) may be braised or used as a pot roast or for boiled beef. It is a flavorful and reasonably tender cut. The drippings

FIGURE 13.10

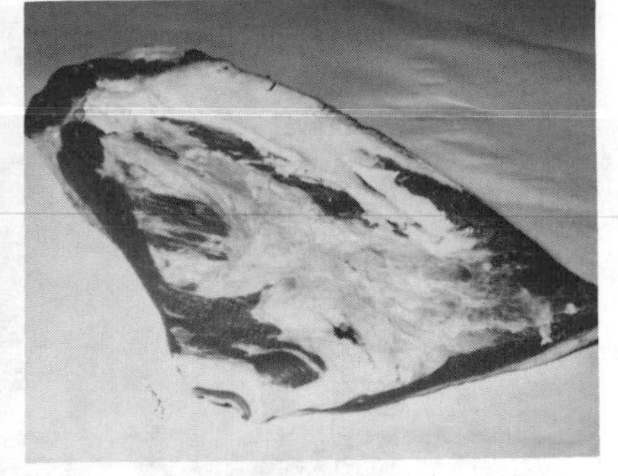

FIGURE 13.11

contain large quantities of beef extract, which make a delicious gravy. A definite slicing plan should be developed. Start at the flat end and (front of picture Fig. 13.11) either slice straight across, perpendicular to the grain or slightly on the bias, which will result in perpendicular to the grain slices at the thick end. About two-thirds of the way through the brisket, three distinct muscles with heavy intermuscular fats present themselves. These muscles may be seamed out, trimmed of excess surface fat, and sliced individually.

Briskets are the most popular cut for corned beef. Though not as lean as round cuts, the brisket is widely accepted because of its flavor.

Foreshank

The foreshank has many uses (Fig. 13.12). Because of the large amount of connective tissue, it is tough, yet flavorful. With the proper

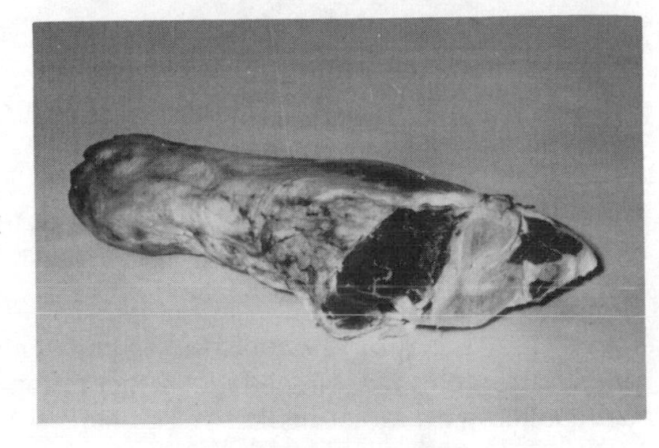

FIG. 13.12. FORESHANK

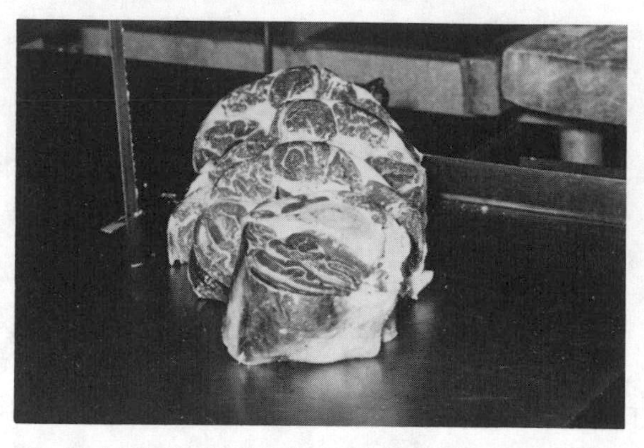

FIG. 13.13. FORESHANK SAWED INTO PORTIONS

cooking, as in braising, it becomes tender. It may be used for the usual by-product items, and is especially good for chili meat and goulash. The foreshank may be sawed into portions (Fig. 13.13) or boned out for soup stock, or used in braised dishes.

Shoulder Clod

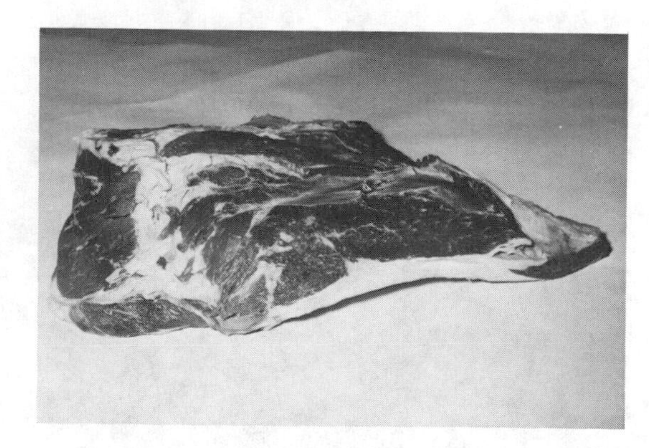

FIG. 13.14. SHOULDER CLOD

The clod (Fig. 13.14) when used as an oven roast is prepared whole, rolled, cut in chunks, or seamed into component muscles. The clod is best cooked with moist heat.

FIG. 13.15. BONELESS CHUCK—CLOD-OUT

Boneless Chuck

The chuck clod-out (Fig. 13.15), as well as the clod, can serve for any "by-product" plan. It makes excellent ground beef. Much or all of the surface and interior fat may be included in the grind for maximum flavor and juiciness. The grind may appear "fatty" and should be carefully cook-tested. Sometimes the "eye of the chuck," a continuation of the rib-eye is shelled out and used for roasting or Swiss steaks. In some institutions, chucks are used for a wide variety of beef items on the menu.

Cutting Variation

As a technique to upgrade the chuck or as a substitute for higher quality "center cuts," the "muscle boning" technique, as developed by the National Live Stock and Meat Board, may be tried (Fig. 13.16).

PRIMAL RIB

The term "prime rib" as it applies to the 7×10 rib or primal cut (Fig. 13.17) is a misnomer. This term is meat packer designed to merchandise as much of the beef carcass as possible at a premium price.

The present trend is to merchandise the roast-ready, export rib, banquet rib, Spencer roll, or rib-eye as the prime rib of beef.

Virtually all operators present the final rib product in the same manner.

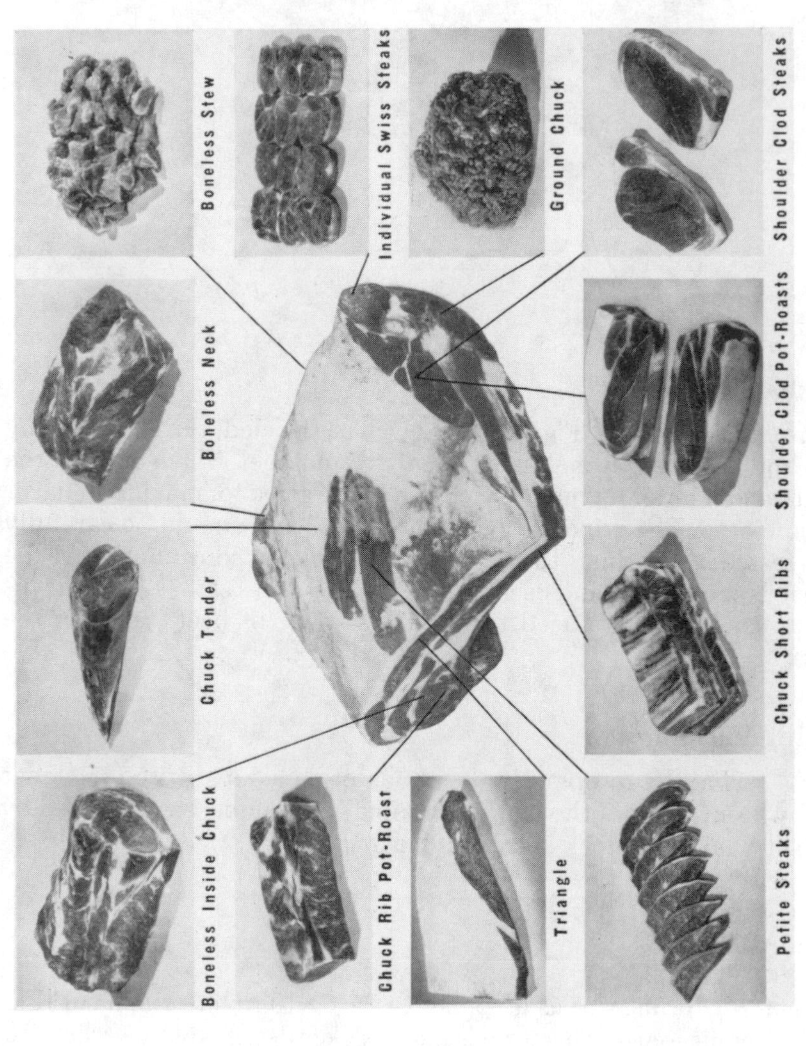

Courtesy of the National Live Stock and Meat Board

FIG. 13.16. CUTS FROM THE MUSCLE BONED CHUCK

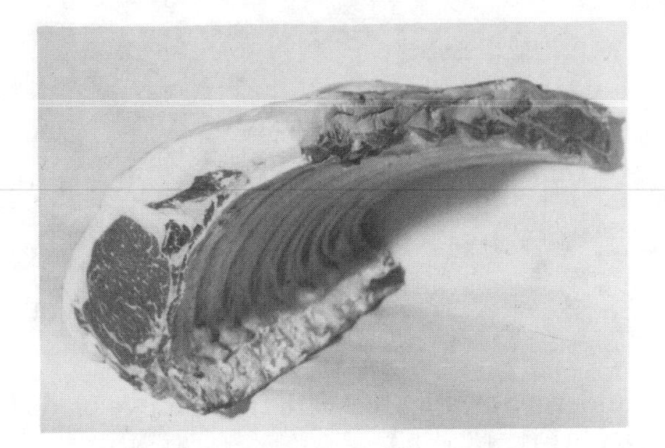

FIG. 13.17. PRIMAL RIB

Oven-prepared Rib Roast

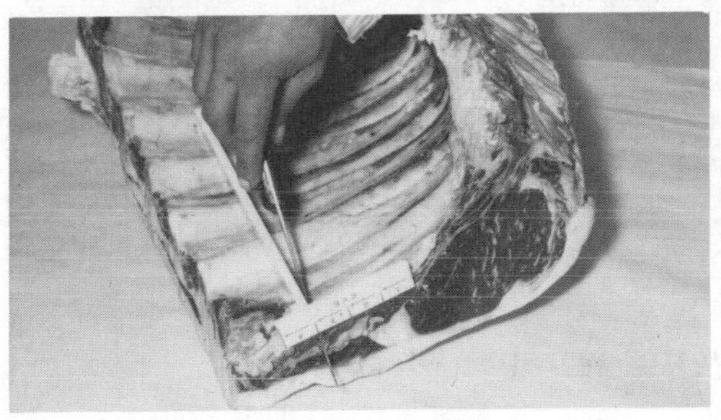

FIGURE 13.18A

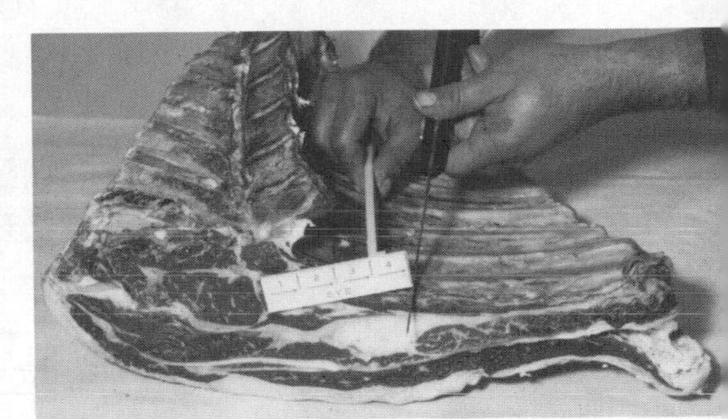

FIGURE 13.18B

(1) Locate a point 3 in. from eye on loin end (Fig. 13.18A). (2) Then locate a point 4 in. from eye on chuck end (Fig. 13.18B). (3) Cut off short

FIGURE 13.18C

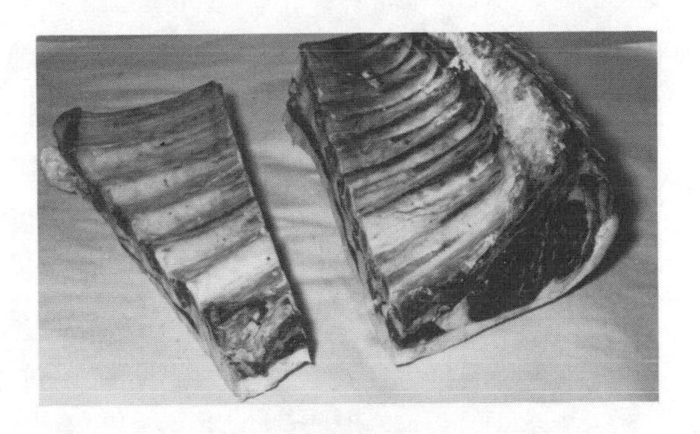

ribs in a straight line between these points (Fig. 13.18C). (4) Remove chine by cutting at juncture of feather bones and rib bones exposing lean

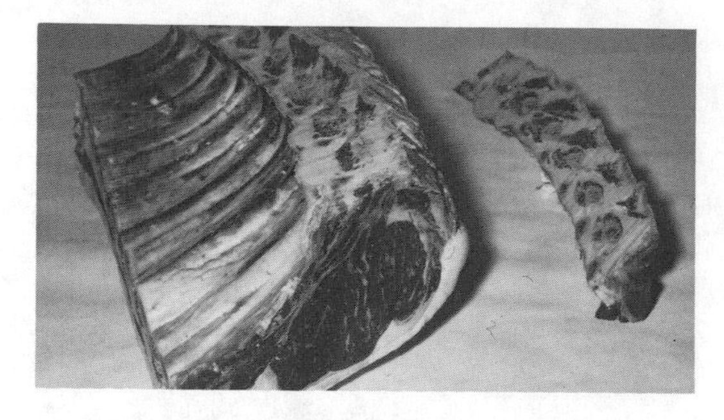

FIGURE 13.18D

underneath (Fig. 13.18D). (**5**) Remove blade bone and cartilage in chuck end (Fig. 13.18E).

FIGURE 13.18E

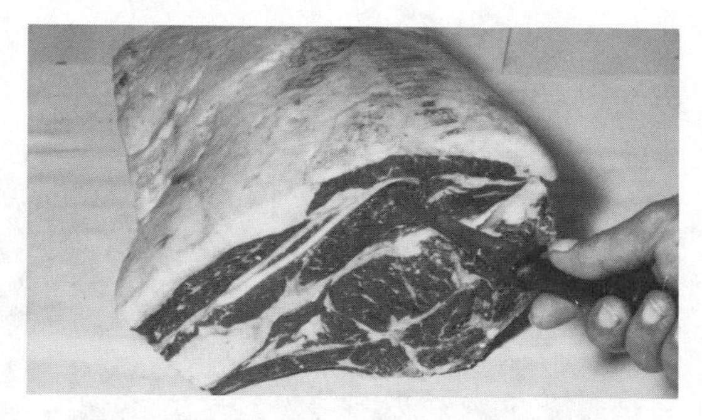

Figures 13.19 and 13.20 show the oven-prepared rib.

Cuts and Yields:

Cut	*Primal Rib Per cent*
Oven-prepared rib	70
Short ribs, fat and bone	30

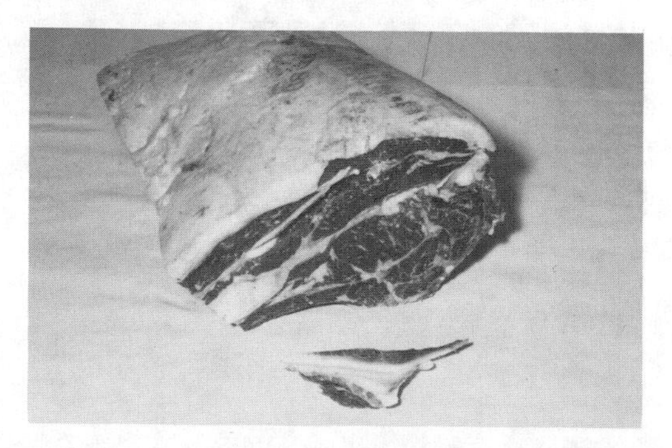

FIG. 13.19. OVEN-PREPARED RIB BLADE BONE OUT—CHUCK END

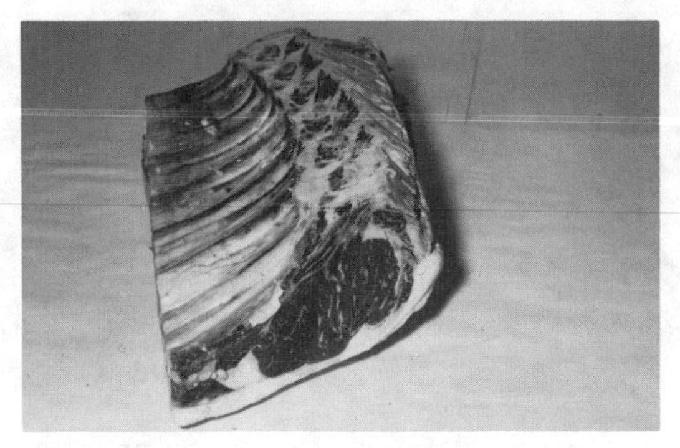

FIG. 13.20. OVEN-PREPARED RIB LOIN END

Most operators will make some further modifications. Some of the variations are presented.

Roast-ready Rib

This is one of the end preparations for the oven-prepared rib. It is an ideal preparation for meat retailers finding it difficult to merchandise the chuck end of the rib.

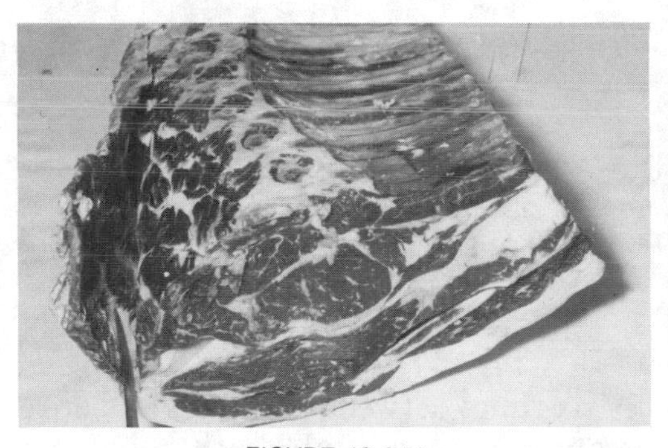

FIGURE 13.21A

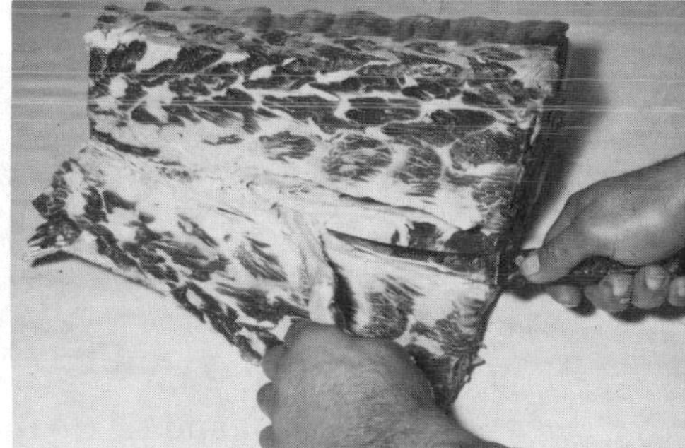

FIGURE 13.21B

(1) Insert knife tip between feather bones and lean at back of rib. Scalp out bones, leaving them "hinged" where they meet fat. The feather bones may be completely removed (Fig. 13.21A). (2) Locate and pull out the backstrap (Fig. 13.21B). (3) Starting at chuck end (large end),

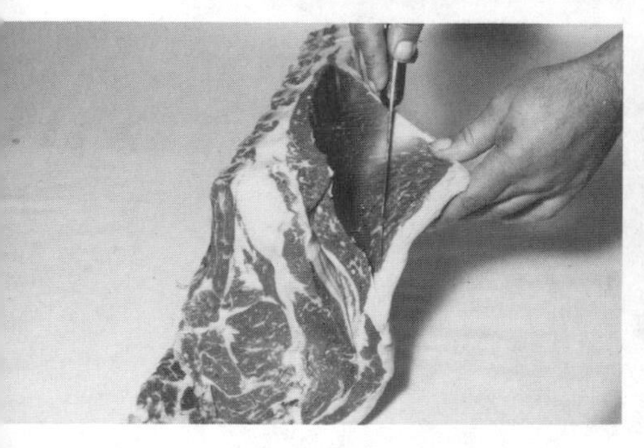

FIGURE 13.21C

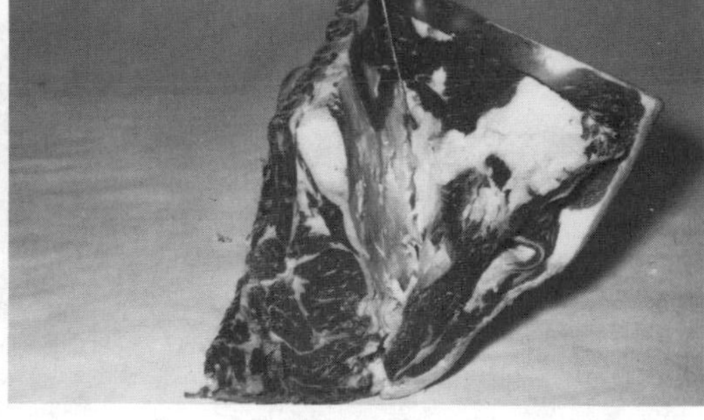

FIGURE 13.21D

loosen cover fat (but leave attached) for first five ribs (Fig. 13.21C). (**4**) Seam carefully between lifter meat above thin fat covering eye of rib (Fig. 13.21D). (**5**) Remove the lean thus seamed out (Fig. 13.21E).

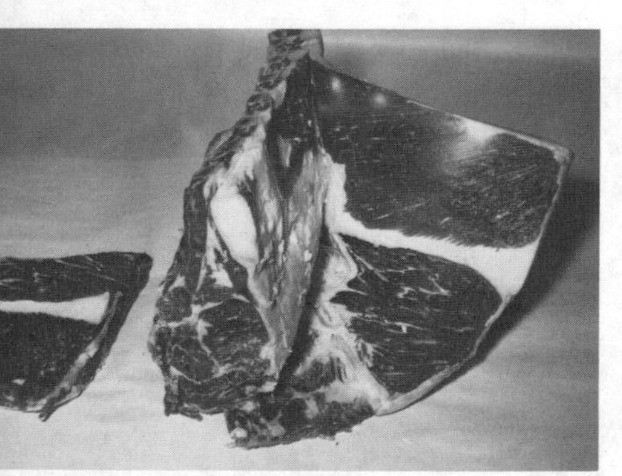

FIGURE 13.21E

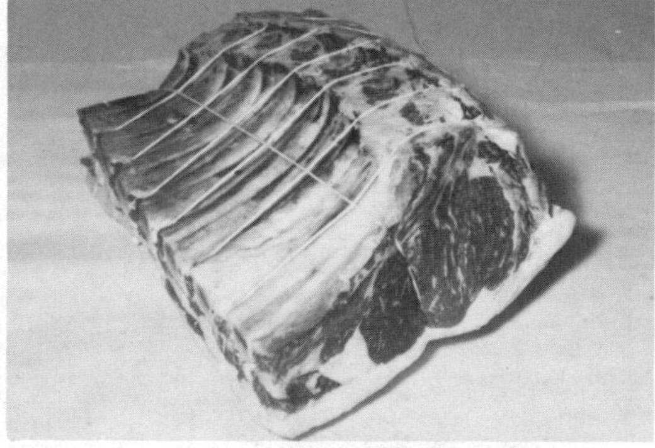

FIGURE 13.21F

(**6**) Replace fat and tie rib firmly six times between rib bones and once lengthwise. Photo shows loin end of roast-ready rib, featherbones on (Fig. 13.21F).

Note that on chuck end, the eye looks very much like the eye on the rib end. This preparation makes for more uniform portions and roasting. In addition, the lifter meat removed, which may contain 2 to 3 lbs of lean, may be used as a by-product, reducing the portion cost of the roast.

Fig. 13.22 shows chuck end of roast-ready rib.

FIG. 13.22. ROAST-READY RIB—CHUCK END

Cuts and Yields:

Cut	*Primal Rib, Per cent*
Roast-ready rib	60
Lifter meat	8
Short ribs, fat and bone	32

Menu Plan

Roasting is the most common method of preparation. To achieve cost control, a weight range should be set for the rib, and a cutting plan determined for portioning. Some plans call for a predetermined cooked weight portion. A simple and commonly used approach is to try to achieve 1, 2, or 3 servings per bone when slicing, or a given number of portions from a given size rib. If the slicing is to be done bone-in, it is important that the chuck end be large enough to rest the roast evenly. Sometimes the loin end is placed in a small pan to steady the roast. Some operators remove the bones before slicing, laying the roast down flat to hand slice or to slice on an electric slicer. The rib bones, meat-on, may be accumulated and used as a low priced entrée—deviled beef bones with horseradish sauce, or barbecued beef bones.

For Steaks.—The oven-ready or roast-ready rib may be cut on the saw (Fig. 13.23). Using the roast-ready rib, first remove the cover fat from the first five ribs, or all seven. Sometimes, a portion of the rib is steaked (usually the first two ribs from the loin end), and the remainder

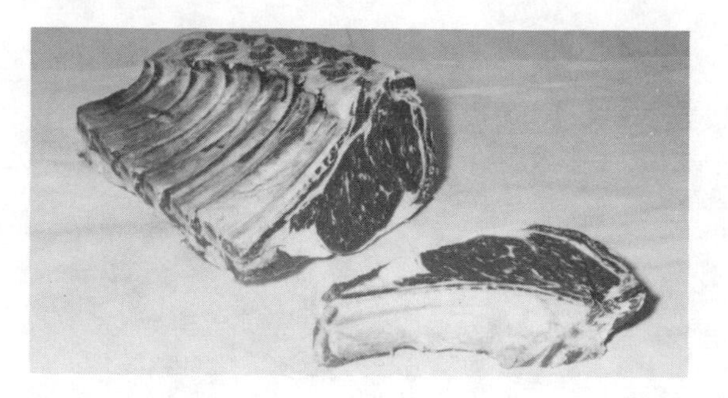

FIG. 13.23. ROAST-READY RIB WITH ONE RIB STEAK REMOVED

used as a roast. Steaks may be cut according to weight or at a given thickness.

Export ribs (Fig. 13.24) is another variation which is receiving broad acceptance. The cover fat and feather bones are removed and excluded from the roast-ready rib. A single grade tab is left on the eye of the rib for identification. This rib has some roasting advantages as it cooks quicker and more evenly; a caramelized crust can be formed on the edible surface which makes a flavor contribution; the probable shelf life is increased by removing the feather bones which tend to puncture the plastic bag, and it need not be tied for roasting.

Economically it is sold competitively with the roast-ready rib; although the selling price is higher, the purchase pounds are less.

FIG. 13.24. EXPORT RIBS

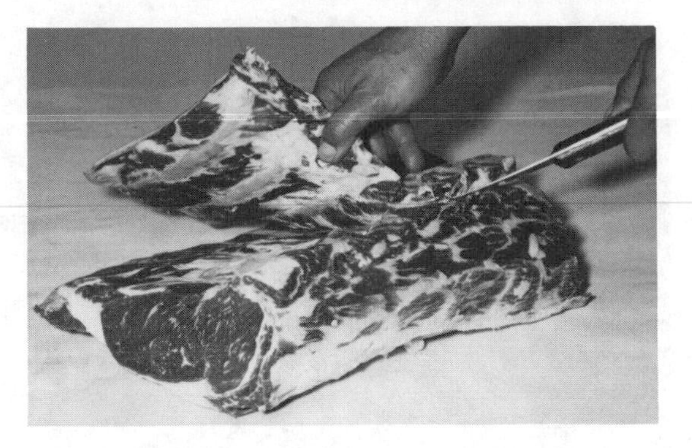

FIG. 13.25. NUDE RIB

A nude rib (Fig. 13.25) is made from the roast-ready rib by seaming off all the cover fat, removing the feather bones, and scalping off the rib bones. The seasoning may be applied close to or on the lean, which crusts while roasting, and contributes to the flavor.

The more by-product that is removed, while not destroying the conformation of the portion, the shorter the roasting time, and the lower the shrink. For the exponents of the slow roasting school, the nude rib might be the most consistent approach. The yield from a 7 × 10 rib is 35 to 40 per cent.

A Boneless Banquet Rib may be made from the nude rib by cutting off the tail in a straight line from point,

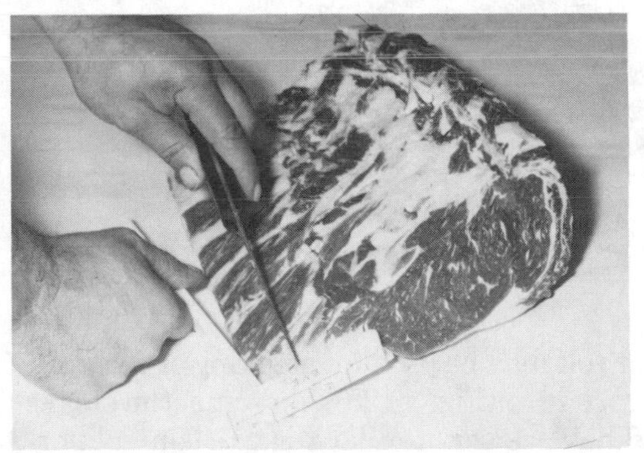

FIGURE 13.26A

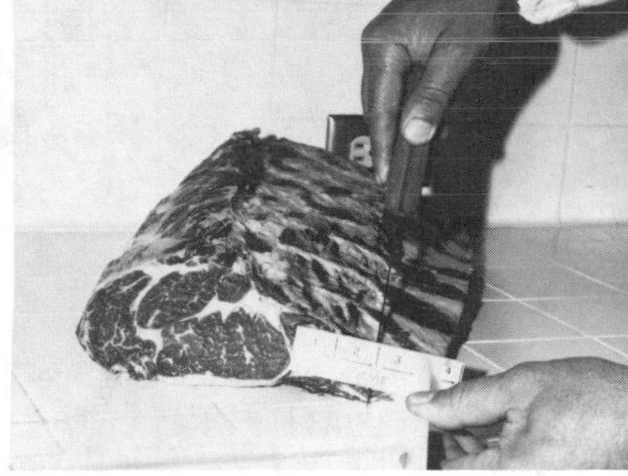

FIGURE 13.26B

(**1**) two inches from eye on the loin end (Fig. 13.26A), (**2**) to a point 2 in. from eye on chuck end (Fig. 13.26B). (**3**) Spencer eye yields 30 to 32 per

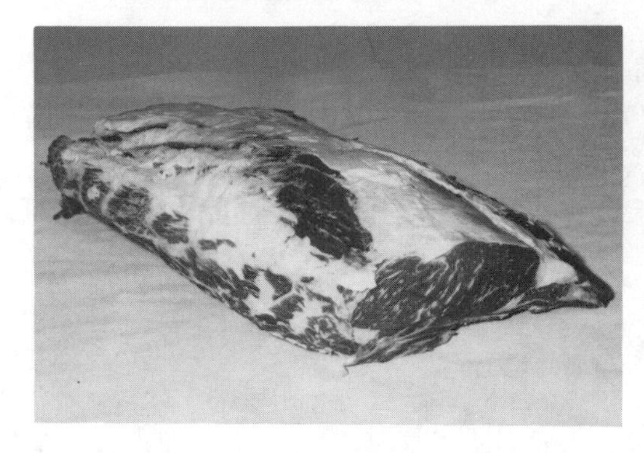

FIGURE 13.26C

cent of primal rib (Figure 13.26C). It is the accepted trade practice to leave a grade stamp firmly attached.

A Rib-eye Roll (Fig. 13.27) may be made from Spencer roll by seaming of all fat and lean that surrounds eye muscle. Yield is 25 to 27 per cent of primal rib.

FIG. 13.27. RIB-EYE ROLL

The rib-eye roll and Spencer roll may be used for steaking or roasting. Steaking is a simple slicing procedure (Fig. 13.28). To some, the steaks have a very desirable flavor. They are very tender, easy to control portion-wise, and economical to use. These boneless roasts are ideal for banquets where speed requires an automatic slicer, and very uniform portions or where the maximum number of portions is desired. The crusty roasted surface is desirable. The eye appeal may not be as desirable as a portion cut from the other styles of rib roasts.

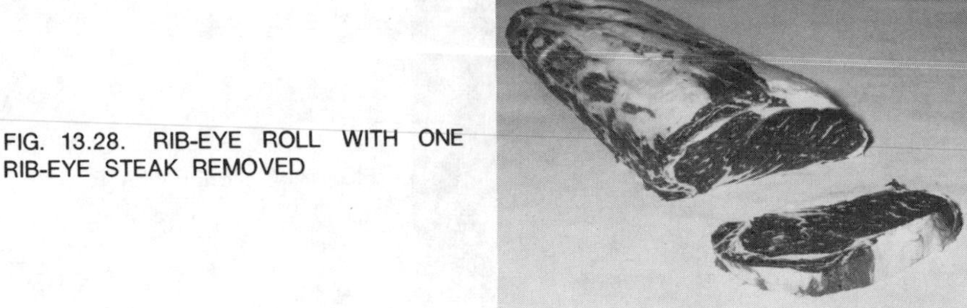

FIG. 13.28. RIB-EYE ROLL WITH ONE RIB-EYE STEAK REMOVED

Short Ribs

There are many different approaches to cutting the short ribs from the rib (Fig. 13.29). There is no general rule among jobbers or restaurants. There are three commonly used methods:

(1) The somewhat outdated method of measuring by placing three or four fingers of the hand next to the eye of the rib and cutting from that point. The cut will vary with the size of the cutter's fingers.

(2) Measuring from the eye with a ruler. This will be varied to suit competition, or to the particular specifications of the operator.

(3) Measuring from the inside of the chine. This is an objective basis, but does not achieve a uniform rib. A rib with either a small eye or from light beef when measured from the chine will have considerably more short ribs than a rib with a large eye or one from a heavy beef.

The term "short rib" is a misnomer from the point of view of end use. It requires some additional preparation. The operator can improvise,

FIG. 13.29. SHORT RIBS—7 BONES

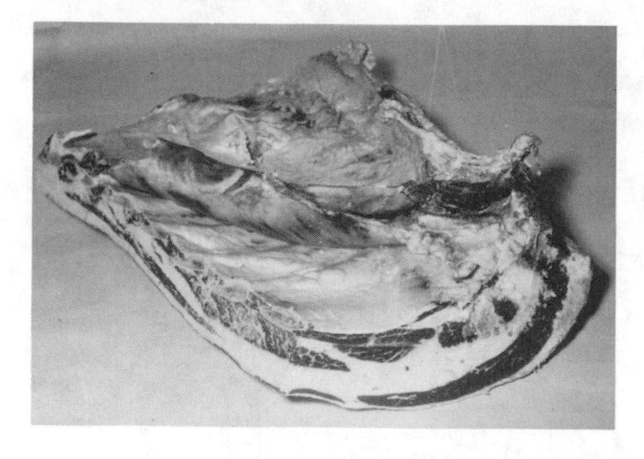

FIG. 13.30. BEEF PLATE

removing the lean portion next to the bone for short ribs, using the middle for boiling beef or short ribs, or other by-products. A careful analysis of the cut may lead to a variety of economical uses.

PLATE

The plate (Fig. 13.30) is a low value beef cut, only reaching the institutional user when the forequarter or beef triangle is purchased. A meat industry boner frequently buys plates for boning, turning out "kosher style" short ribs, boneless pastrami navels, and trimmings.

REFERENCES

ANON. 1945. Meat Handbook of the U.S. Navy. U.S. Navy Dept., Washington, D.C.

ANON. 1960. Institutional meat purchase specifications for fresh beef. (Series *100*). U.S. Dept. Agr., Washington, D.C.

ANON. 1960. Merchandising beef, muscle boning the chuck. National Live Stock and Meat Board, Chicago, Ill.

ANON. 1961. Meat Buyer's Guide to Standardized Meat Cuts. National Association of Hotel and Restaurant Meat Purveyors, Chicago, Ill.

ANON. 1962. Facts about beef. National Live Stock and Meat Board, Chicago, Ill.

BULL, S. 1951. Meat for the Table. McGraw-Hill Book Co., New York.

ROMANS, J. R. and ZIEGLER, P. T. 1977. The Meat We Eat, 11th Edition. Interstate Printers and Publishers, Danville, Illinois.

WANDERSTOCK, J. J., and WELLINGTON, G. H. 1961. Let's Cut Meat. Cornell Extension Bull. *1053.* Ithaca, N.Y.

14

Veal

PRIMAL CUTS

Veal (Fig. 14.1–14.2), being a young bovine, has the identical structure as presented in the chapter on beef. Like beef the forequarter may be

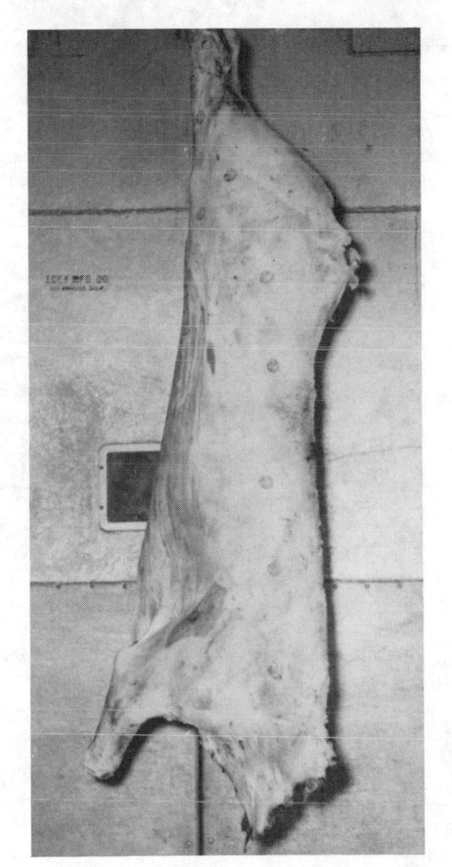

FIG. 14.1. VEAL SIDE

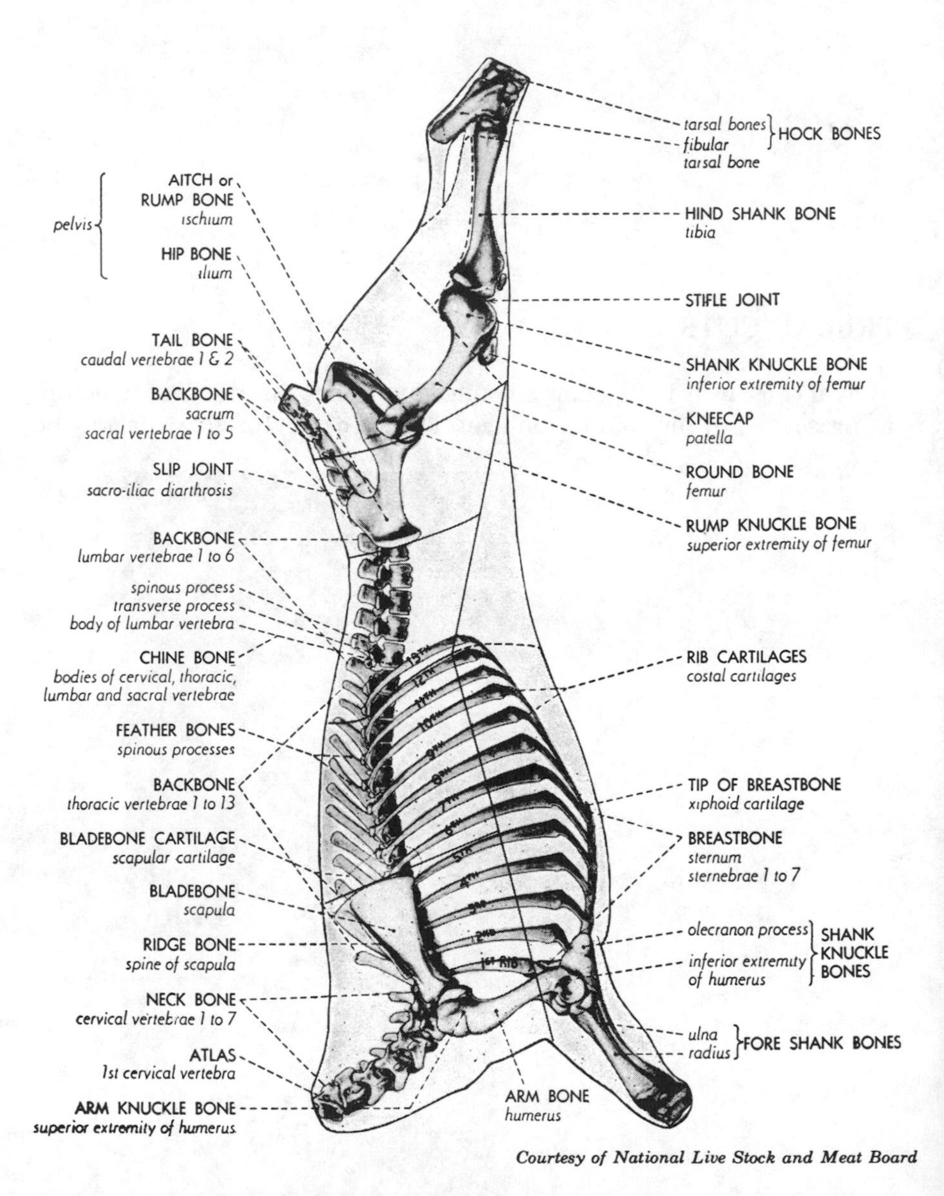

Courtesy of National Live Stock and Meat Board

FIG. 14.2. VEAL SIDE—SKELETAL STRUCTURE

separated from the hindquarter, with one rib bone on the hind. In practice, especially among Kosher packers, where a considerable portion of high quality veal is produced, two or three ribs are left on the hind. This practice is followed in that the hindquarter commands a considerably higher price than the forequarter. This creates a favorable economic break for the packer. The short loin user gets a couple of extra rib chops.

The term hindsaddle and foresaddle are used for both veal and lamb. These carcasses are not split lengthwise like beef. When saddles are made, the two forequarters or hindquarters are left joined at the backbone.

Yield for a two rib hindsaddle (Figs. 14.3–14.5) is about 50 per cent and foresaddle 50 per cent.

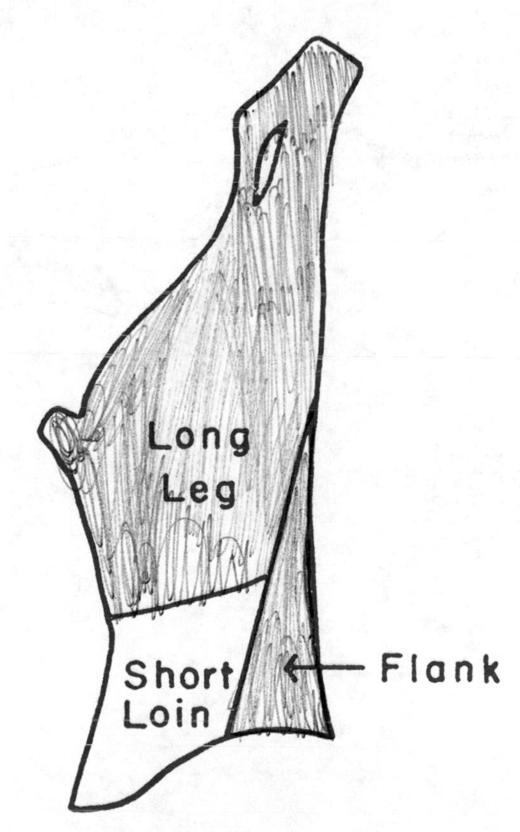

FIG. 14.3. VEAL HINDQUARTER—PRIMAL CUTS

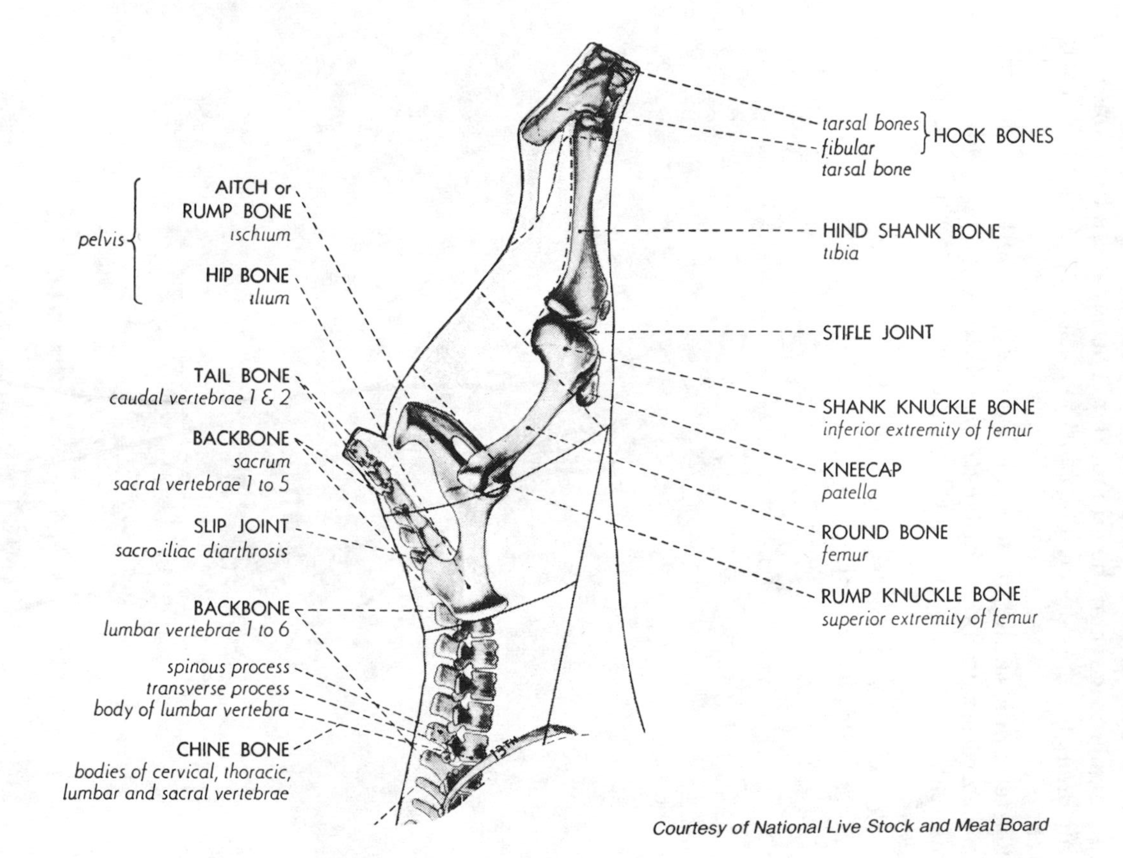

FIG. 14.4. VEAL HINDQUARTER—SKELETAL STRUCTURE

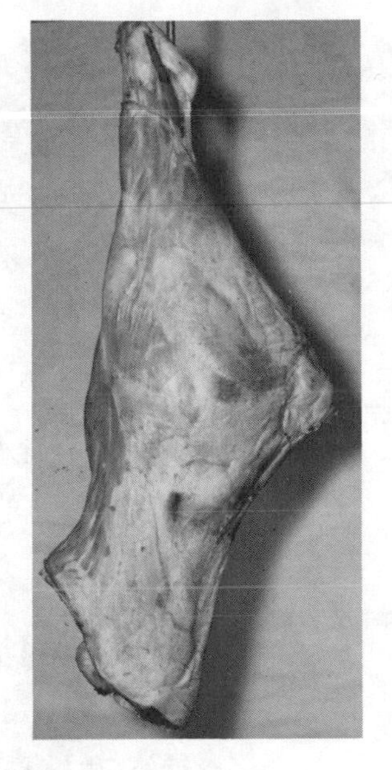

FIG. 14.5. VEAL HINDQUARTER

BREAKING HINDQUARTER

Primal Cuts and Yields:

Cut	*Percent*
Long leg	77
Trimmed short loin	14
Flank, kidney, and trim	9
	100

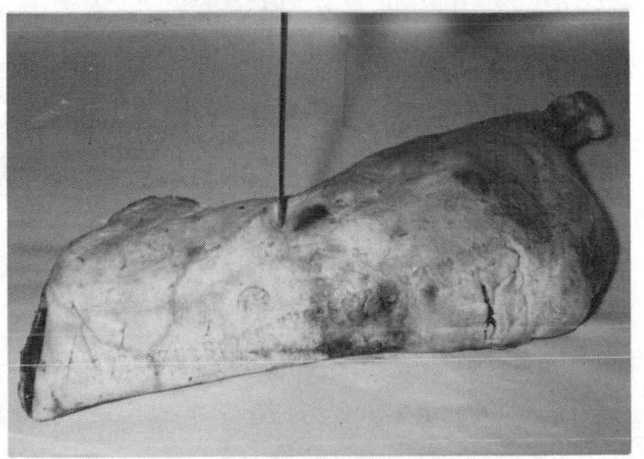

FIGURE 14.6A

(1) Locate anterior end of hip bone with knife tip (Fig. 14.6A).

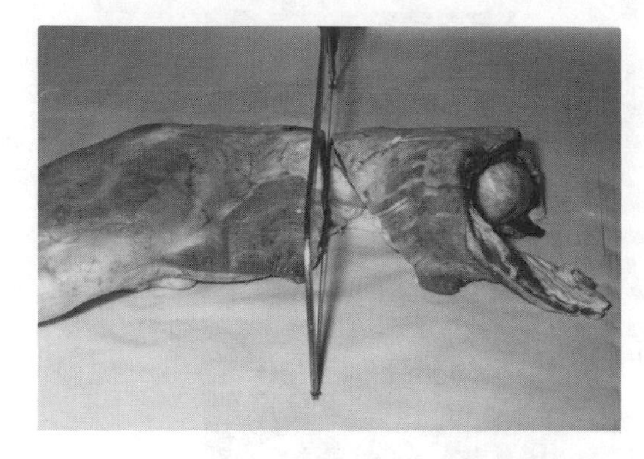

FIGURE 14.6B

FIGURE 14.6C

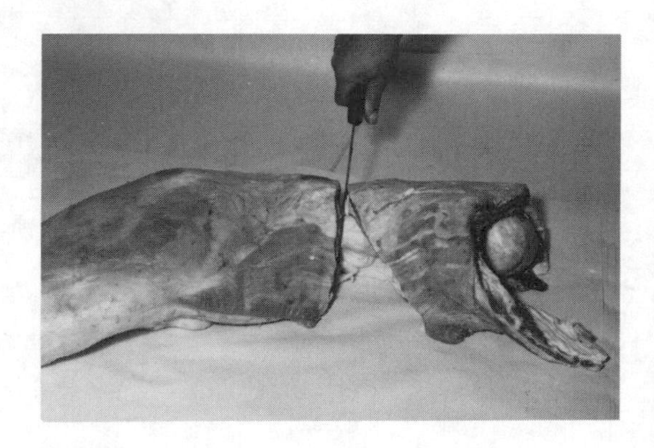

(**2**) Cut perpendicular to lean through to backbone (Fig. 14.6B).
(**3**) Saw through bone, leaving hip bone in long leg (Fig. 14.6C).

Cuts and Yield from Hindquarter:

Cut	*Per cent*
Long leg	77
Drop short loin	23
	100

MUSCLE BONING THE LONG VEAL LEG

Among restaurant users, the popular approach is to bone the long leg (Fig. 14.7), and to tie it for roasts, or muscle seam it for cutlets, roasts, etc., using each cut according to the quality standards set by the house.

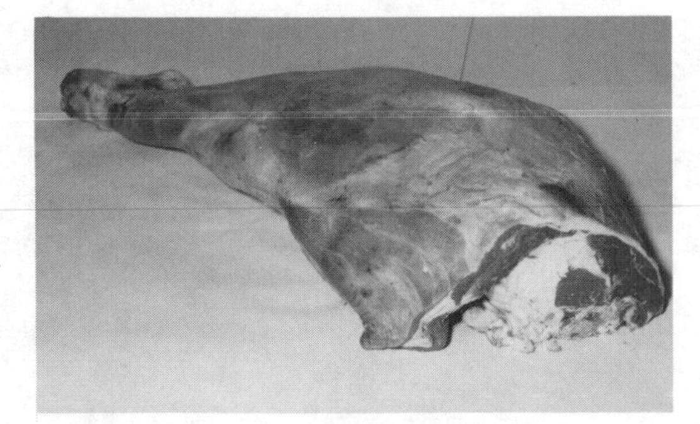

FIG. 14.7. LONG LEG OF VEAL

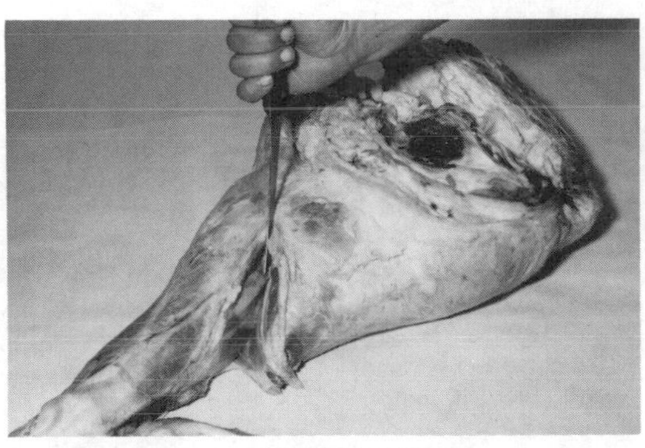

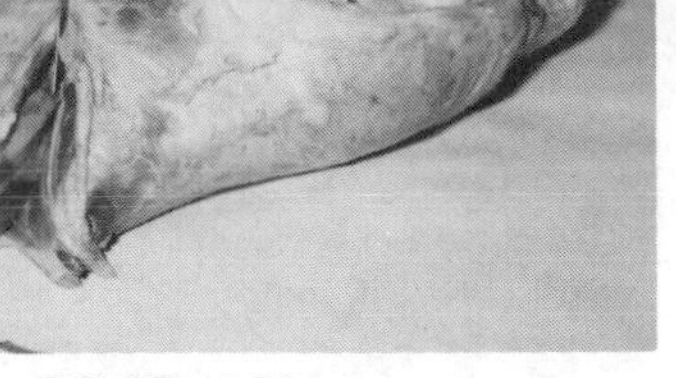

FIGURE 14.8A

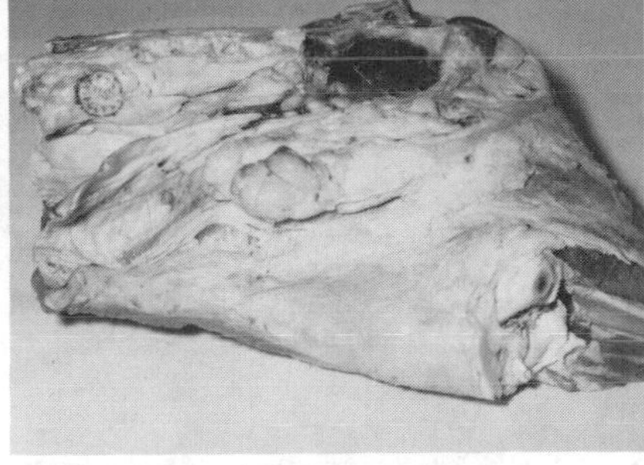

FIGURE 14.8B

(1) As in the case of the beef round, cut gambrel cord, follow natural seam along shank to stifle joint (Fig. 14.8A). (2) Cut through joint, crack off as on a beef round, making a long leg, shank off (Fig. 14.8B).

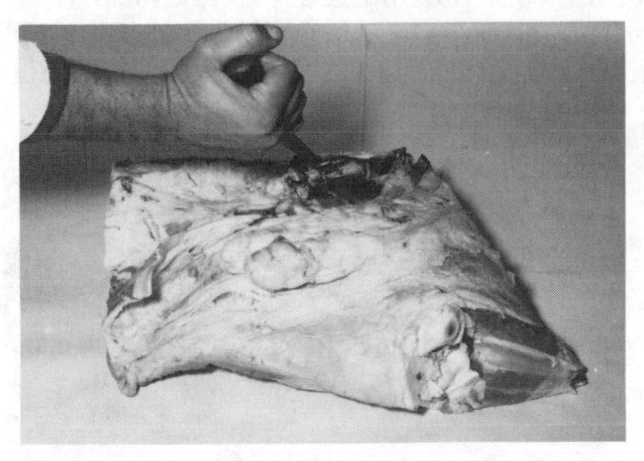

FIGURE 14.8C

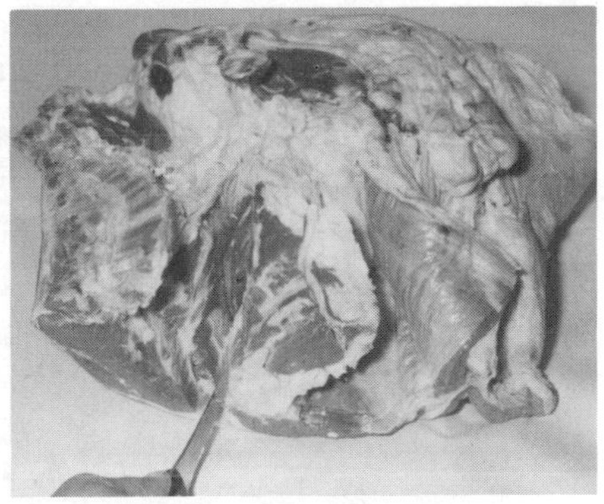

FIGURE 14.8D

(3) Score along aitchbone to anterior end of leg (Fig. 14.8C). (4) Extend cut along bone (Fig. 14.8D). (5) Cut through ball and joint at extremity

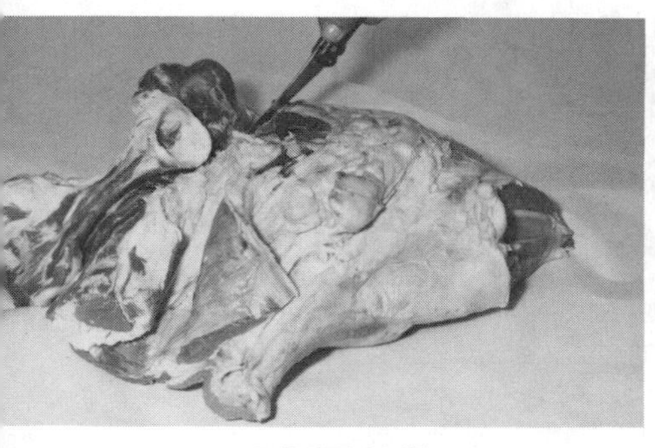

FIGURE 14.8E

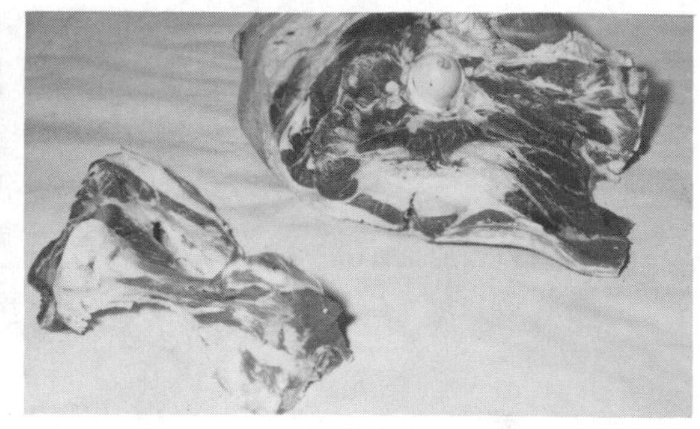

FIGURE 14.8F

of round bone (Fig. 14.8E). (6) Cut until back bone, hip bone, and tail bones come off in one piece (Fig. 14.8F).

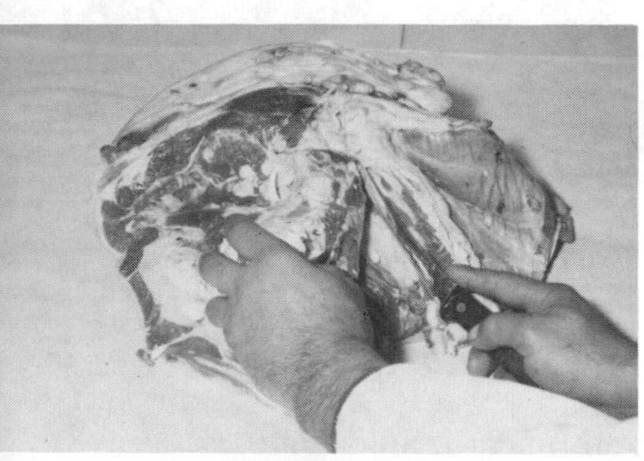

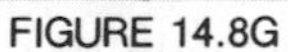

FIGURE 14.8G

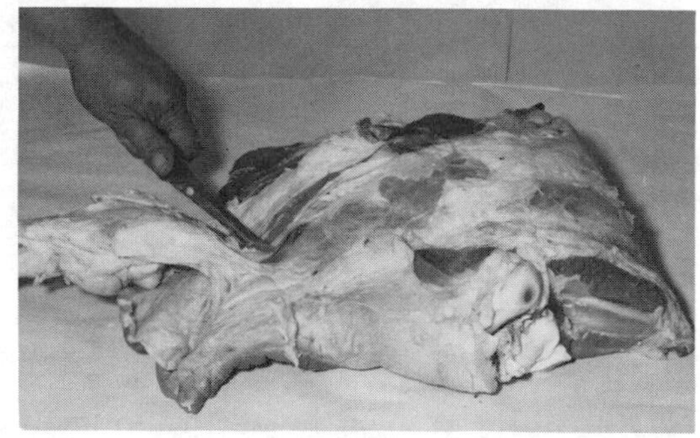

FIGURE 14.8H

(7) Locate and seam out tenderloin portion (butt tenderloin) (Fig. 14.8G). (8) Trim off udder or cod fat (Fig. 14.8H). (9) Score along round bone between top round and knuckle (Fig. 14.8I). (10) Extend cut (Fig. 14.8J).

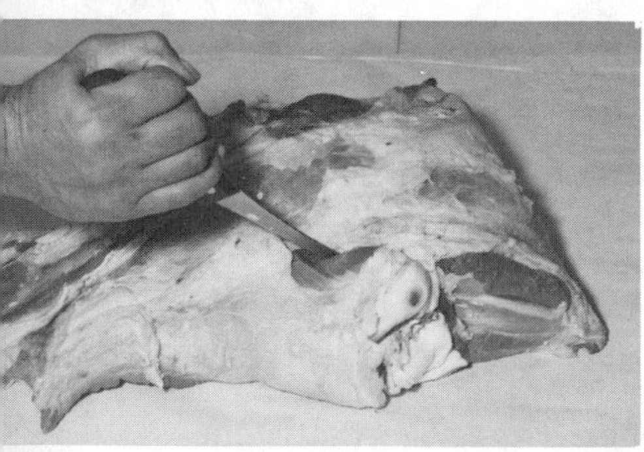

FIGURE 14.8I

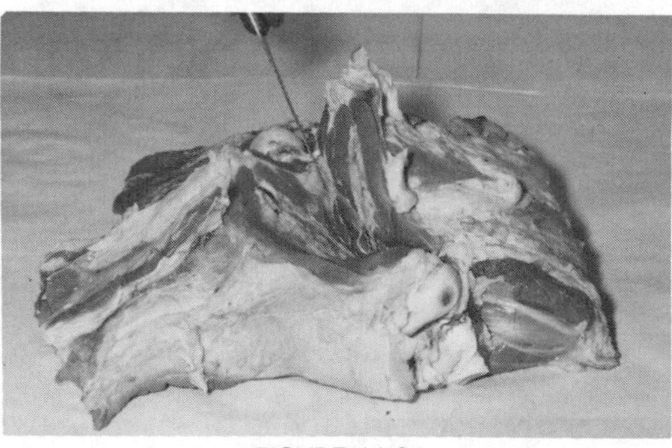

FIGURE 14.8J

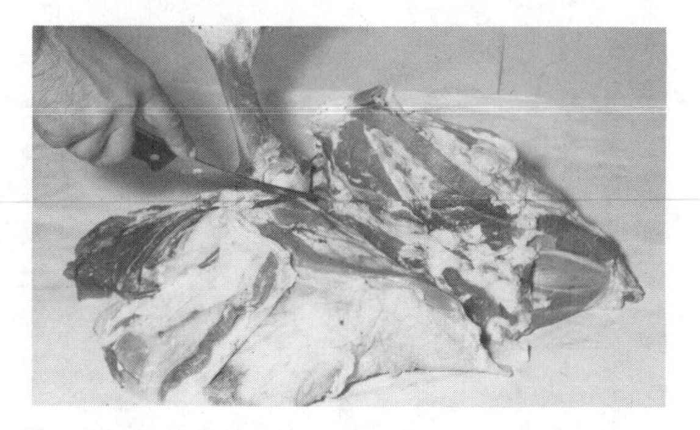

FIGURE 14.8K

(11) Chisel out bone (Fig. 14.8K). Seam out primal cuts, as in a beef round, making top round, knuckle, and bottom round. Separate loin end from bottom round. Remove kneecap from knuckle and trim all cuts. The sequence of quality and cuts are illustrated in Figs. 14.9A–E, the filet butt being most desirable.

FIG. 14.9A. FILET BUTT

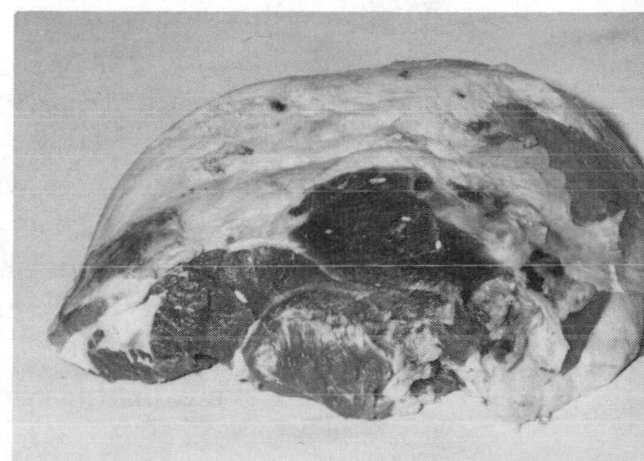

FIG. 14.9B. TOP ROUND

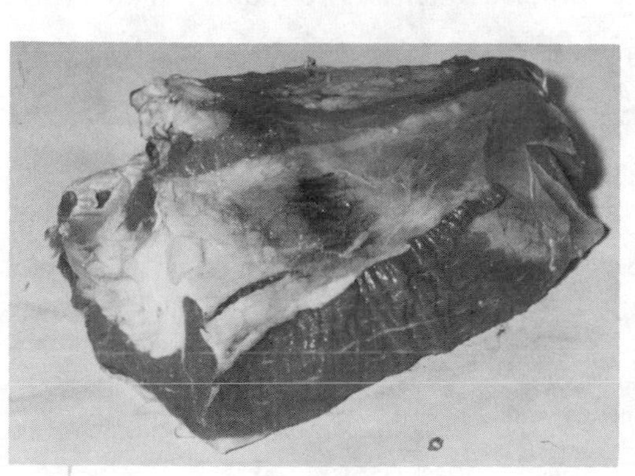

FIG. 14.9C. SIRLOIN BUTT

FIG. 14.9D. KNUCKLE

FIG. 14.9E. BOTTOM ROUND

Yield of lean meat from long leg is 60 per cent, excluding shank meat.

MENU PLAN

Veal dishes may be prepared successfully in any manner, except broiling. Almost any cut can be interchanged for any dish. The quality and tenderness will vary with the cut selected. The quality standard, and the economic factor should determine the cut used.

In some instances, for sautéed and fried items, it may be desirable to tenderize the cut mechanically. Usually the raw serving is put through a cubing machine which cuts the fibers. Another technique is to use a pounding device, or the side of a cleaver. Veal bones are especially good for sauce bases.

The veal shank (Fig. 14.10) cannot be used like the other cuts. However, it is most tasty for braised dishes. Frequently it is cut off

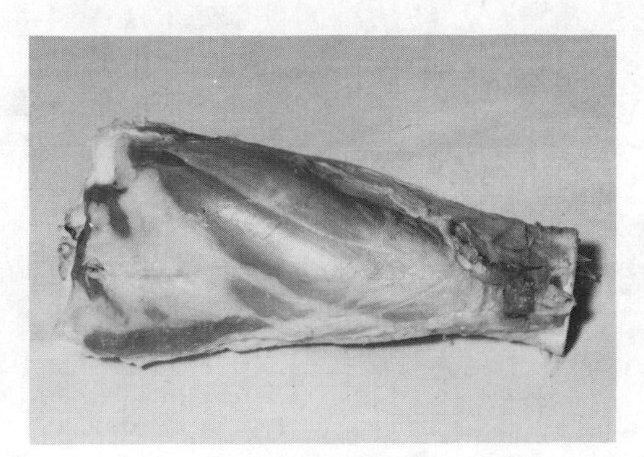

FIG. 14.10. VEAL SHANK

FIG. 14.11. VEAL SHANK—CENTER CUT

inside of both joints to make bone-in entrées (Fig. 14.11). Sometimes it is cut into two portions (Fig. 14.12). The meat may be removed from the bone and braised or ground. Osso buco is a most popular bone-in preparation.

FIG. 14.12. VEAL SHANK—PORTION CUT

Short loin. The regular (drop) short loin (Fig. 14.13) is usually made into a trimmed loin.

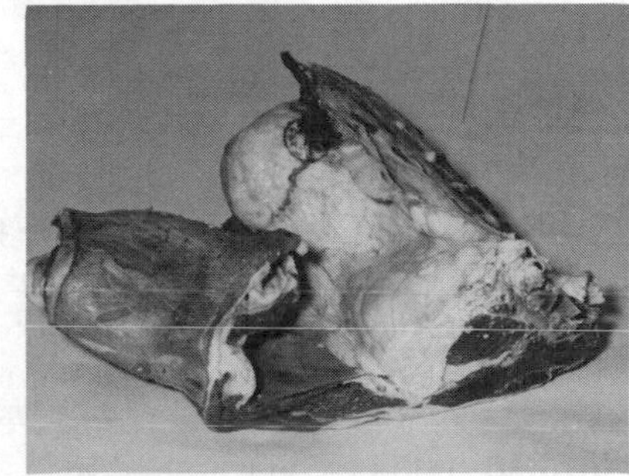

FIG. 14.13. VEAL SHORT LOIN—REGULAR

Making Trimmed Veal Short Loin

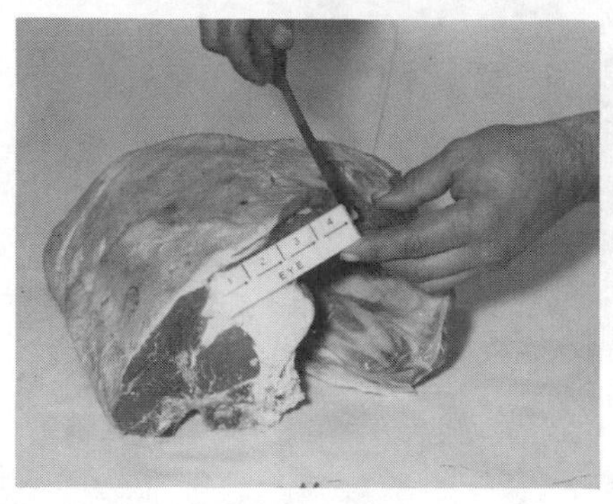

FIGURE 14.14A

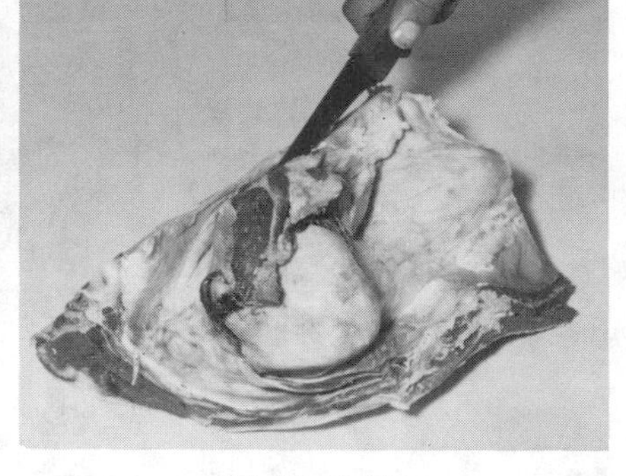

FIGURE 14.14B

(1) Locate a point 4 in. from eye on both ends of short loin. Cut off flank in a straight line through these points. Saw through rib bones (Fig. 14.14A). (2) Cut out kidney knob and trim off fat in excess of ½ in. at butt end, tapering to rib end (Fig. 14.14B). Trimmed loin yields 60 per cent of drop short loin, 14 per cent of veal hindquarter. (3) The short loin may be cut or sawed into chops (Fig. 14.15).

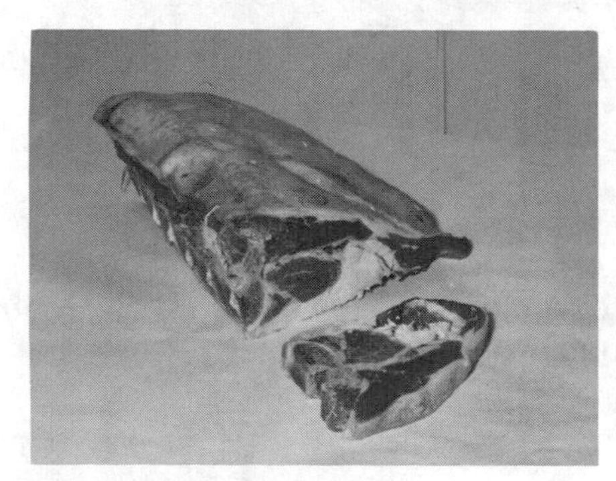

FIGURE 14.15

The flank may be left on where the plan is to muscle bone and roll the short loin to roast, using the flank as the outside layer. Otherwise trim out flank for by-product and take kidney from knob.

Veal short loin steaks are sometimes featured as "baby T-bones." These are usually made from kip calves. The finished product when broiled tends to be tough. This is at best a marginal approach.

Muscle Boning Veal Short Loin

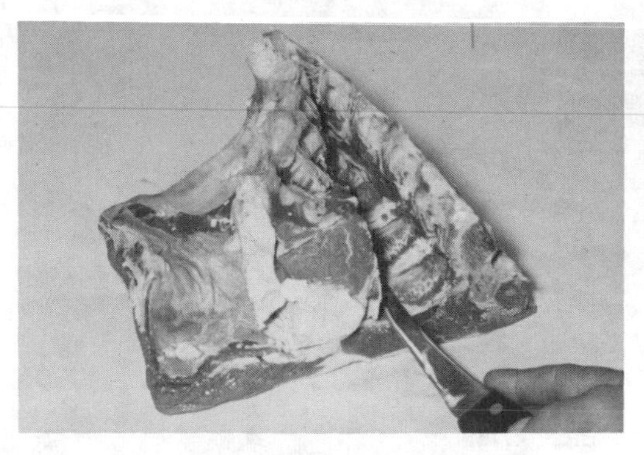

FIGURE 14.16A

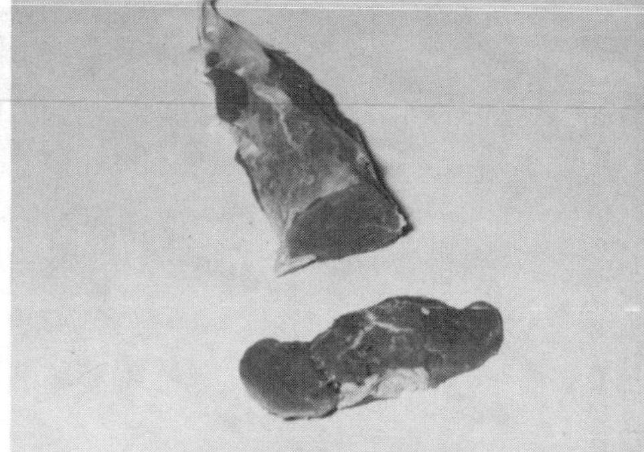

FIGURE 14.16B

(1) Tenderloin should be shelled out from inside of chine bones (Fig. 14.16A). (2) Remove all surface fat and the attached small side strip muscle from the tenderloin. Steaks or cutlets may be butterflied (Fig. 14.16B). (3) Strip loin should be boned like beef, and flank should be

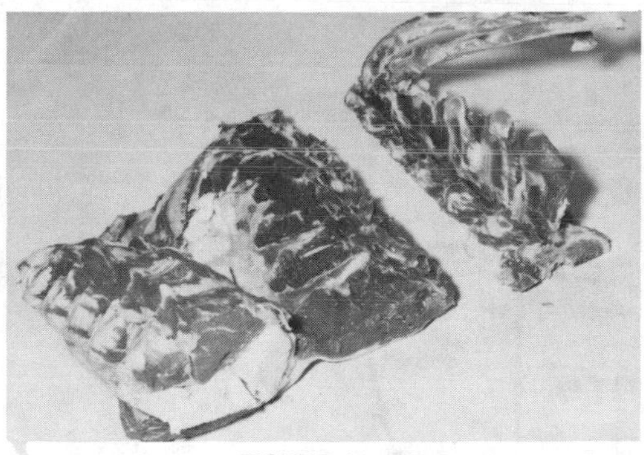

FIGURE 14.16C

FIGURE 14.16D

trimmed off at eye (Fig. 14.16C). (4) Trim off skin, silver seam below it, and back strap (Fig. 14.16D).

The strip loin (Fig. 14.16E) may be sliced thin or butterflied. Slices may be flattened carefully or cooked as is. This most delicate piece of veal makes for the finest sautéed dishes.

FIGURE 14.16E

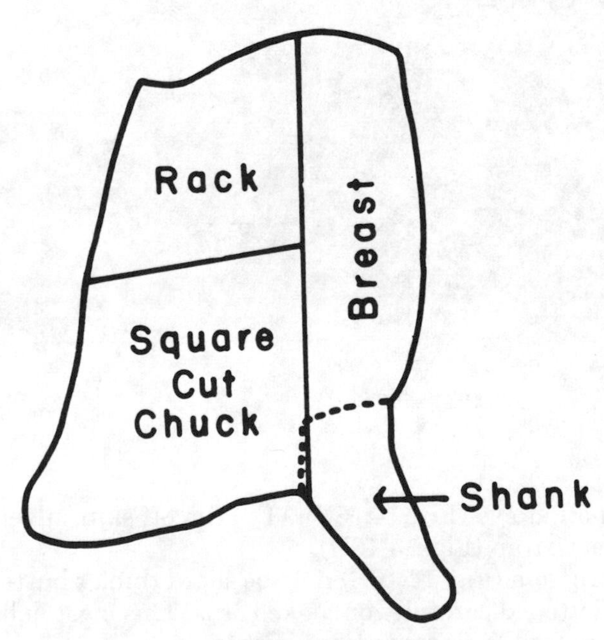

FIG. 14.17. VEAL FOREQUARTER—PRIMAL CUTS

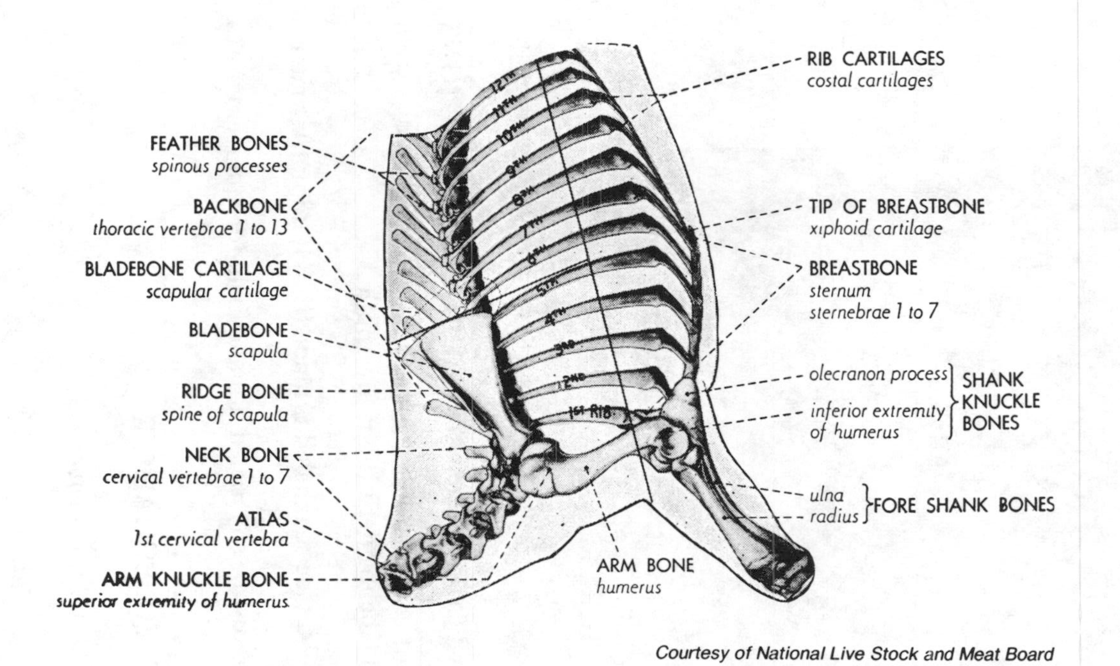

Courtesy of National Live Stock and Meat Board

FIG. 14.18. VEAL FOREQUARTER—SKELETAL STRUCTURE

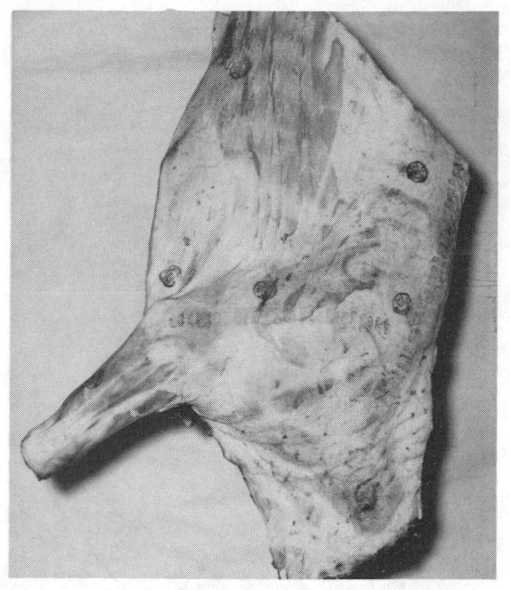

FIG. 14.19. VEAL FOREQUARTER

U.S. DEPARTMENT OF AGRICULTURE DEFINITIONS

For pre-cut veal steaks and cutlets produced under Federal inspection, special labeling definitions are applied.

Cutlets.—May be so labeled if made from hindquarter cuts.

Steaks.—Must be so labeled if made from forequarter cuts.

BREAKING FOREQUARTER

Cuts and Yields (11 rib forequarter):

Cut	*Per cent*
Breast	22
Rack	18
Shank	6
Square cut chuck	54
	100

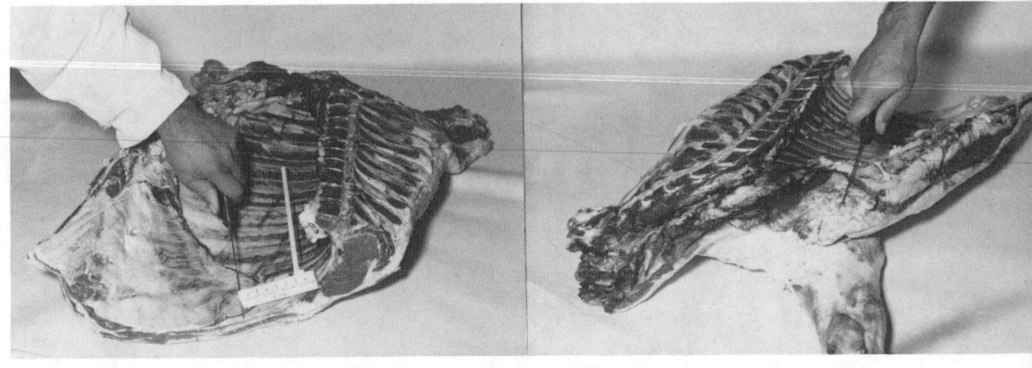

FIGURE 14.20A FIGURE 14.20B

(1) From a point 4 in. from eye on loin end (Fig. 14.20A), (2) to a point at juncture of the first rib and breastbone (Fig. 14.20B), (3) saw off breast (plate and brisket) in a straight line (Fig. 14.20C). (4) Insert tip of knife between 5th and 6th rib and cut through to bone. The number of ribs in rack will vary with ribs left on hindquarter (Fig. 14.20D).

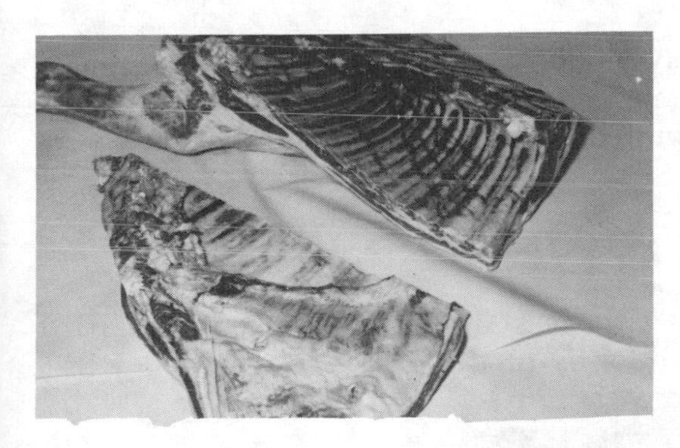

FIGURE 14.20C

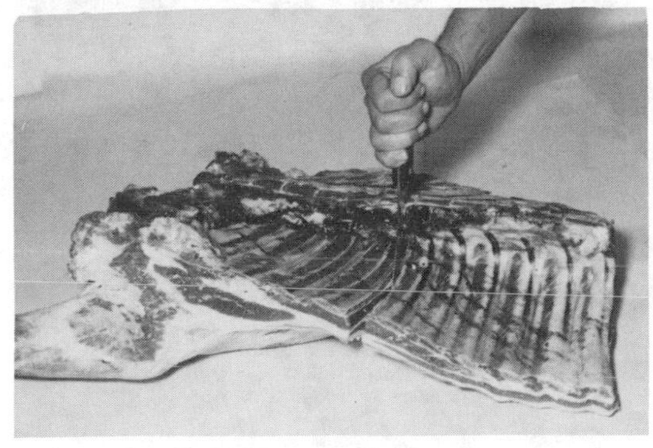

FIGURE 14.20D

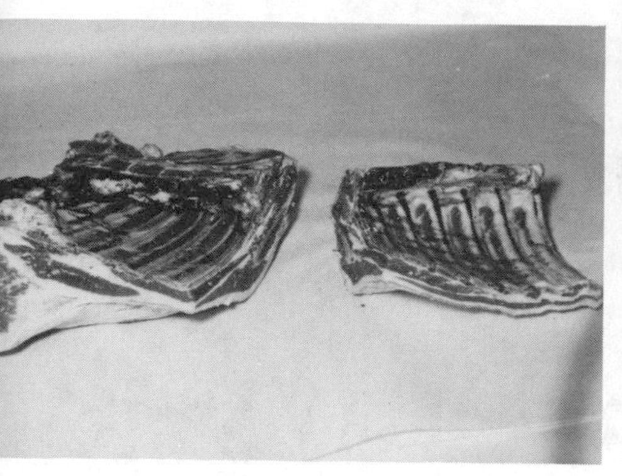

FIGURE 14.20E

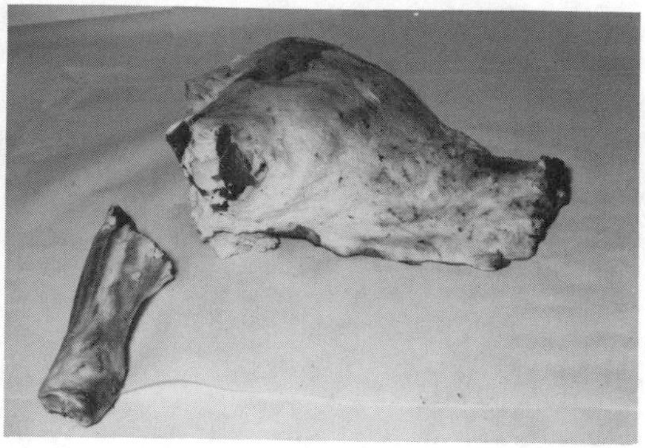

FIGURE 14.20F

(**5**) Saw through to separate rack (on right) (Fig. 14.20E). (**6**) Knife through knuckle at inferior extremity of the humerus (see skeletal chart Fig. 14.18) and remove foreshank meat (left) making square cut chuck (right) (Fig. 14.20F).

MENU PLAN

Most veal cuts are relatively tender due to the youthfulness of the animal. Consequently, the forequarter cuts may be used in most types of preparation. This is only limited by the quality standard of the operator.

Veal Breast

The veal breast (Fig. 14.21) is the least desirable forequarter cut. It may be muscle boned and used for by-product.

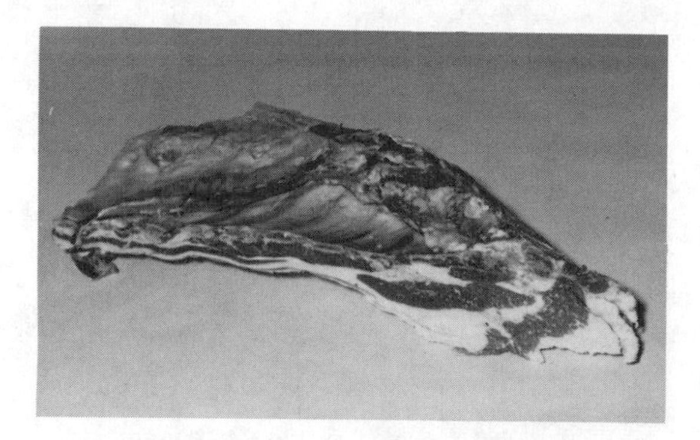

FIG. 14.21. VEAL BREAST

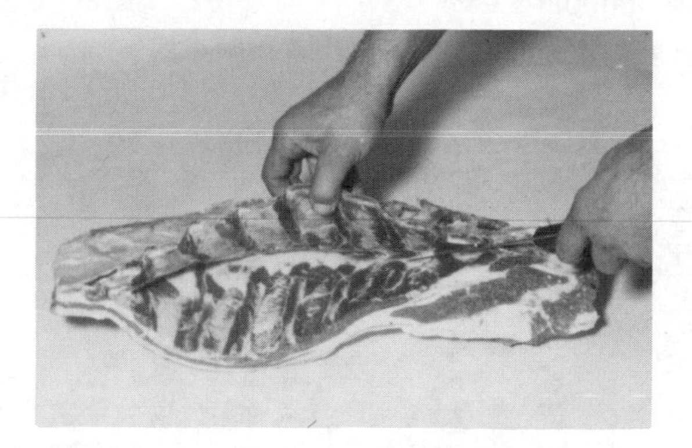

FIG. 14.22. VEAL BREAST WITH POCKET

Sometimes a pocket is made by shelling along the rib bones (Fig. 14.22). This may be stuffed and roasted or braised. Sometimes the stuffed breast is frozen and band sawed into "veal steaks" for a low priced entrée.

Single Rack

The veal rack (Fig. 14.23) is the most desirable forequarter cut, comparable to the prime rib of beef. It is sometimes roasted, and sometimes muscle boned to make a veal rib-eye which is used for sautéed dishes.

FIG. 14.23. SINGLE VEAL RACK

(1) To ready the rack, chine bone should be removed (right) and short ribs cut off (left) at some suitable point (Fig. 14.24A). (2) Very desirable

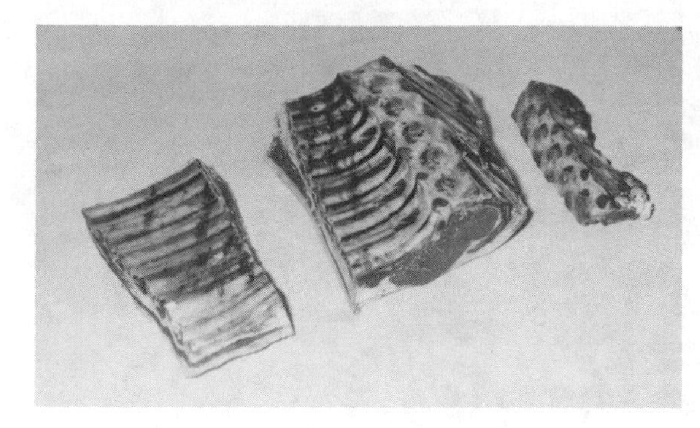

FIGURE 14.24A

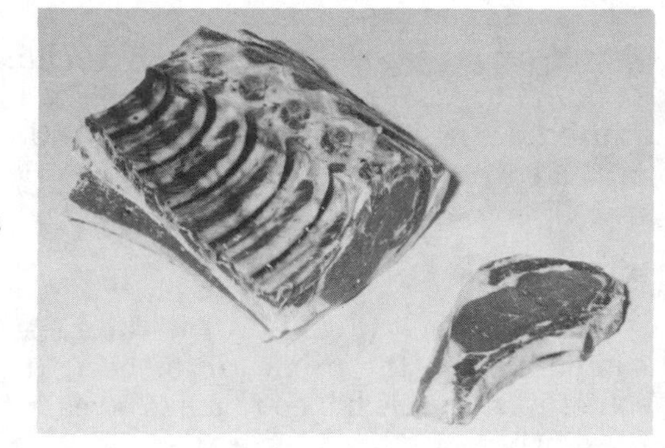

FIGURE 14.24B

chops may be cut with a saw. Double chops are sometimes butterflied and stuffed. These chops may be Frenched (Fig. 14.24B).

Short ribs (Fig. 14.25) may be cut up with bone-in for stew or boned out for by-product.

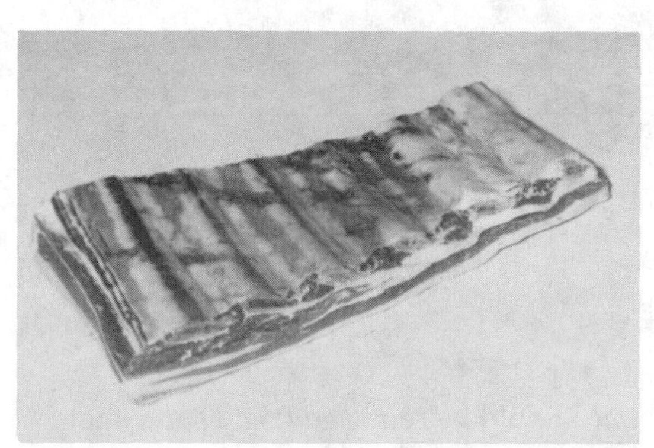

FIG. 14.25. VEAL SHORT RIBS

Foreshank

Veal foreshank (Fig. 14.26) may be treated the same as hindshank.

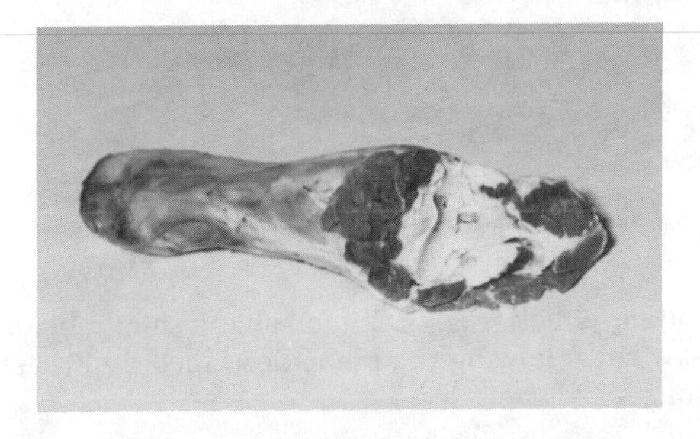

FIG. 14.26. VEAL FORESHANK

Square Cut Chuck

The chuck (Fig. 14.27) is usually muscle boned. Cutting procedure is the same as beef. The clod and boneless chuck may be used for steaks, which are usually cubed or pounded, roasts, braised dishes and by-product.

Retail butchers frequently merchandise the chuck cut as bone-in steaks. This approach might have merit for some institutions with low priced entrées.

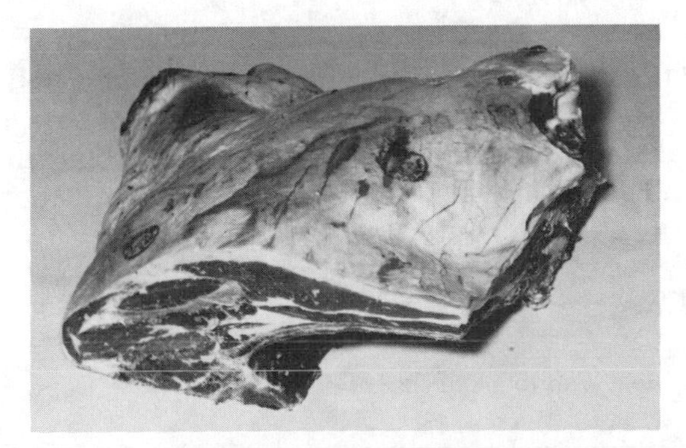

FIG. 14.27. SQUARE CUT VEAL CHUCK

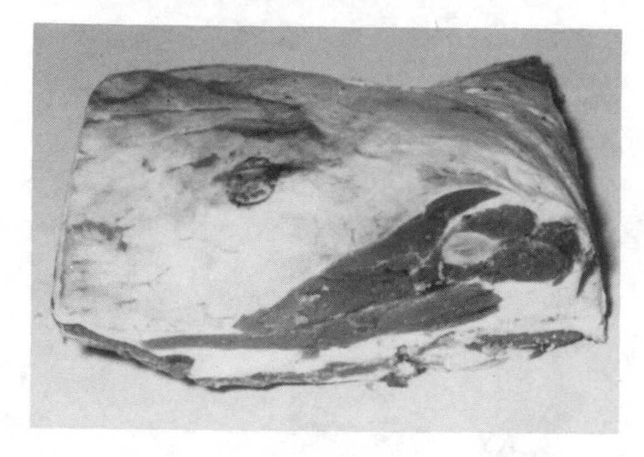

FIG. 14.28. SQUARE CUT VEAL CHUCK—STEAKING STYLE

When steaking chuck (Fig. 14.28) foreshank should be taken off with a straight saw cut following the plane described by breast, instead of following bone structure.

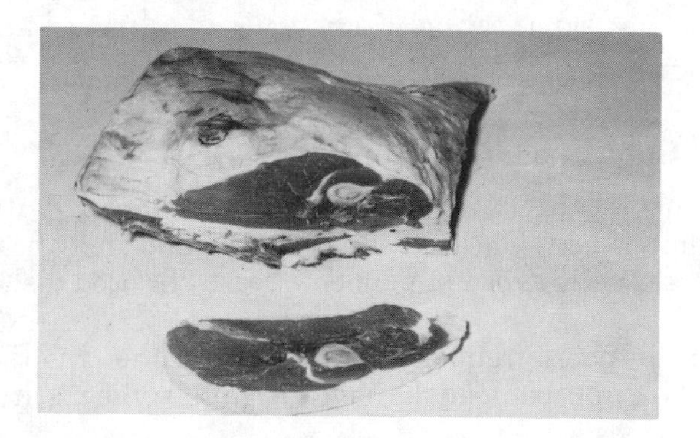

FIG. 14.29. SQUARE CUT VEAL CHUCK—WITH ONE VEAL STEAK REMOVED

It is a good plan to freeze chuck before sawing. Cut arm bone steaks first (Fig. 14.29). Cut to point where blade steaks are a desirable size. Cut blade steaks, cutting parallel with the cut made removing the rack. Chine bone and excess fat should be trimmed off of all chops.

REFERENCES

ANON. 1945. Meat Handbook of the U.S. Navy. U.S. Navy Dept., Washington, D.C.

ANON. 1960. Institutional meat purchase specifications for fresh veal and calf Series *300*. U.S. Dept. Agr., Washington, D.C.

ANON. 1961. Meat Buyer's Guide to Standardized Meat Cuts. National Association of Hotel and Restaurant Meat Purveyors. Chicago, Ill.

BULL, S. 1951. Meat for the Table. McGraw-Hill Book Co., New York.

ROMANS, J. R. and ZIEGLER, P. T. 1977. The Meat We Eat, 11th Edition. Interstate Printers and Publishers, Danville, Illinois.

WANDERSTOCK, J. J., and WELLINGTON, G. H. 1961. Let's cut meat. Cornell Extension Bull. *1053.* Ithaca, N.Y.

15

Lamb

PRIMAL CUTS

The terms hindsaddle and foresaddle are used in describing the two basic divisions. Lamb carcasses are not split lengthwise into sides like

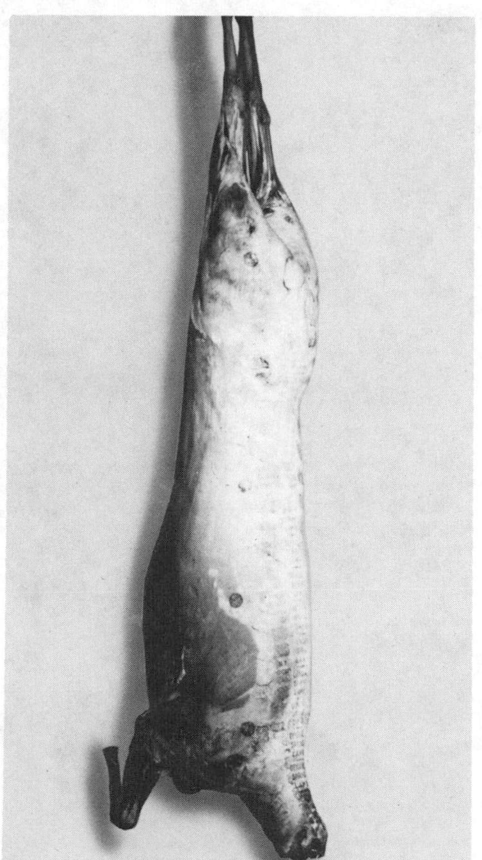

FIG. 15.1. LAMB CARCASS

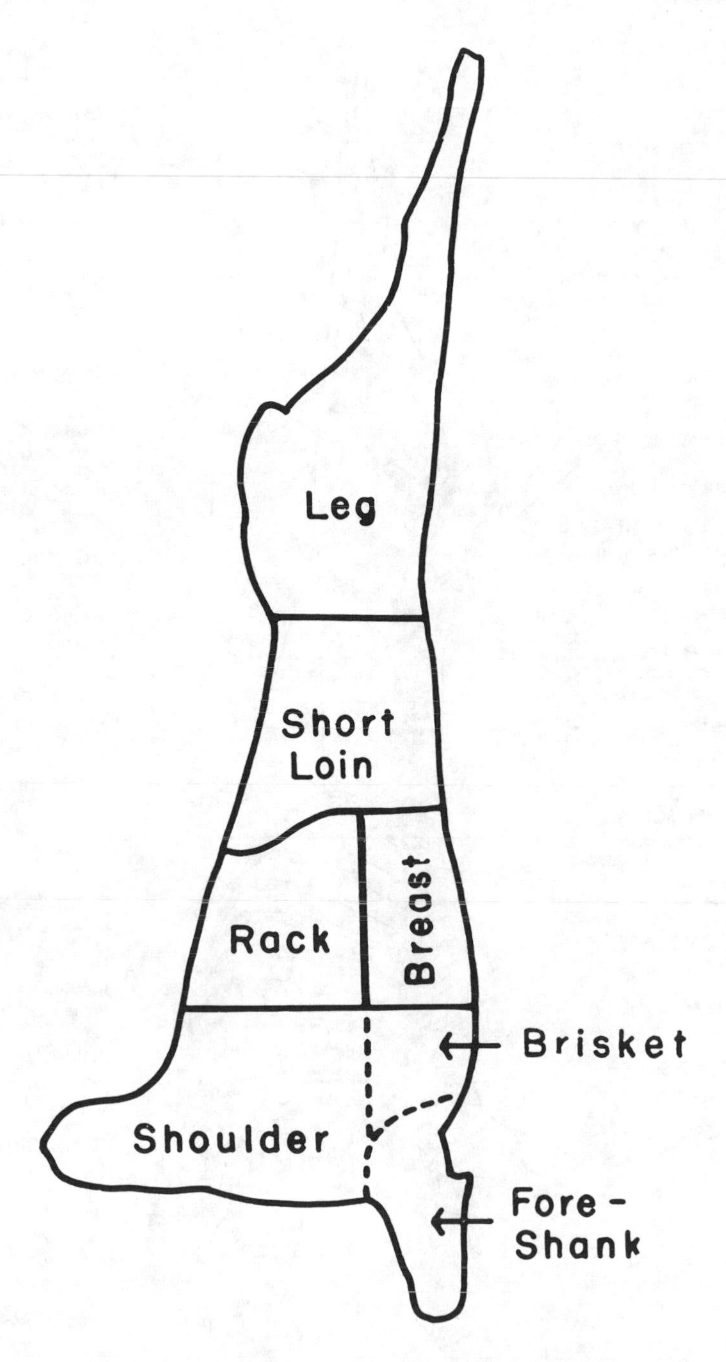

FIG. 15.2. LAMB—PRIMAL CUTS

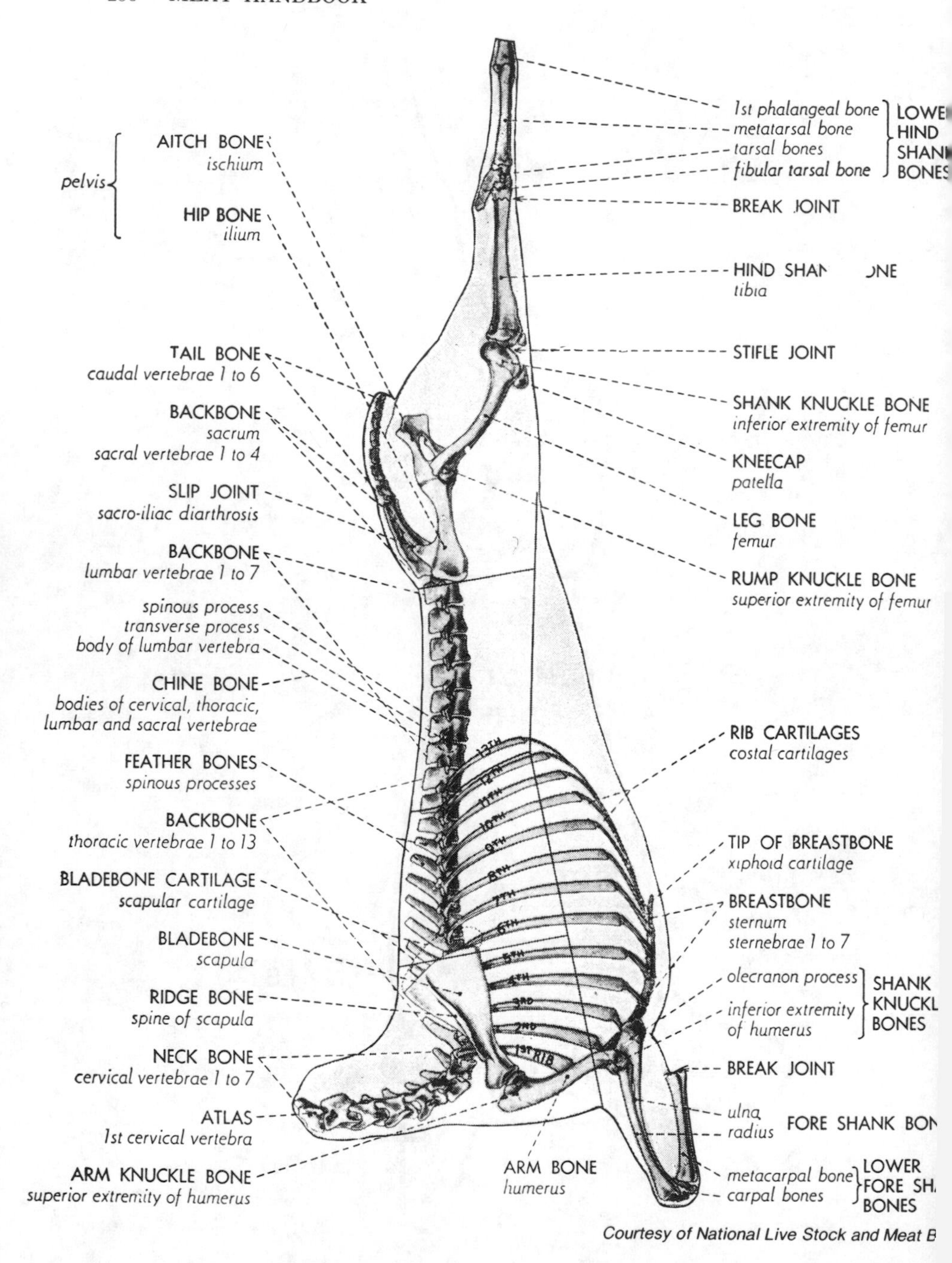

FIG. 15.3. LAMB SIDE—SKELETAL STRUCTURE

beef. When saddles are made, two forequarters or hindquarters are left joined at the backbone.

The whole carcass is seen in Fig. 15.1, and the primal cuts and skeletal structure in Fig. 15.2 and 15.3.

	Per cent
Foresaddle	50
Hindsaddle	50
	100

Cuts and Yields:

Cut	
Rack	12
Plate	8
Chuck	30
Drop loin	17
Leg	33
	100

Breaking the Carcass

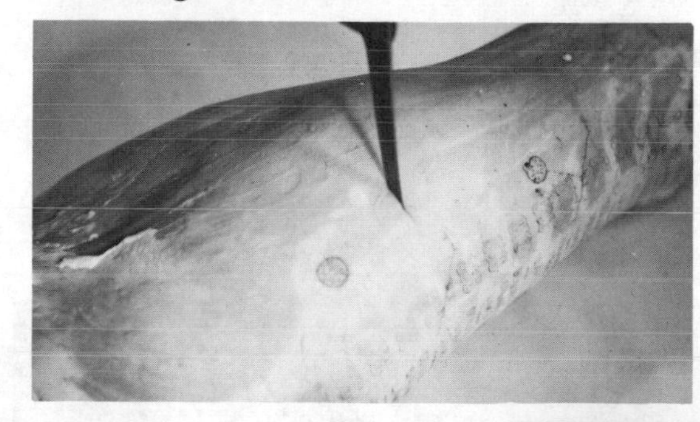

FIGURE 15.4A

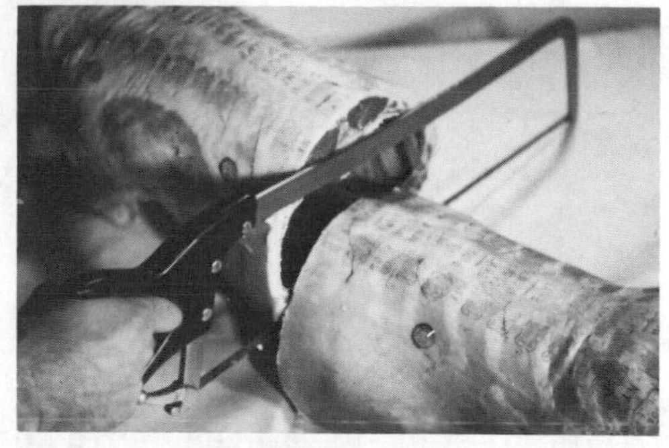

FIGURE 15.4B

(1) Insert knife tip between 12th and 13th rib (Fig. 15.4A). (2) Saw through carcass, cutting perpendicular to back (Fig. 15.4B), separating foresaddle (Fig. 15.5) and hindsaddle (Fig. 15.6).

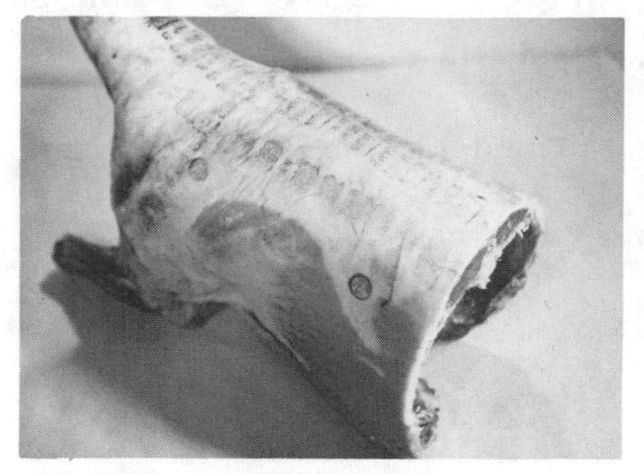

FIG. 15.5. LAMB FORESADDLE

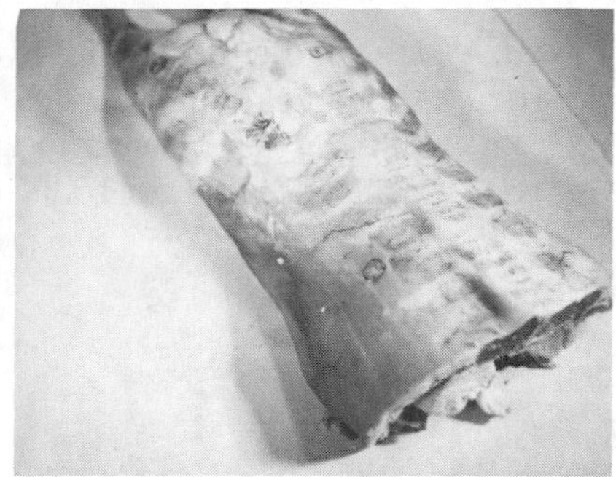

FIG. 15.6. LAMB HINDSADDLE

Breaking Foresaddle

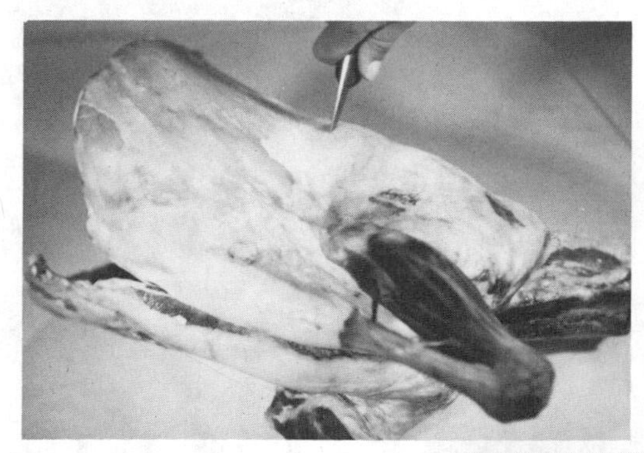

FIGURE 15.7A

FIGURE 15.7B

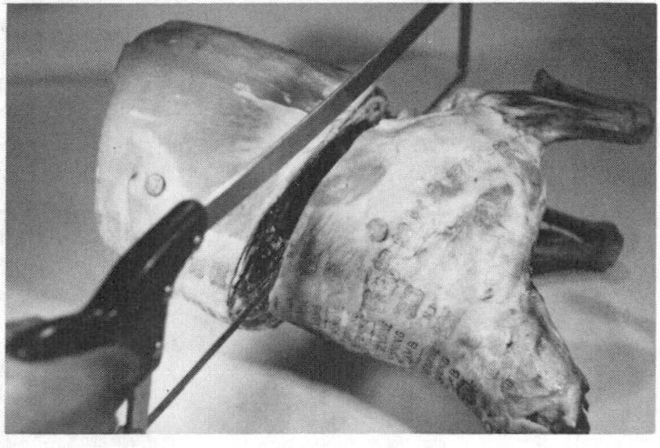

(1)Insert knife tip between fourth and fifth ribs (Fig. 15.7A). (2) Saw perpendicular to spine, separating double chuck (right) and bracelet (left) (Fig. 15.7B). (3) Measure 4 in. from eye on loin end and chuck end of bracelet (Fig. 15.7C). (4) Saw in a straight line between two points,

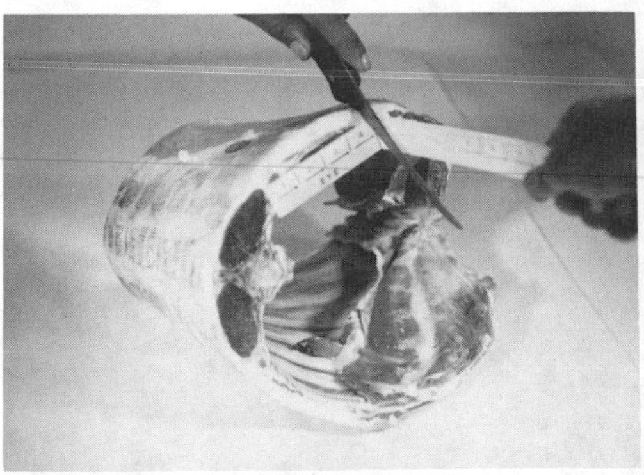

FIGURE 15.7C

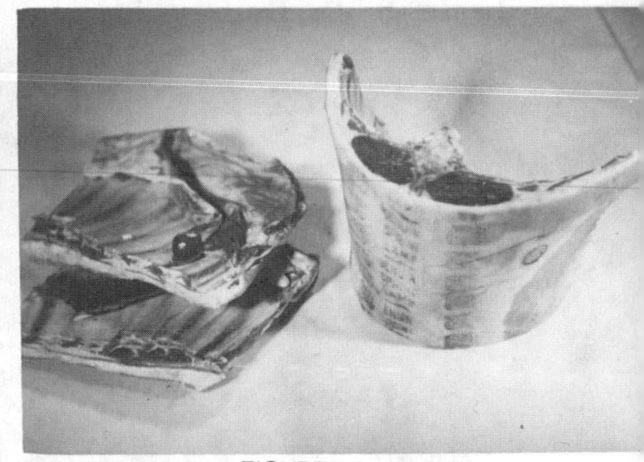

FIGURE 15.7D

making a double hotel rack (right) and two plates (left) (Fig. 15.7D). (**5**) Locate a point at juncture of the first rib and anterior extremity of

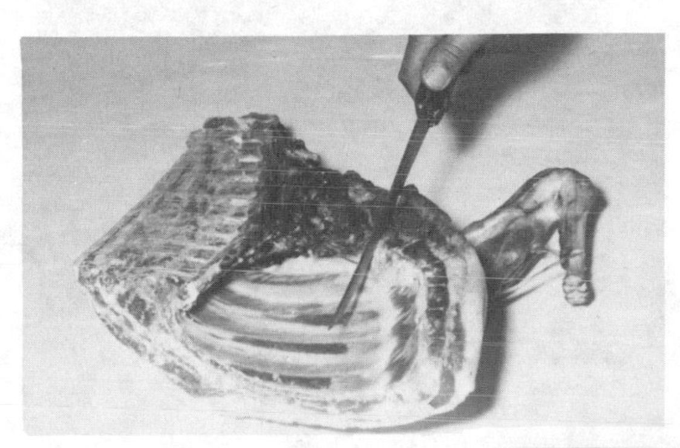

FIGURE 15.7E

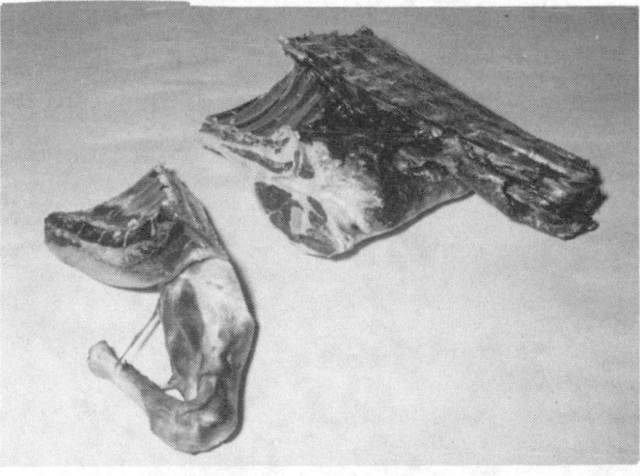

FIGURE 15.7F

the breastbone (Fig. 15.7E). (**6**) Through this point, and in a line perpendicular to cut on chuck, saw and separate shanks and briskets from shoulder (Fig. 15.7F).

Breaking Hindsaddle

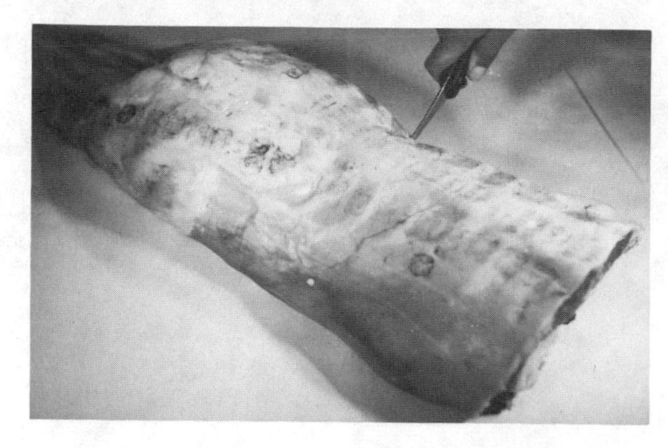

FIGURE 15.8A

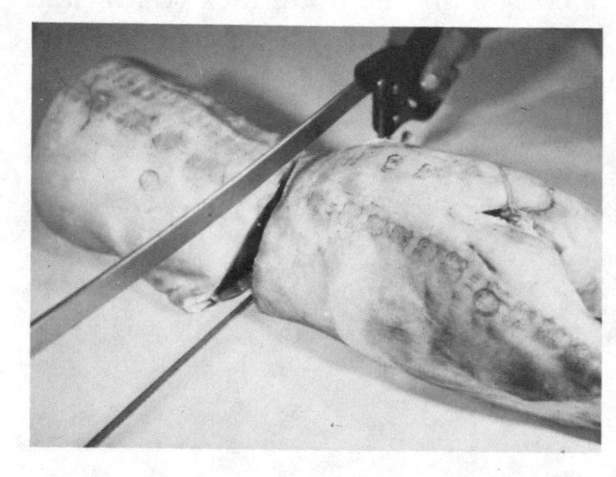

FIGURE 15.8B

(**1**) With a knife tip, locate anterior tip of hip bone (Fig. 15.8A). (**2**) Saw through perpendicular to back (Fig. 15.8B). (**3**) Separate legs (pair) (Fig. 15.8C—right) and double drop (regular) lamb loin (Fig. 15.8C—left).

FIGURE 15.8C

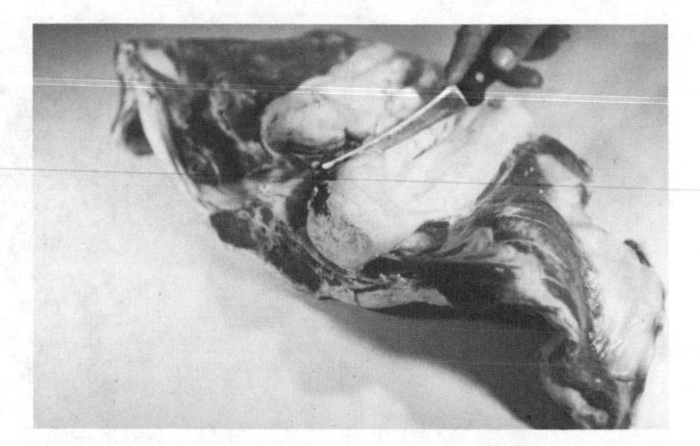

FIGURE 15.8D

(4) Cut out kidney knobs and trim lumbar fat to a maximum of ½ in. tapered in thickness to the lean not over three-fourths of the length of loin (Fig. 15.8D). (5) Measure 4 in. from eye on both ends. Cut off flank in a straight line (Fig. 15.8E).

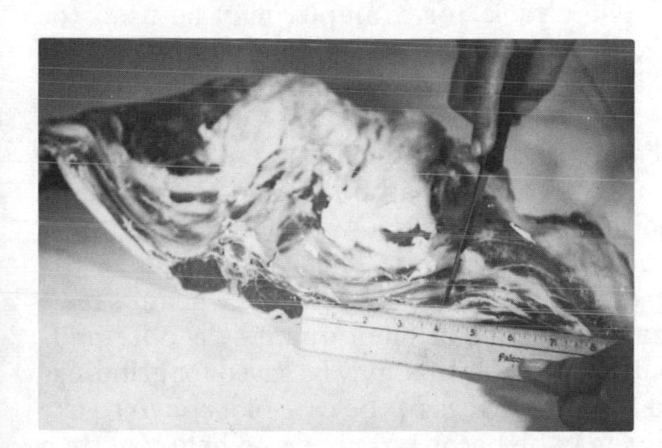

FIGURE 15.8E

Trimmed short loins are 12 per cent of the lamb, or about 65 per cent of the drop lamb loin.

Optional Break

Sometimes a lamb back, or "saddle of lamb" (Fig. 15.9), is broken by cutting at posterior end of the short loin and anterior end of the rack, leaving loin and rack in one piece. This is the untrimmed or regular back. A trimmed back has the kidneys, flanks, and plates cut off.

The breast is sometimes taken as a full cut starting at cod or udder, through the flank, plate, and including the brisket.

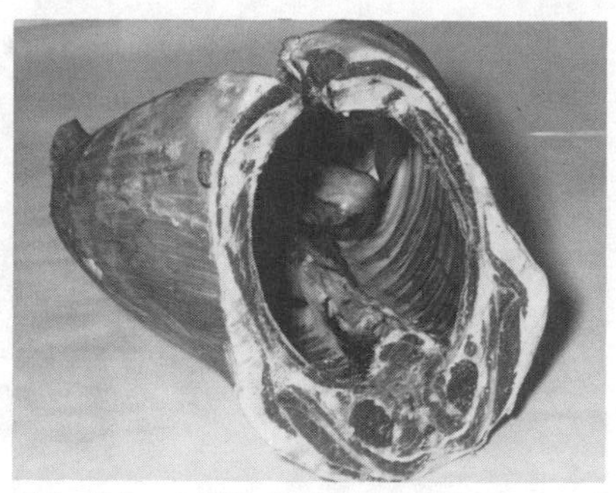

FIG. 15.9. LAMB BACK—REGULAR

MENU PLAN

Relative to the degree of quality to be achieved, almost any portion of a USDA Choice or USDA Prime carcass, except the shank meat, may be used for any purpose. The chuck, for example, may be used for broiling, roasting, braising, and stewing. The flavor and tenderness of chuck cuts are adequate for economical preparation. Other cuts will be better.

Table 15.1 may be used as a guide for lamb cuts.

The shank, flank, neck, and breast may all be treated as secondary or by-product. Though their relative value is low, they may be utilized in the same manner in almost any operation. The shanks (Fig. 15.10) make a tasty bone-in braised dish or stew. As a by-product, shanks may be accumulated in the freezer until a sufficient quantity is obtained. The neck and flank have portions of lean that may be saved for grinding. Ground lamb may be used for stuffing, as in the case of a crown roast, or prepared as patties which may be bacon wrapped and held together with a skewer or toothpick.

The fell is the tough, parchment-like, cover on the surface of the meat. To increase the attractiveness of the rack and loin cuts, the fell should

TABLE 15.1. MENU PLAN FOR LAMB CUTS

Cut	Chops	Oven Roasting
Rack	Excellent	Gourmet item
Shoulder	Fair	Fair
Short loin	Excellent	Gourmet item
Legs	Good	Excellent

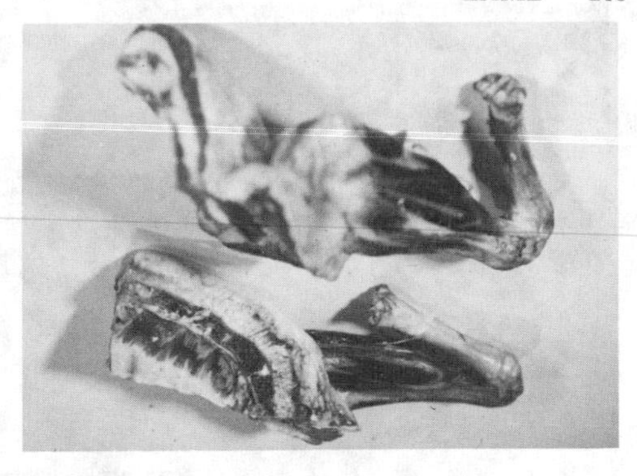

FIG. 15.10. LAMB FORESHANKS

be removed. It need not be removed from the other cuts, as it has been proved experimentally that it helps preserve the shape during cooking, shortens roasting time, and retains juices.

Lamb Breast

The excess fat should be trimmed, and the breast may be barbecued, stewed, or braised.

For best acceptance, breast bone portion may be cut off (Fig. 15.11).

A pocket may be provided by shelling along bones with knife tip. Breast may be stuffed for baking, or stuffed with ground lamb, frozen, and sawed into "chopettes" (Fig. 15.12).

Hotel Rack (Double)

The hotel rack (Fig. 15.13) is generally used for rib lamb chops or for roast rack of lamb. In either case, the basic preparation is the same. Among the better restaurants, the rack is generally selected for chops.

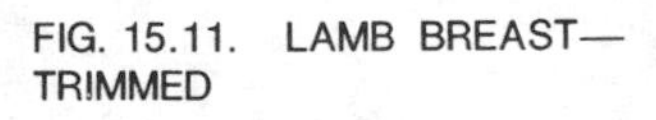

FIG. 15.11. LAMB BREAST—TRIMMED

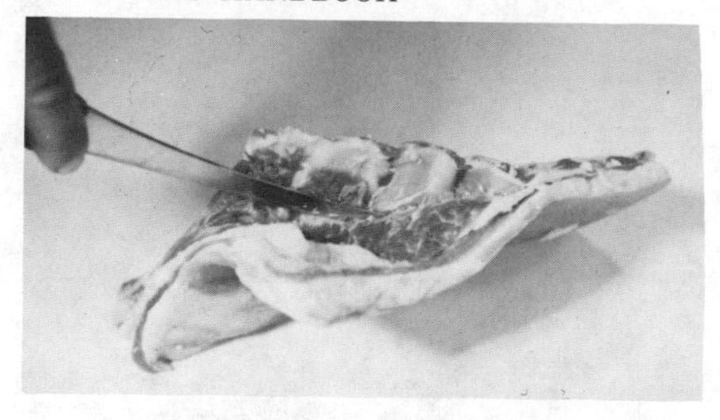

FIG. 15.12. LAMB BREAST—WITH POCKET

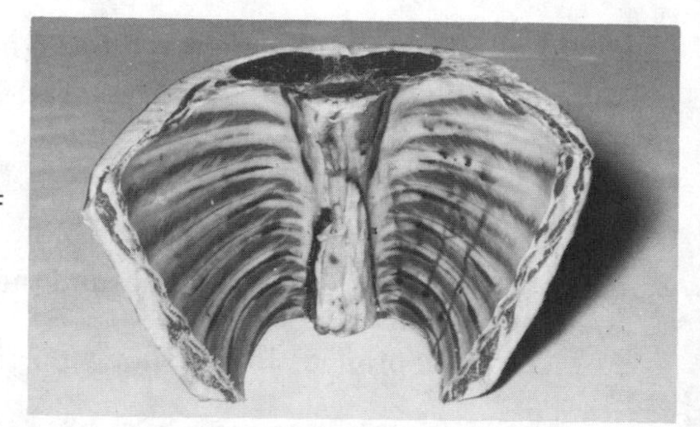

FIG. 15.13. HOTEL RACK OF LAMB (DOUBLE)

For roasting, a crown roast is the most elegant presentation. The crown may be stuffed with ground lamb, beef, or a variety of other suitable stuffings.

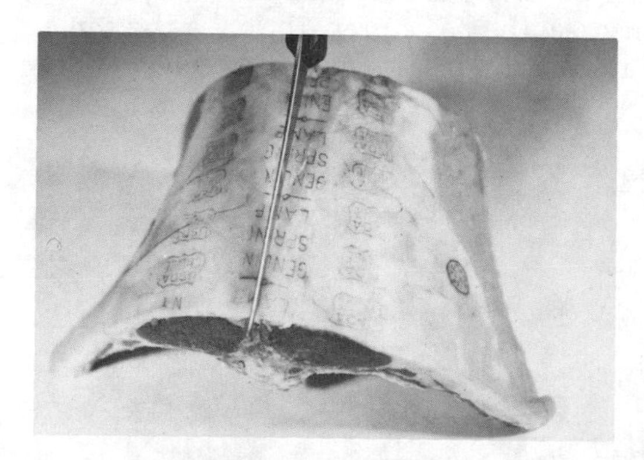

FIGURE 15.14A

(1) Knife along both sides of chine bone (Fig. 15.14A). (2) Mark rack $3\frac{1}{2}$ in. from eye on rib end (Fig. 15.14B). (3) Saw through both sides of

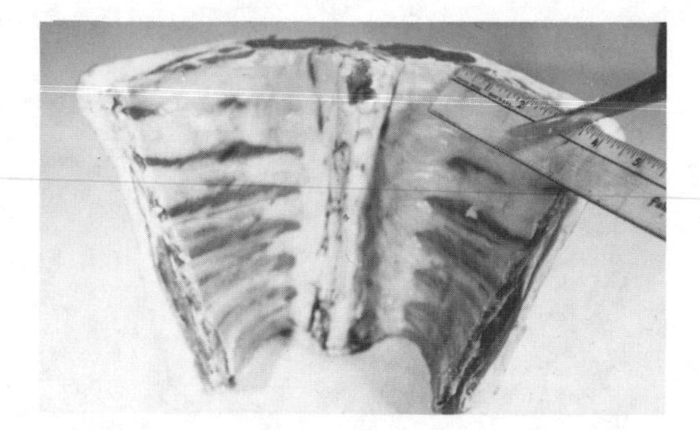

FIGURE 15.14B

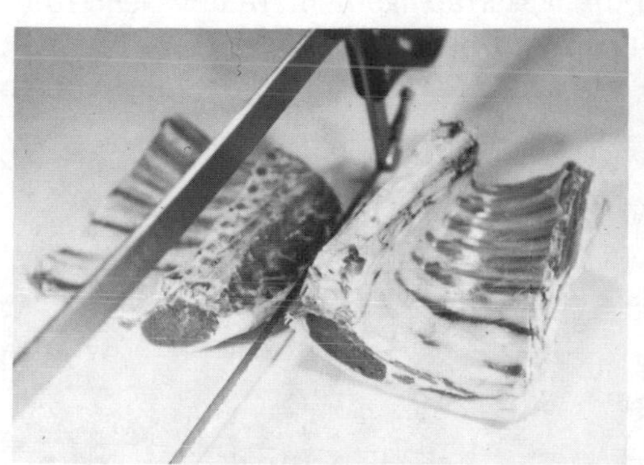

FIGURE 15.14C

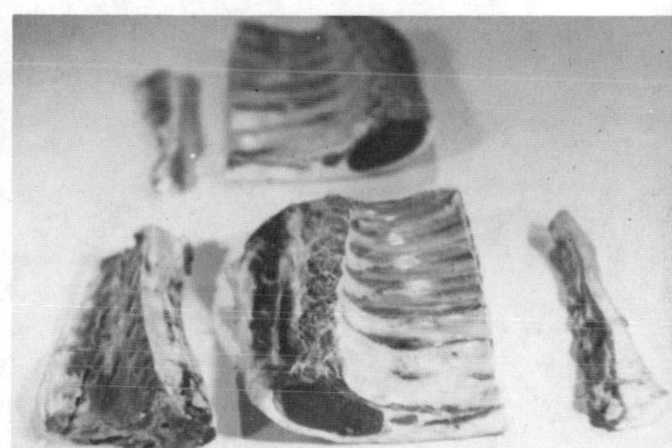

FIGURE 15.14D

chine on inside of rack to knife cuts, freeing chine bone in one piece (Fig. 15.14C). (**4**) Saw off rib bones from 3½ in. point parallel with back of rack (Fig. 15.14D). (**5**) Make a knife cut 2 in. from rib bone end (Fig. 15.14E). (**6**) Remove blade cartilage and meat above it. Seam carefully

FIGURE 15.14E

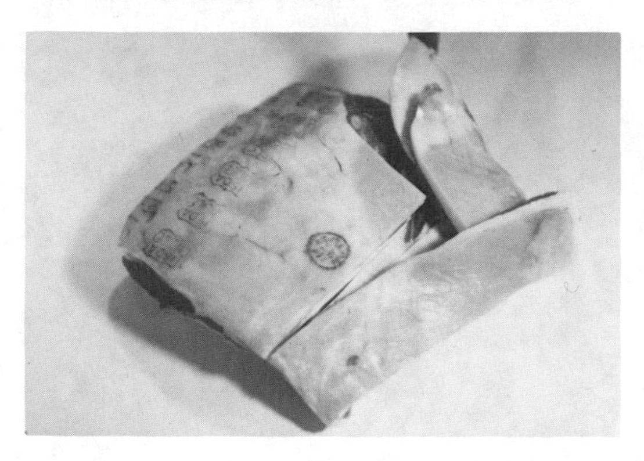

FIGURE 15.14F

(Fig. 15.14F). (7) Remove fell, pull backstrap ligament, remove lean from between exposed rib bones. Trim and bevel excess fat (Fig. 15.14G). (8) The Frenched ray may be left whole for roasting. To make a crown

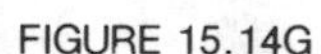

FIGURE 15.14G

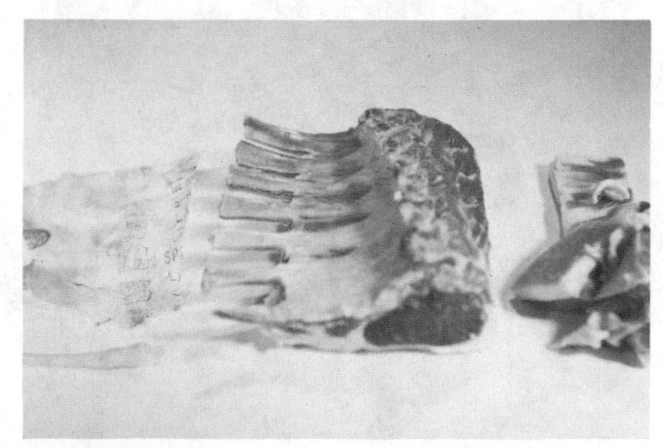

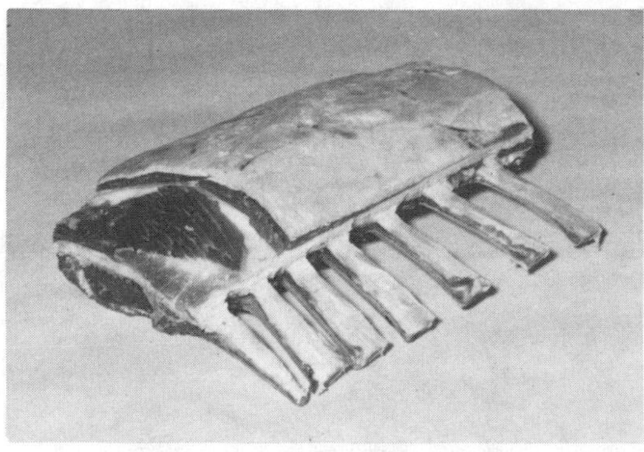

FIGURE 15.14H

roast, sew two or more single racks together. Chop holders on each bone, placed there after roasting, add to the general presentation. A head of cauliflower may be placed in the center instead of stuffing. Yield of chop ready rack is 50 per cent (Fig. 15.14H). (9) Fig. 15.14I illustrates double

FIGURE 15.14I

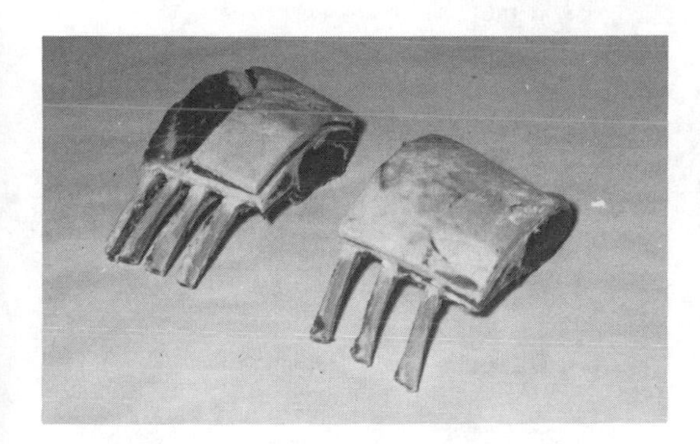

FIGURE 15.14J

chops with both rib bones. Frequently, only one of the rib bones is left on. (**10**) Uniform double chops may be cut by locating mid point of rack and cutting in half (Fig. 15.14J). (**11**) Locate mid point of "half," and

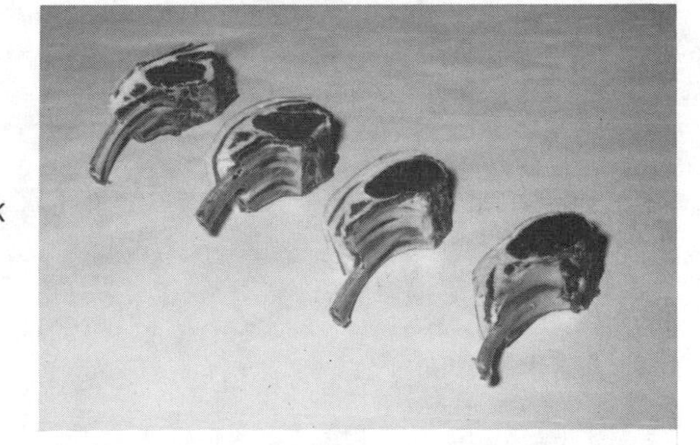

FIGURE 15.14K

divide again, making four equally thick "double" chops. This avoids cutting at angle of rib bones and may take care of odd portion of ninth rib. Double rib chops have only one bone left on (Fig. 15.14K).

Shoulder (Double)

Two approaches are presented. Shoulder (Fig. 15.15) may be cut on the saw for chops, or bone-in braising meat. It may be muscle boned and

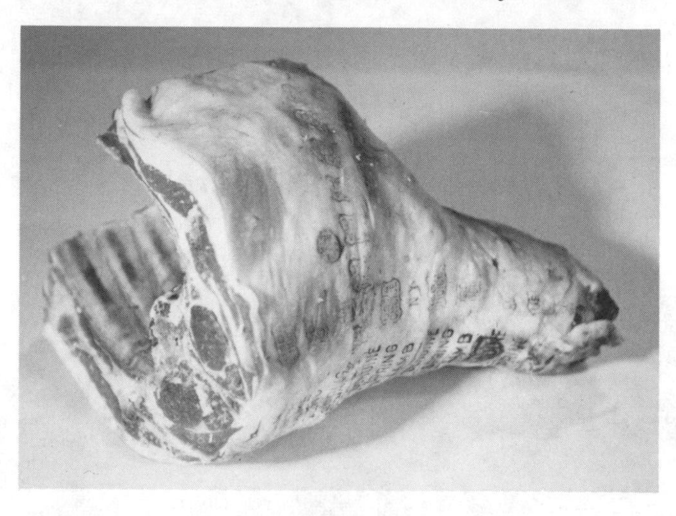

FIG. 15.15. LAMB SHOULDER (DOUBLE)

used for braising, for roasting, or divided and used partially for roasting, partially for boneless chops.

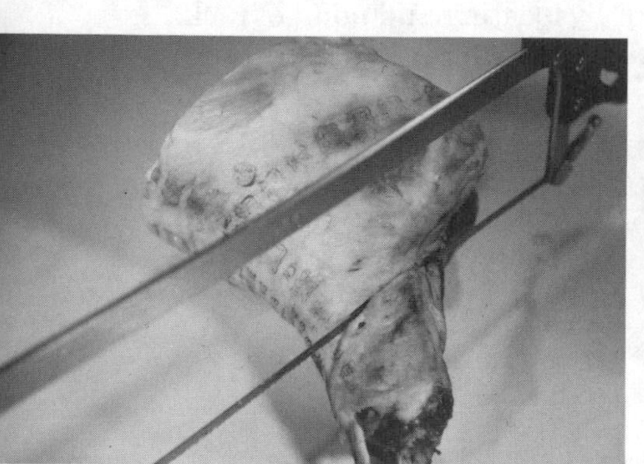

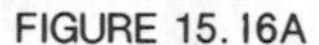

FIGURE 15.16A

FIGURE 15.16B

(1) Cut off neck, within 1 in. of shoulder (Fig. 15.16A). (2) Split shoulder with saw (Fig. 15.16B). (3) Cut arm bone chops first. Stop cutting where the remaining portion of shoulder makes desirable blade chops (Fig.

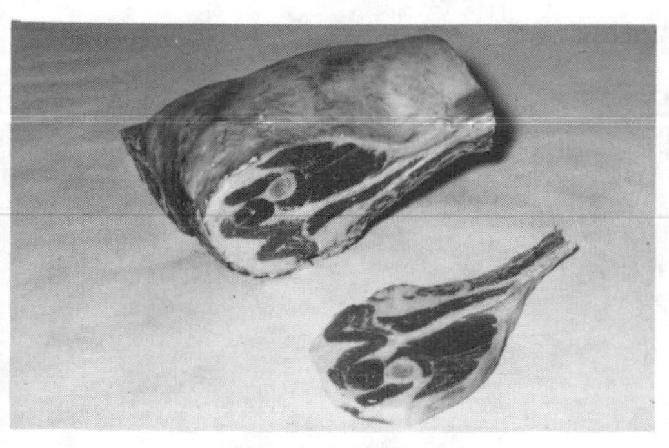

FIGURE 15.16C

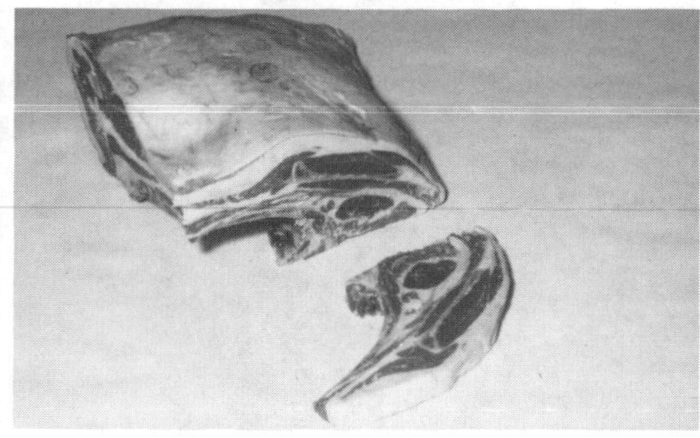

FIGURE 15.16D

15.16C). (4) Cut blade chops (Fig. 15.16D). Remove feather and chine bones from all chops for best presentation (Fig. 15.17).

FIG. 15.17. LAMB SHOULDER—BLADE AND ARM CHOPS

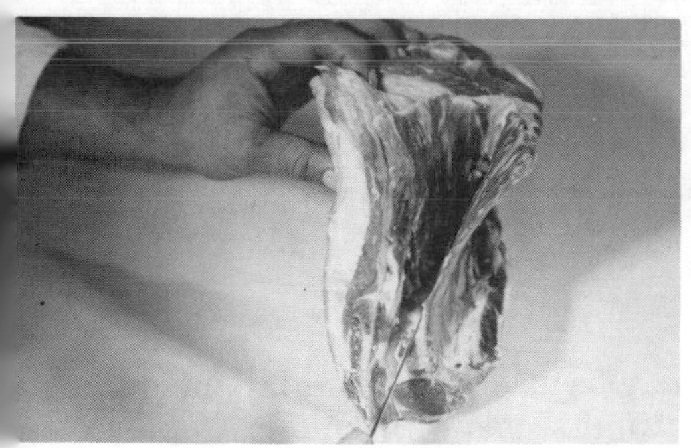

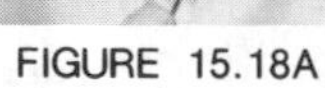

FIGURE 15.18A

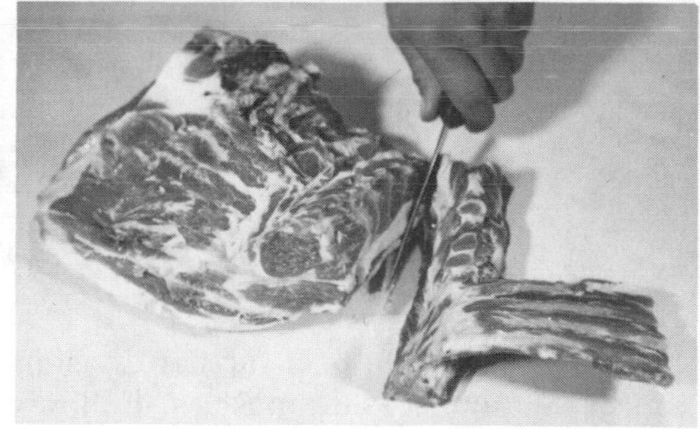

FIGURE 15.18B

Muscle Boning—Lamb Shoulder

(1) Insert knife at rib bones and shart to shell along bones (Fig. 15.18A). (2) Continue shelling behind neck bones and chine bone (Fig. 15.18B).

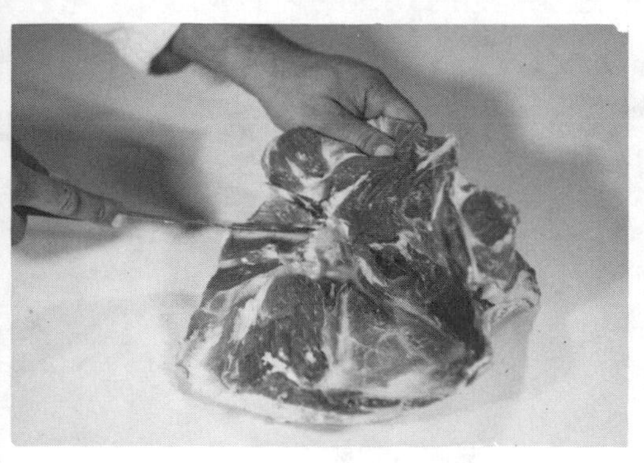

FIGURE 15.18C

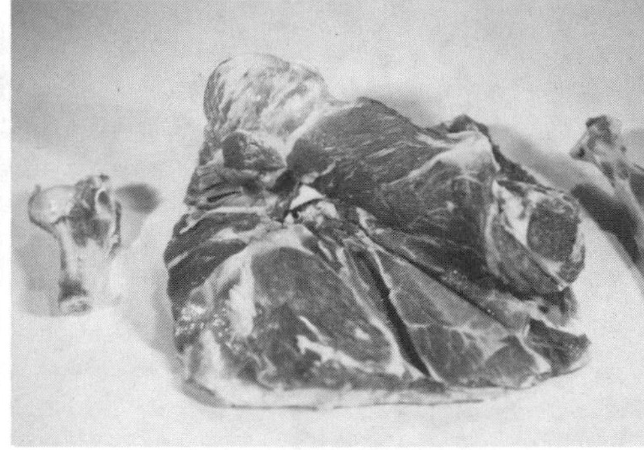

FIGURE 15.18D

(**3**) Insert tip of knife above blade bone. Follow natural seam to round bone (Fig. 15.18C). (**4**) Score on both sides of blade bone and round bone. Pull blade bone. Cut out round bone (Fig. 15.18D). (**5**) Remove back strap and trim off excess fat. The boneless shoulder may be tied to roast, cut into cubes for stewing or braising, or Shish Kebab (Fig.

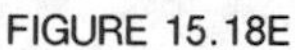

FIGURE 15.18E

FIGURE 15.18F

15.18E). (**6**) A further division may be made by separating the inside and outside muscles. Follow natural seams with knife to separate. Fashion the inside piece into a roll and fasten with skewers at intervals equal to the intended thickness of chops. Slice the rib eye muscle into chops (Fig. 15.18F). "Saratoga chops," a suggestive merchandising term, is generally used for any cut of lamb that is boned, rolled, tied, and sliced into broiling portions. The outside muscle can be used like the whole shoulder. Boneless yield is about 65 per cent.

Trimmed Short Loin (Double)

The short loin (Fig. 15.19), because of its relatively high portion cost, is not as popular as the rack, though high in quality for either chops or roasting. It is prepared with bone-in as chops or boneless as chops or

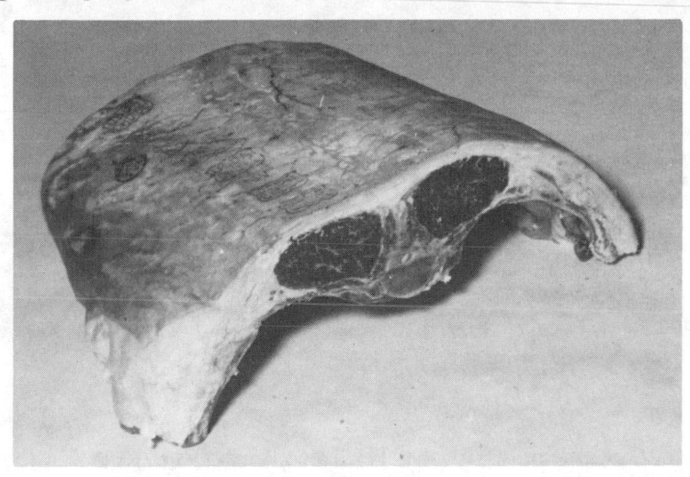

FIG. 15.19. TRIMMED LAMB SHORT LOIN (DOUBLE)

roast. As a boneless chop, the loin is sometimes rolled with the kidney inside. The flank is sometimes left on the boneless roast as part of the roll.

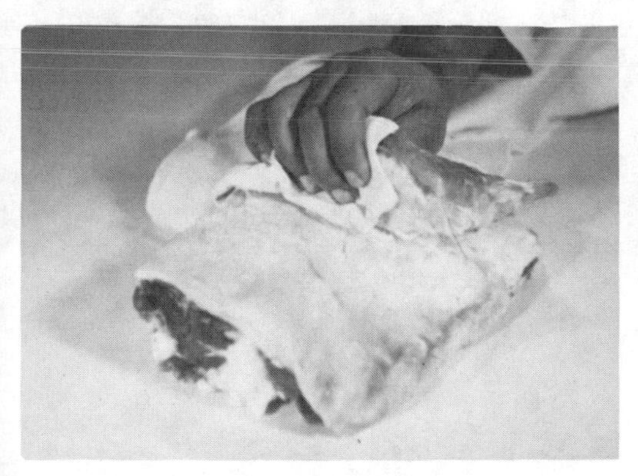

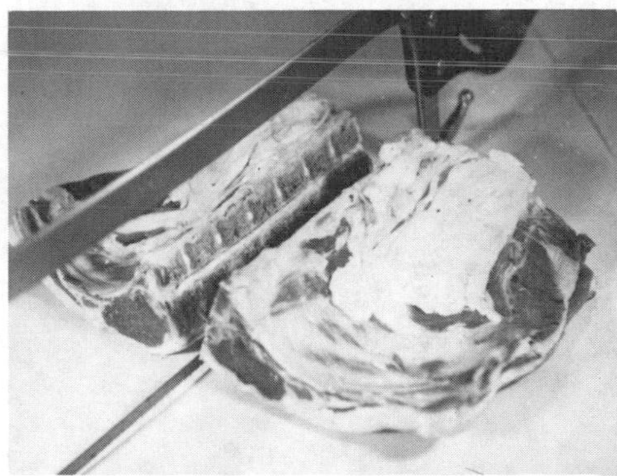

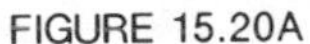

FIGURE 15.20A

FIGURE 15.20B

(1) Remove fell by scoring lightly pushing up with knife tip and pulling with cloth (Fig. 15.20A). (2) Split with saw (Fig. 15.20B). The whole double loin is sometimes cut into chops or boned out and rolled into a

roast, not split. This is not a common practice. (**3**) Pre-trim flank to chop ready length. This will vary with each operator's idea of the best presentation. Trim off excess fat. Score loin as per cutting plan for uniform

FIGURE 15.20C

FIGURE 15.20D

appearance, thickness, or weight. Knife to bone (Fig. 15.20C). (**4**) Saw chops. Trim off excess fat. A maximum and uniform thickness for the fat should be maintained (Fig. 15.20D).

Boning Lamb Short Loin

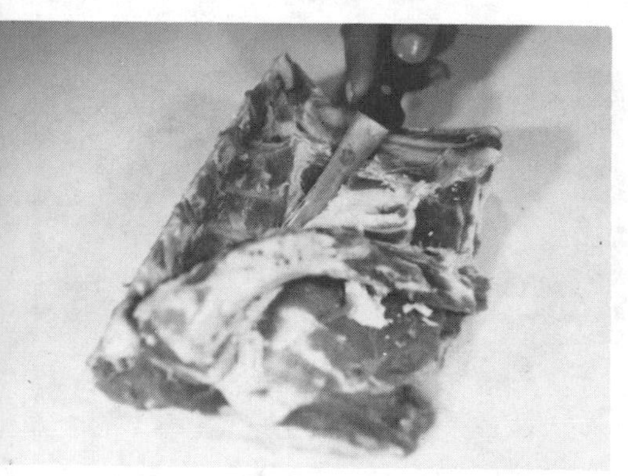

FIGURE 15.21A

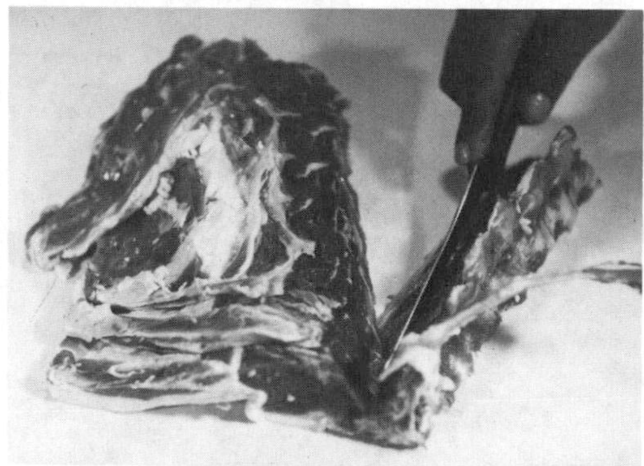

FIGURE 15.21B

(**1**) Remove all channel fat and loosen tenderloin (similar to beef tenderloin) (Fig. 15.21A). (**2**) Proceed as in boning beef strip loin. Loosen rib bone and cut strip loin muscle away from back bone (Fig. 15.21B).

FIGURE 15.21C

(3) Trim off excess fat, roll, and tie to roast (Fig. 15.21C). If boneless loin chops or Saratoga chops are intended, place skewers or large toothpicks at approximately the midpoint of each chop before slicing. The skewer should be visible to the patron or removed before serving.

Legs (Double or Pair)

Legs (Fig. 15.22) are the most popular moderate cost lamb item for roasting. In retail meat operations, generally the head loin (hip) is sawed off for bone-in chops or steaks, or boned out for a roast, Saratoga chops, or other secondary lamb products. The short cut leg, or "round," is sometimes steaked with a saw. Generally, each leg is prepared for a roast without boning. The cooked portion yield increases when the leg is first boned and rolled. The shank portion should be separated from the roast and treated as a secondary product for braising or stewing.

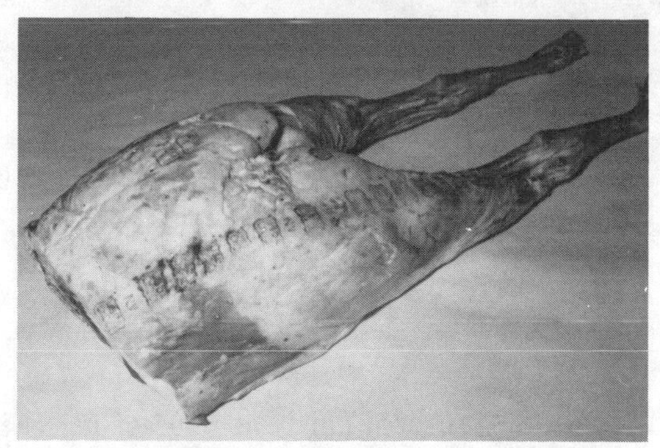
FIG. 15.22. LAMB LEGS (DOUBLE)

FIGURE 15.23A

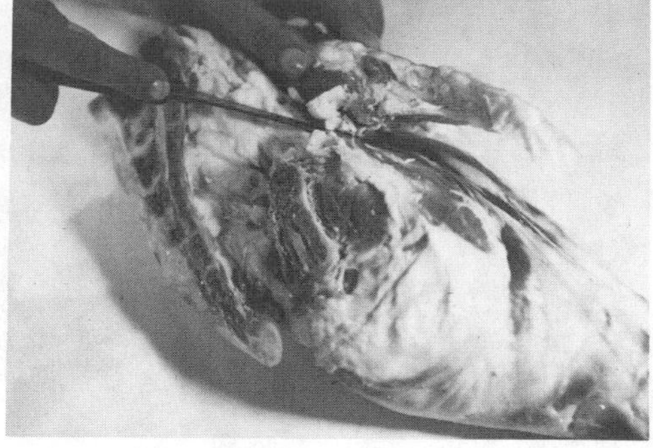

FIGURE 15.23B

(**1**) Split into single legs (Fig. 15.23A). (**2**) Trim off flank and excess cod or udder fat (Fig. 15.23B). (**3**) Seam out pelvic bone and tail section (Fig.

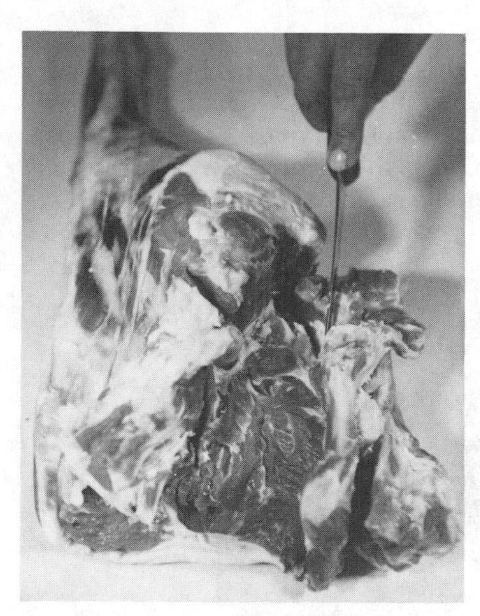

FIGURE 15.23C

15.23C). (**4**) Cut tendon (gambrel cord) at base of leg (Fig. 15.23D). (**5**) Score and crack hindshank at break joint (Fig. 15.23E). (**6**) Cut along

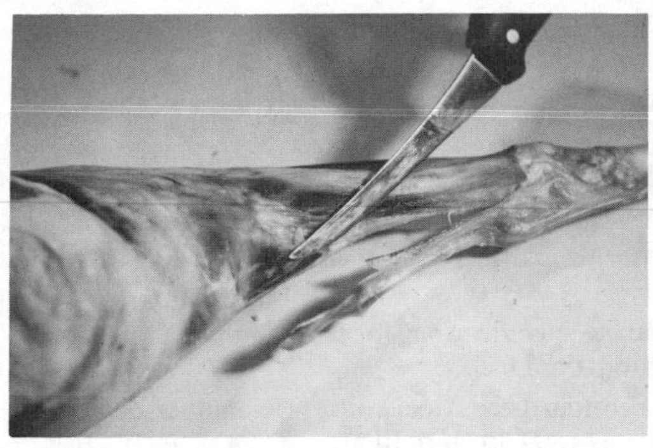

FIGURE 15.23D

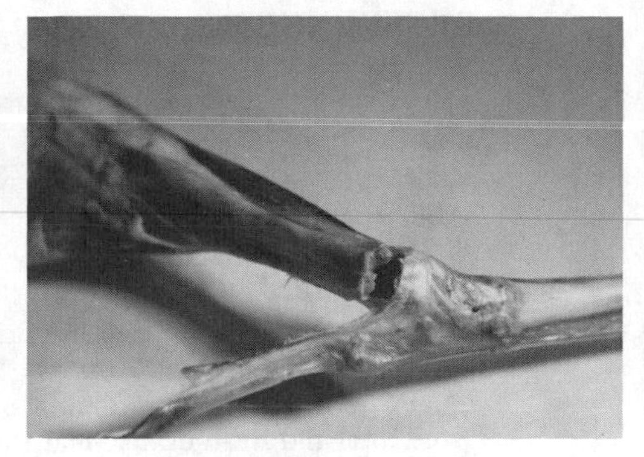

FIGURE 15.23E

natural shank seam to stifle joint. Knife through joint and separate meat-on-shank (Fig. 15.23F). (7) Tie leg with leg bone-in. This is a fully

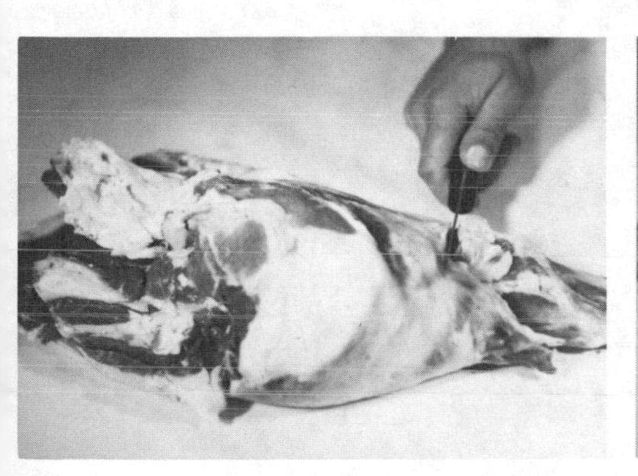

FIGURE 15.23F

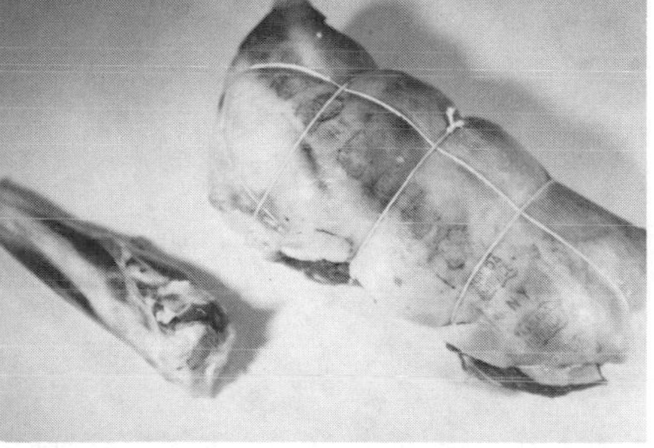

FIGURE 15.23G

oven prepared lamb leg. Yield is 75 per cent leg and 7 per cent shank. Leg bone may be chiseled or drilled out, making a boneless roast (Fig. 15.23G).

Lamb Hams

Legs of lamb are sometimes cured in the same manner as a ham, and then smoked. This is not a popular item. Legs of mutton are never used because the combination of the mutton flavor and the smoking process makes for an unpalatable product.

REFERENCES

ANON. 1945. Meat Handbook of the U. S. Navy. U. S. Navy Dept., Washington, D. C.

ANON. 1957. Cashing in on lamb. National Live Stock and Meat Board, Chicago, Ill.

ANON. 1959. Merchandising heavy lamb. National Live Stock and Meat Board, Chicago, Ill.

ANON. 1960. Institutional meat purchase specifications for fresh lamb and mutton, Series *200.* U. S. Dept. Agr., Washington, D. C.

ANON. 1961. Meat Buyer's Guide to Standardized Meat Cuts. National Association of Hotel and Restaurant Meat Purveyors. Chicago, Ill.

ANON. 1962. Something different! Starring today's lamb. Tech. Bull. *170,* American Lamb Council, Denver, Col.

BULL, S. 1951. Meat for the Table. McGraw-Hill Book Co., New York.

ROMANS, J. R. and ZIEGLER, P. T. 1977. The Meat We Eat, 11th Edition. Interstate Printers and Publishers, Danville, Illinois.

WANDERSTOCK, J. J., and WELLINGTON, G. H. 1961. Let's cut meat. Cornell Extension Bull. *1053.* Ithaca, N. Y.

16

Fresh Pork

USES

Pork packing is the best example of the meat industry described as a "disassembly operation." Few packer sales of pork are made as carcass pork, except to special processors, who function as subpackers. The carcass is fabricated into a wide variety of cuts at the packing house level. Some 30 per cent of the product is sold fresh, the rest is rendered into lard, cured, smoked, and processed into sausage. The industry realized early the advantage of this type of merchandising. Consequently, there will be no attempt to develop the carcass breakdown on pork in this chapter (Fig. 16.1–16.2).

A discussion of specifications also will be omitted, in that packer produced fresh or fresh frozen pork cuts, in any given category, are relatively uniform. Trim and quality variations will have to be determined by use and testing.

APPROXIMATE YIELDS*

NAME OF CUT	PERCENT
Fresh Hams, Skinned	18.5
Loins Blade on	15.0
Boston Butts	6.5
Picnics, Regular	8.5
Bacon, Square Cut	17.5
Spareribs	3.0
Jowl, Trimmed	3.0
Feet, Tail, Neckbones	5.0
Fat Back Clear Plate and all Fat Trimmings	18.0
Sausages Trimmings	5.0
Total	100

*Packer Dressed Hog, Head off, Leaf out
No allowance for cutting shrink

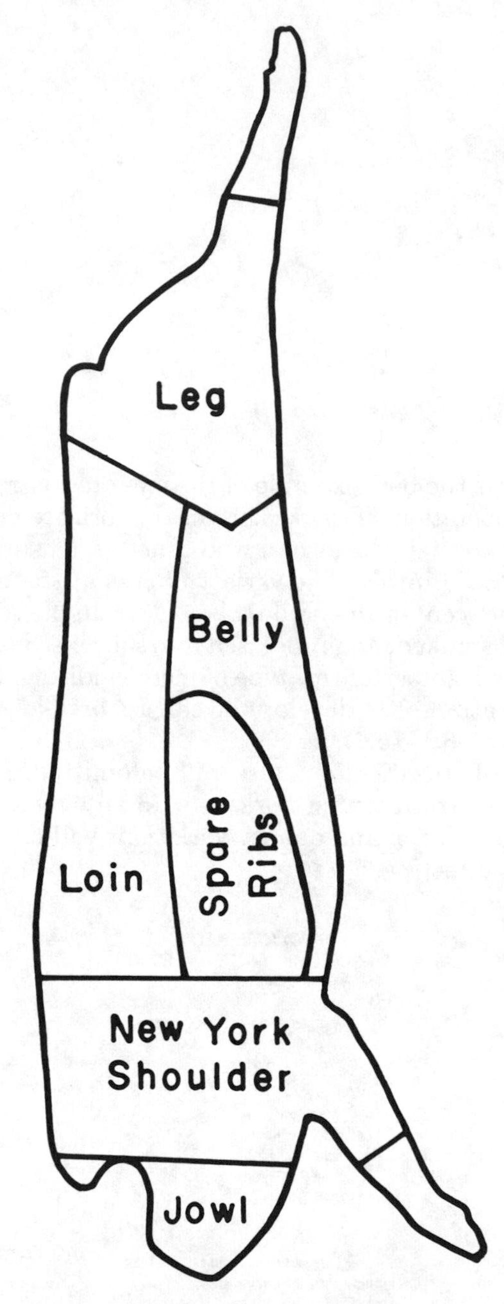

FIG. 16.1. PORK—PRIMAL CUTS

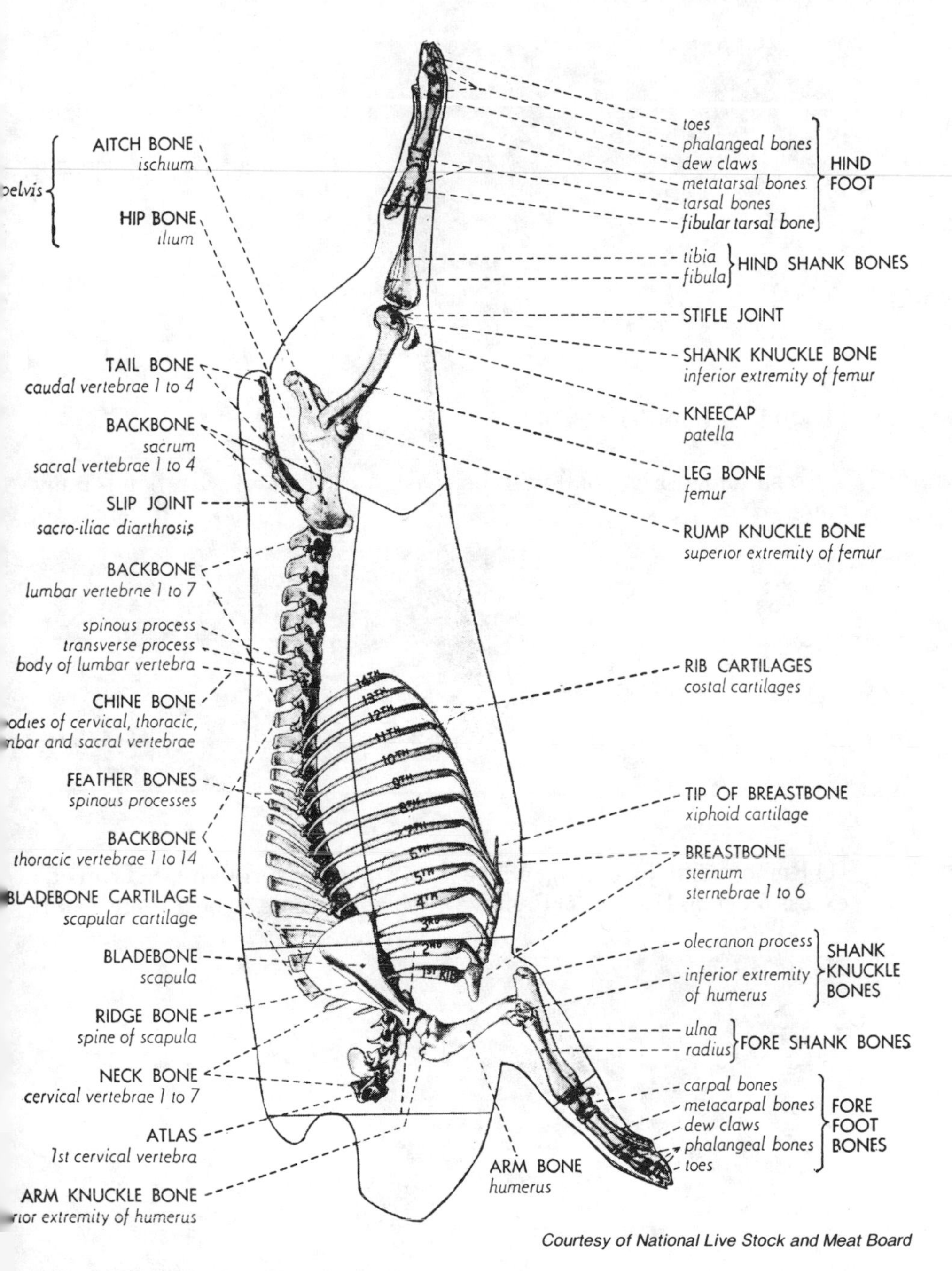

Courtesy of National Live Stock and Meat Board

16.2. PORK SIDE—SKELETAL STRUCTURE

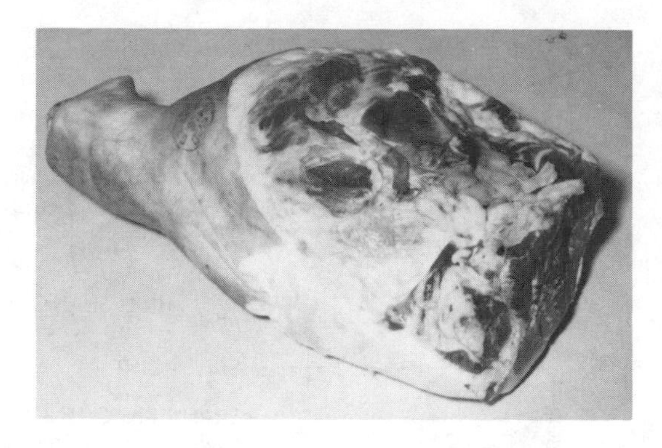

FIG. 16.3. FRESH SKINNED HAM

FRESH SKINNED HAM

Fresh hams (Fig. 16.3) may be roasted or steaked with bone-in or boneless.

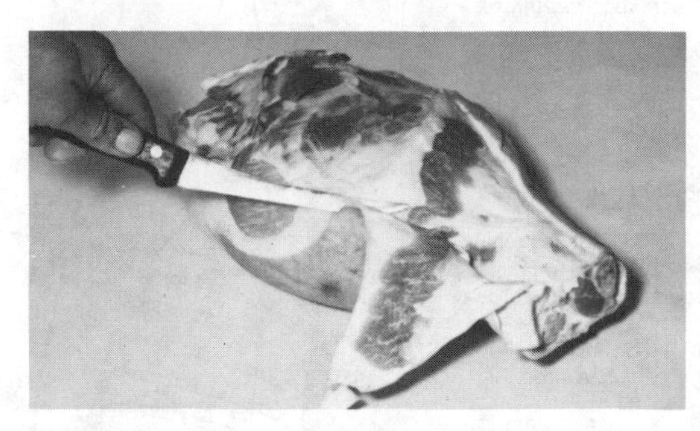

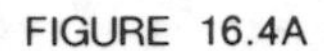

FIGURE 16.4A

(**1**) Remove skin by cutting above major muscles through fat. Trim off excess fat (Fig. 16.4A). (**2**) Remove aitch bone (same as beef round) and

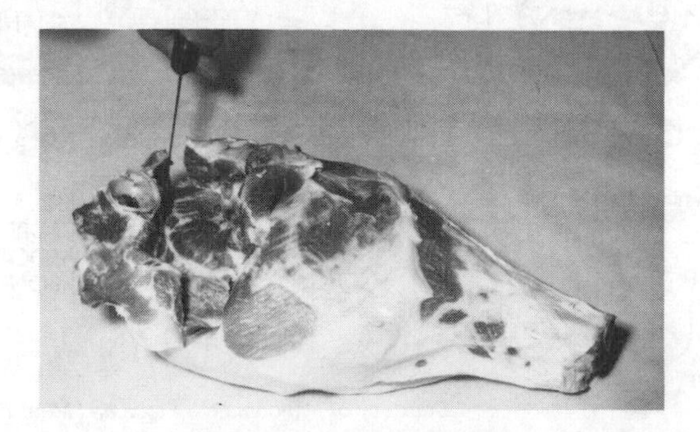

ham is ready to steak, roast, or bone-out completely (Fig. 16.4B). (**3**) Cut boneless steaks off butt end. Cut should be started perpendicular

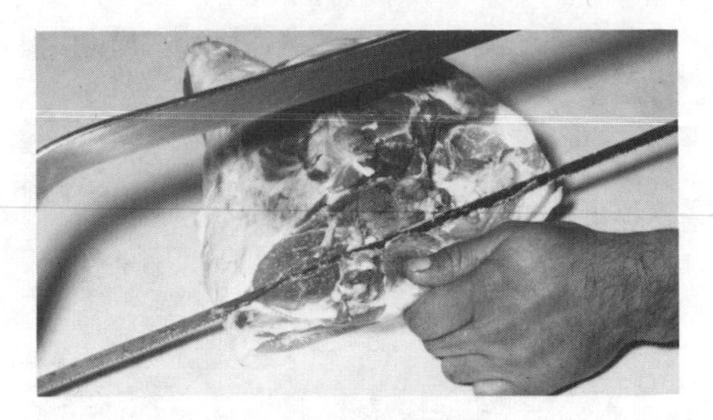

FIGURE 16.4C

to leg bone. Trim off excess fat. The leg bone portion may be roasted or cut into center cut steaks with saw (Fig. 16.4C).

When steaking, continue to a point in shank where steaks no longer have a desirable conformation. Treat shank as by-product.

To prepare a boneless roast, follow steps 1 and 2, cut off shank at stifle joint, open ham exposing leg bone and seam out. Tie into roast, or

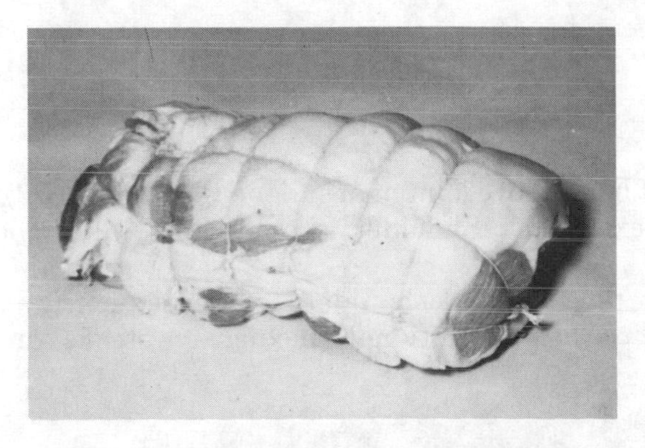

FIG. 16.5. FRESH SKINNED HAM—BONED AND ROLLED

muscle bone leg like beef round (Fig. 16.5). Boneless ham may be made into cubed, diced, or ground items.

NEW YORK PORK SHOULDER

Pork shoulder (Fig. 16.6) may be roasted or steaked like fresh ham. The shoulder is not equal in quality to the ham and much fatter.

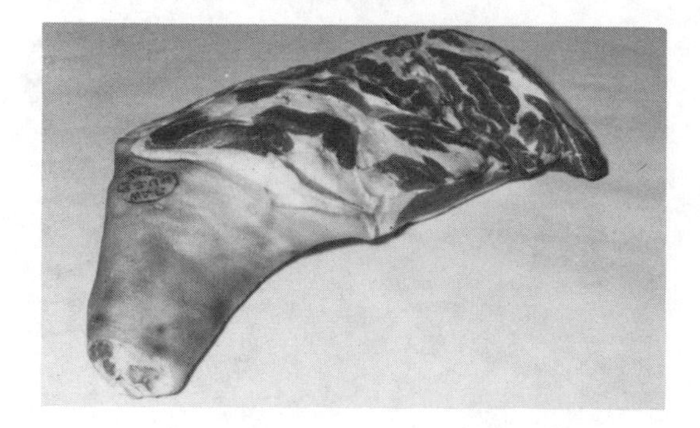

FIG. 16.6. NEW YORK PORK SHOULDER

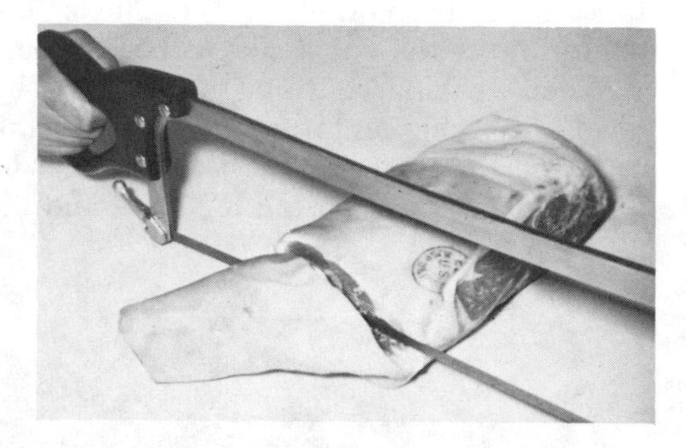

FIGURE 16.7A

(1) Saw off hock above joint of arm bone (Fig. 16.7A). (2) Remove skin and trim excess fat. Shoulder is ready to steak or bone out (Fig. 16.7B).

Steaking the New York shoulder is more frequently a retail butcher shop practice than institutional. Inexpensive steaks can be cut on the

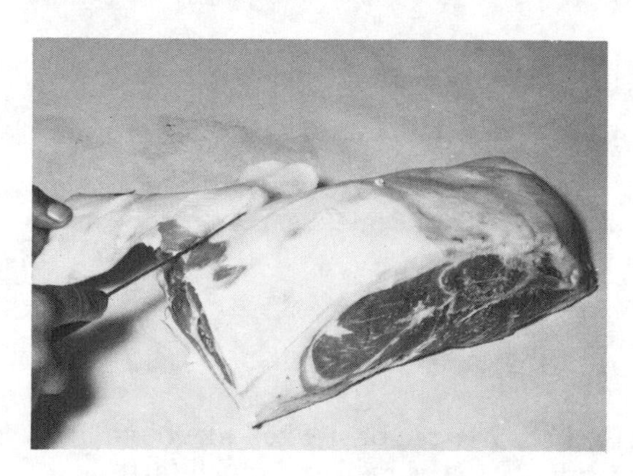

FIGURE 16.7B

saw. Saw off arm bone steaks first, cutting perpendicular to the arm bone. Continue to steak to the desired width of the blade bone steaks, and cut blade steaks.

This steaking is best done on an electric saw with a frozen product. It can be done manually by steaking to the bone with a knife, and sawing through.

Muscle Boning New York Pork Shoulder

Follow steps 1 and 2, and place shoulder fat side down on work table.

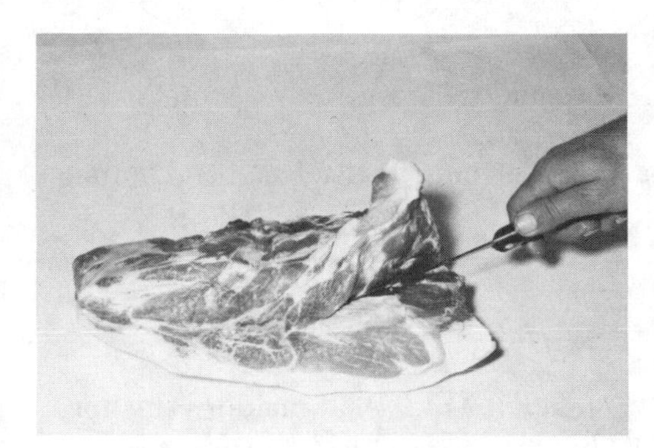

FIGURE 16.8A

(**1**) Starting from shank end, knife through fat below lean and mark to natural seam (Fig. 16.8A). (**2**) Seam to the tip of blade bone, cut down

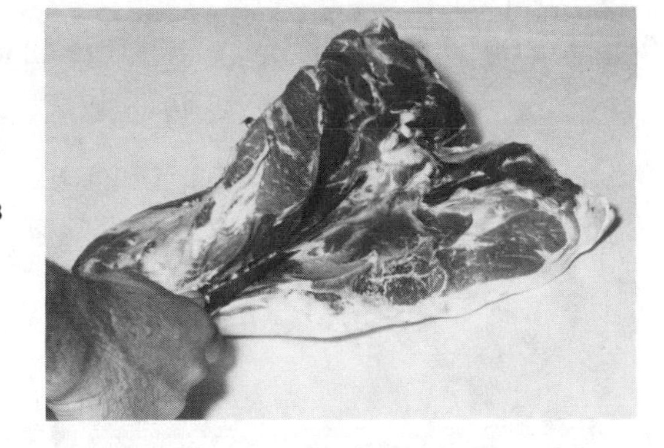

FIGURE 16.8B

and through shoulder to separate inside cut (Fig. 16.8B). Trim off excess fat. Roll inside cut into roast, starting with thick ends. Tie to hold in shape. (**3**) Seam out both bones from outside cut. Roll shoulder

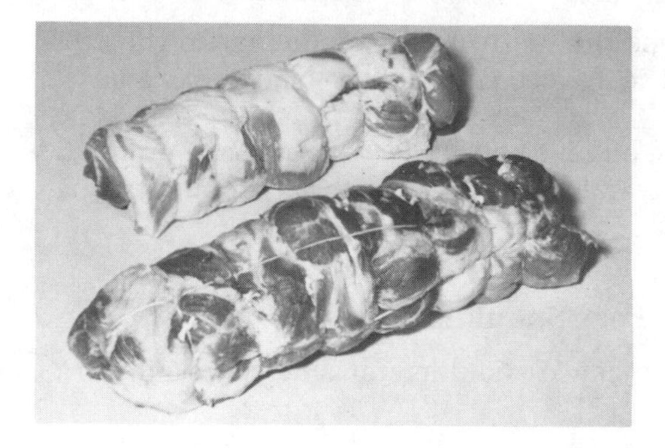

FIGURE 16.8C

lengthwise and tie (Fig. 16.8C). To hold roast together, a tying needle may be used.

Any portion of the boneless shoulder may be used as diced, cubed, or ground items.

PORK LOIN

The term is a misnomer. The pork loin (Fig. 16.9) includes the bone and muscle structure extending along the greater part of the backbone. The cut extends from about the third rib, which is a shoulder cut, through the rib area, through the loin. Because of the varied structure, a wide variety of cutting plans can be developed. Pork loins under 16 lbs are generally flavorful and tender.

A simple plan is to cut chops of uniform weight straight through the loin. On the rib end, the meat above the blade may be cut off for by-

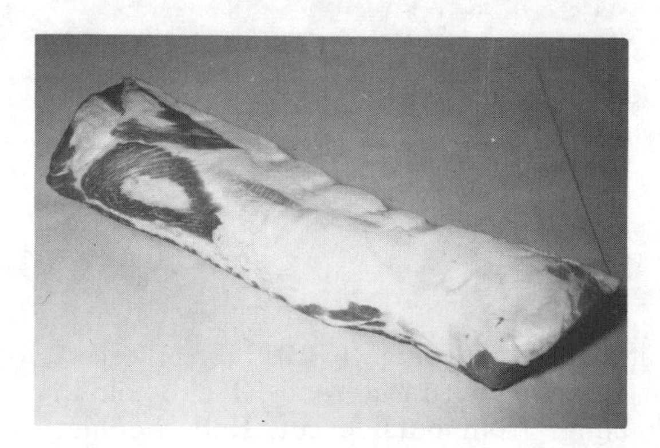

FIG. 16.9. PORK LOIN

products. The obvious disadvantages are that the conformation and quality of the chops will vary widely. Uniform portions will be difficult to achieve by weight unless some chops are cut very thin (the butt section), and some cut very thick (the rib section). In addition, though the loin appears relatively inexpensive, a simple inventory of the chops cut, divided into the cost of the loin, will generally demonstrate a relatively high unit cost.

Making Center Cut Pork Loin

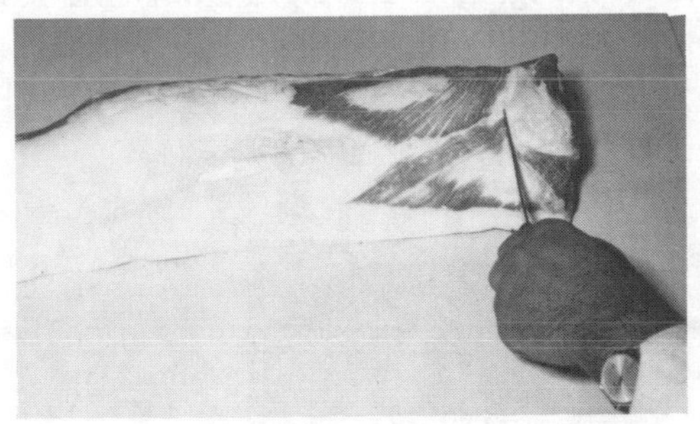

FIGURE 16.10A

FIGURE 16.10B

(1) Locate tip of blade bone with knife tip and mark (Fig. 16.10A). (2) Cut through and saw off shoulder end (Fig. 16.10B). (3) Locate anterior

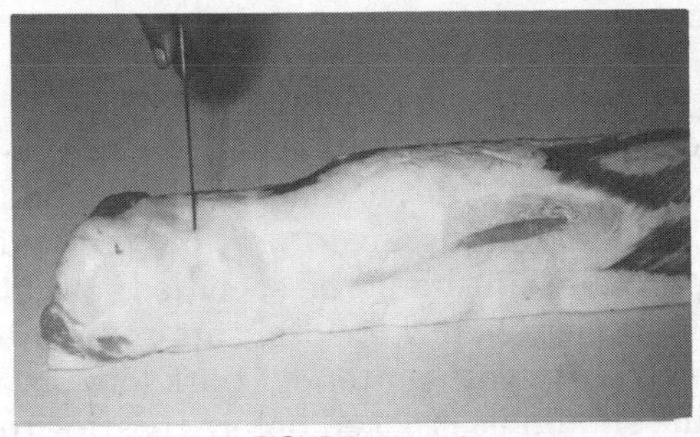

FIGURE 16.10C

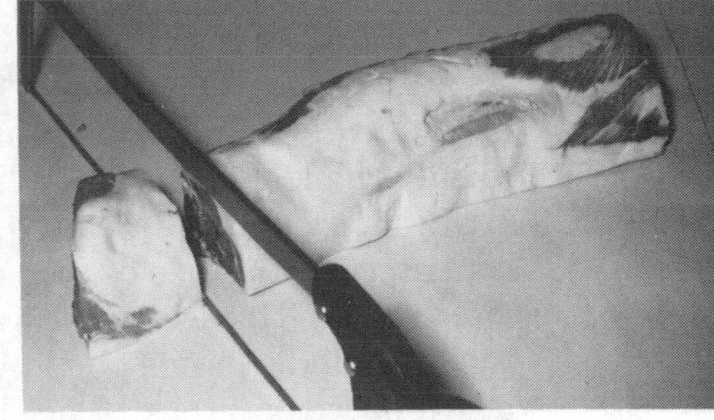

FIGURE 16.10D

tip of aitch bone at first lumbar vertebra (Fig. 16.10C). (4) Knife through and saw off (Fig. 16.10D). (5) The center cut pork loin (Fig. 16.10E—top)

FIGURE 16.10E

is the primary product. The loin end (Fig. 16.10E—right front) and rib end (Fig. 16.10E—left front) are secondary products.

The center cut pork loin may be prepared in several different ways.

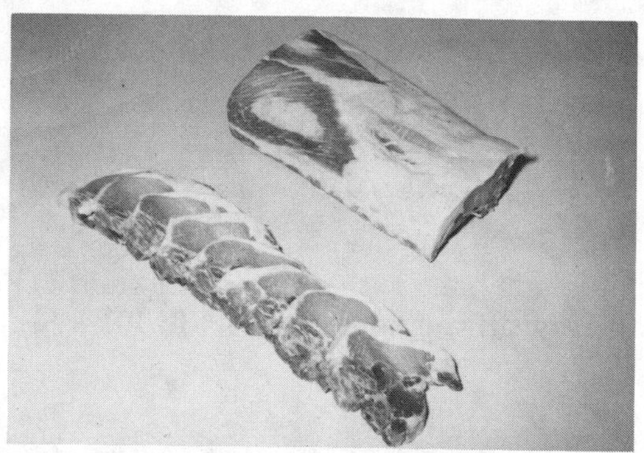

FIG. 16.11. CENTER CUT PORK LOIN—PARTIALLY CUT INTO CHOPS

(1) To achieve uniformly appearing portions, chops may be cut uniformly thick (Fig. 16.11). When two chops are served, a loin chop and a rib chop should be paired. To achieve any given portion size and control cost, the size of the center cut loin should be specified so that at any given thickness, the average size chop will closely approximate the intended average portion. (2) The whole center cut pork loin may be prepared as a roast. The chine should be cracked before roasting to facilitate slicing. The rib portion may be cut off at the 14th rib. The fat between the bones may be Frenched, making a crown pork roast (Fig. 16.12). The loin portion is then used for chops (step 1). Rib chops may be cut thick enough for a pocket and stuffing (Fig. 16.13).

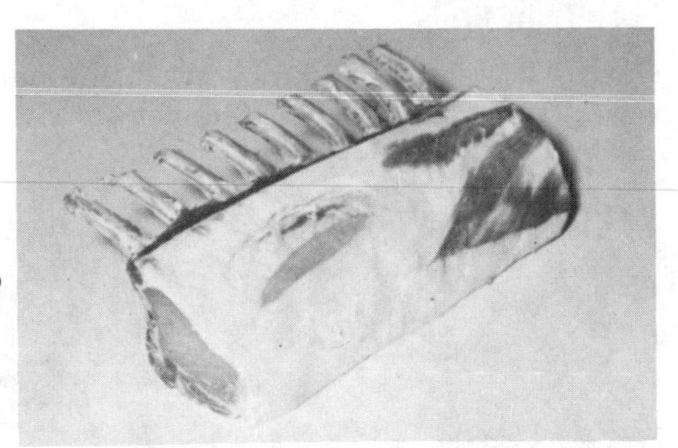

FIG. 16.12. RIB PORTION PORK LOIN ROAST—FRENCHED

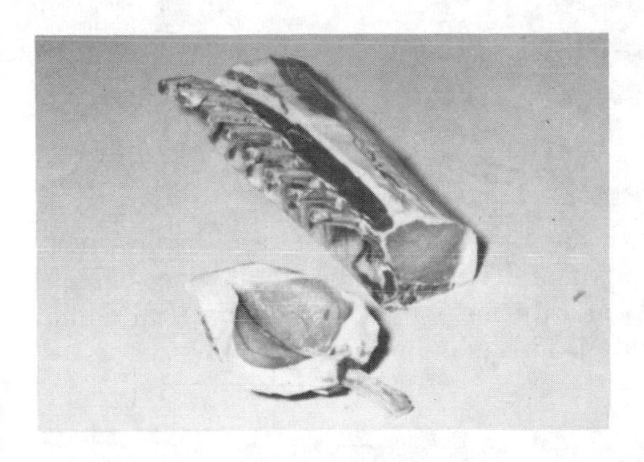

FIG. 16.13. FRENCHED PORK LOIN ROAST—ONE CHOP SEPARATED WITH POCKET

(3) The tenderloin may be pulled from the center cut loin before cutting chops (Fig. 16.14). The loin chops (tenderloin out) and rib chops have similar conformation. When the center cut loin, tenderloin out, is made from the whole loin, tenderloin should be pulled, before the hip end is cut off.

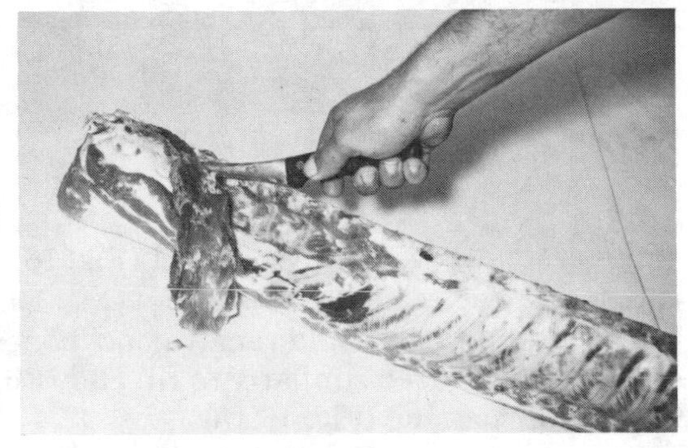

FIG. 16.14. PULLING PORK TENDERLOIN

Pork Loin Ends

(1) On rib end, remove meat above blade, and remove blade bone. Lean is by-product (Fig. 16.15A). (2) Shell out rib bones and back bone. Trim

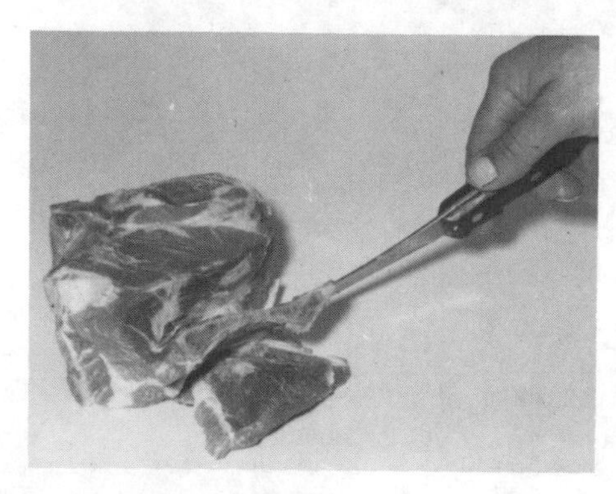

FIGURE 16.15A

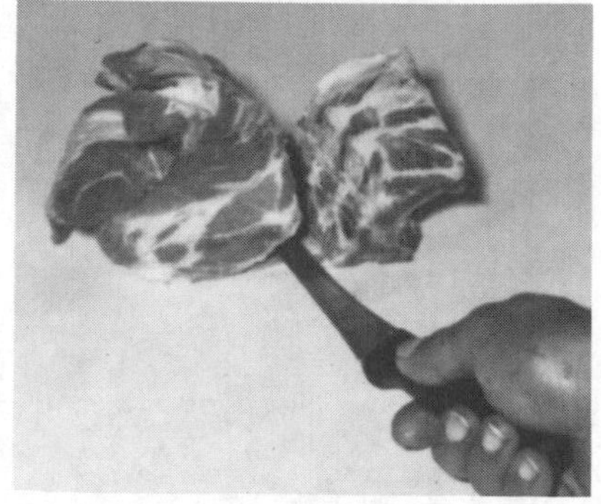

FIGURE 16.15B

closely (Fig. 16.15B). Cut with saw between rib bones and back bones. Rib bones will make a portion of back ribs. (3) The whole piece of lean may be used for roasting. As an alternate, by following natural seam, the eye muscle may be separated. This may be used for cutlets, boneless

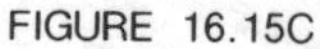

FIGURE 16.15C

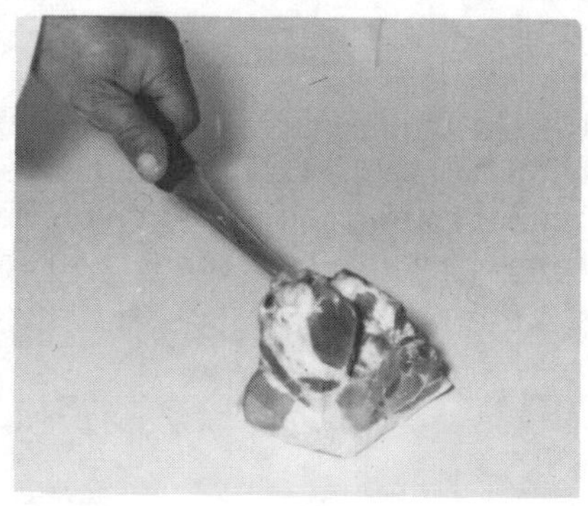

FIGURE 16.15D

chops, a small roast, or by-product (Fig. 16.15C). (4) On the hip or loin end, locate and seam out tenderloin (Fig. 16.15D). Shell out aitch bone (like beef top sirloin) and remove backbones. (5) The two large pieces of lean may be used similarly to rib end (Fig. 16.15E), (6) or tied lean to lean for roasting (Fig. 16.15F).

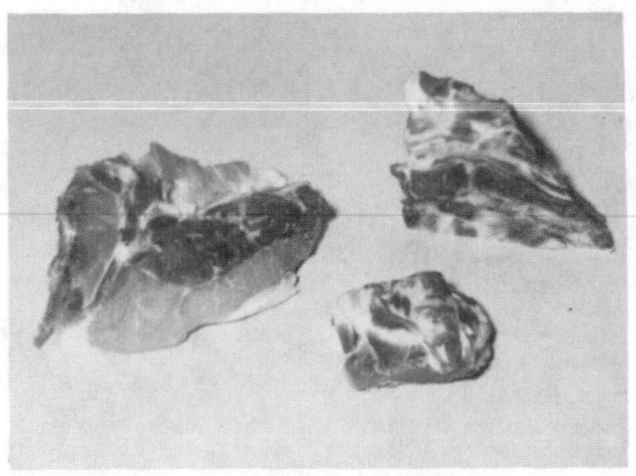

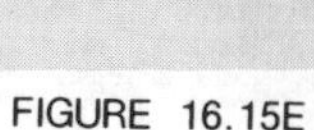

FIGURE 16.15E

FIGURE 16.15F

Muscle Boning the Pork Loin

(1) Locate and seam out (Fig. 16.14) full tenderloin (Fig. 16.16).

FIG. 16.16. PORK TENDERLOIN

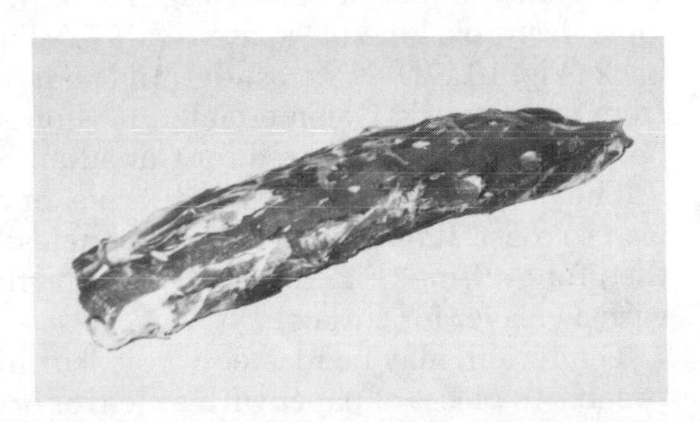

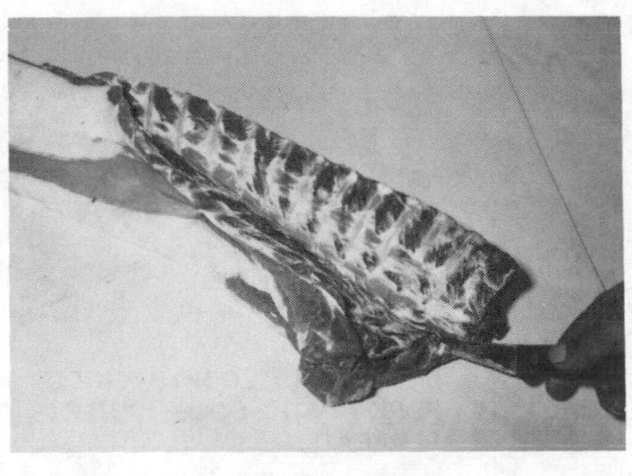

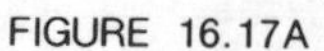

FIGURE 16.17A

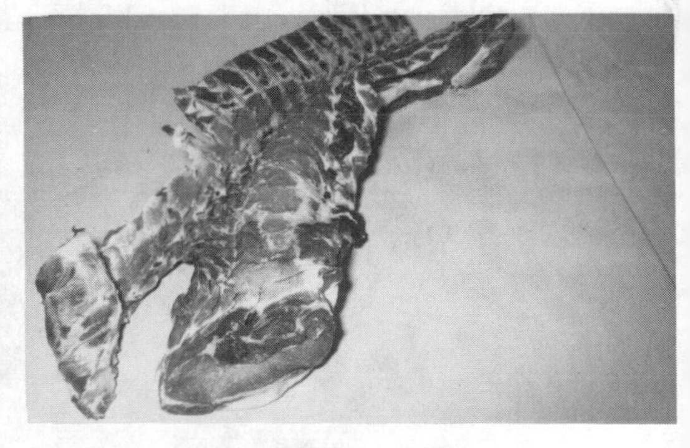

FIGURE 16.17B

(2) Seam along rib bones (Fig. 16.17A). (3) Continue through the loin portions until lean and bones are separated (Fig. 16.17B). Remove blade

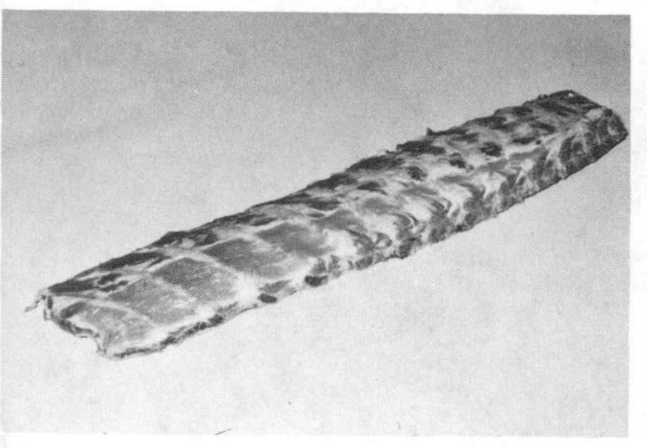

FIGURE 16.17C

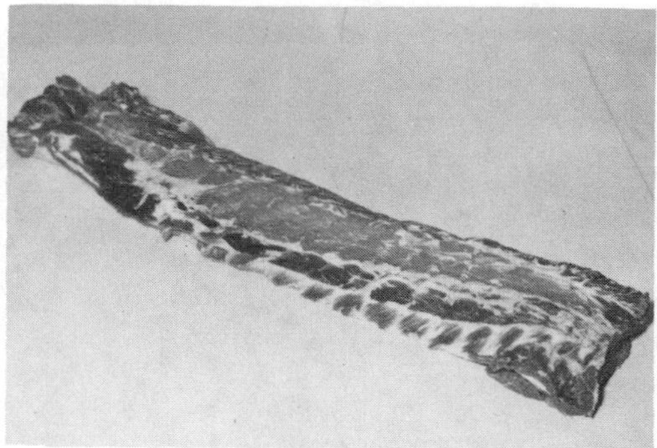

FIGURE 16.17D

cartilage and lean above it on rib end. (**4**) Cut off rib bones in a slab where they join back bones (Fig. 16.17C). These are loin back ribs, a most desirable kind of "spare ribs." (**5**) The boneless pork loin or loin back (Fig. 16.17D) is a most desirable cut of fresh pork when produced from young pork. Commercially, it is only available from older type pork, and usually prepared for Canadian bacon.

The loin back may be roasted whole, or cut in half and tied lean-to-lean to roast. It may be sliced into boneless chops or cutlets, or butterflied for stuffing (Fig. 16.18). The butterflied chops may be flattened with a cleaver for cutlets.

Tenderloin may be roasted whole, cut into slices, and flattened between two pieces of paper with a cleaver or a similar device, or cut into uniform lengths and split lengthwise (butterflied) and flattened with cleaver. This cut is ideal for making stuffed pork birds.

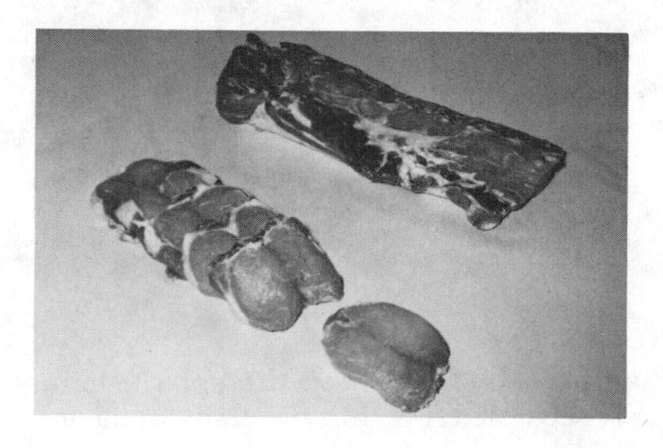

FIG. 16.18. LOIN BACK OF PORK—WITH SOME CUTLETS SEPARATED

SPARE RIBS

This is the whole rib section removed from the belly portion of the hog. Spare ribs (Fig. 16.19) are usually classified by weight as under 2 lbs, under 3 lbs, 3 to 5 lbs, 5 lbs and up. There is also a special pack "right side rib" which some find more desirable than the unmatched left side. Palatability and weight are inversely proportional.

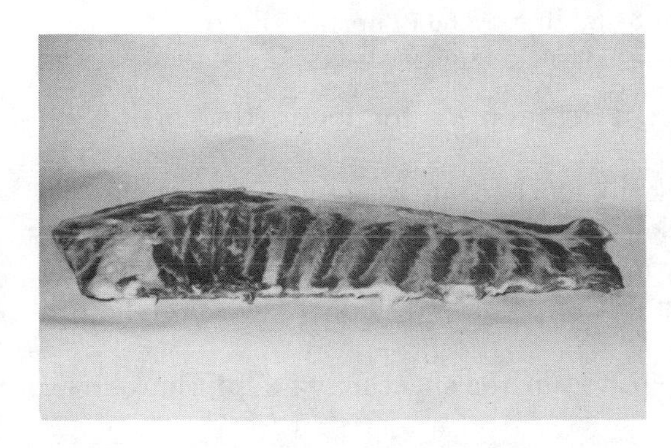

FIG. 16.19. SPARE RIBS

Menu Plan

The whole sheet may be prepared or it may be split between each bone, or split lengthwise, depending upon use. The plate and brisket portion are little more than fat and bone. By knifing along the tip of the breast bones in the soft cartilage, the brisket and plate may be removed (Fig. 16.20). This type of spare rib is available from the packers under a variety of names, such as special trimmed or barbecue ribs. Although

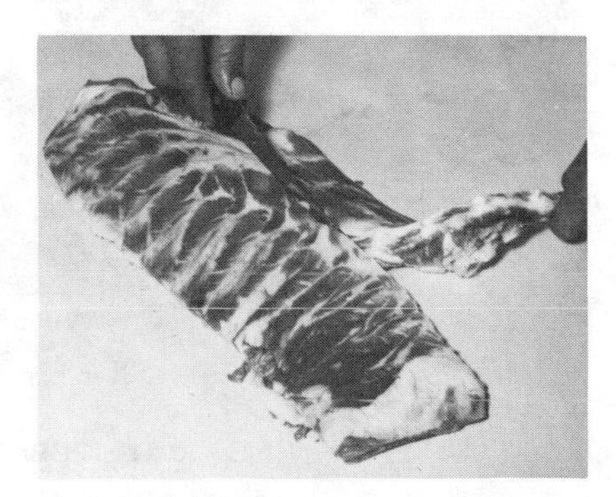

FIG. 16.20. REMOVING BRISKET AND PLATE PORTION OF SPARE RIBS

the price per pound will be higher, the total delivered cost of the sheet will be no greater because it will weigh commensurately less. For example:

3 lb sheet at $1.20 lb = $3.60 (2 portions)
portion cost = $1.80 each

Same sheet with special trim:

2 lb sheet at $1.80 lb = $3.60 (2 portions)
portion cost = $1.80 each

Note that with a 50 per cent higher price per pound, the portion cost is the same.

The loin back rib, or back rib, has become as popular as the spare rib. This is a by-product of pork loins cut for Canadian bacon. This is a very meaty-type rib. It is usually made from older hogs. Being a little tougher, it requires moist heat preparation. This cut makes exceptionally uniform portions.

Single bones may be separated or the sheet may be split lengthwise for hors d'oeuvres.

MISCELLANEOUS CUTS

Hocks

Pork hocks (Fig. 16.21) are the shank portions of hams and shoulders. They may be skinned, defatted, and cut into bone-in pieces or boned for by-product.

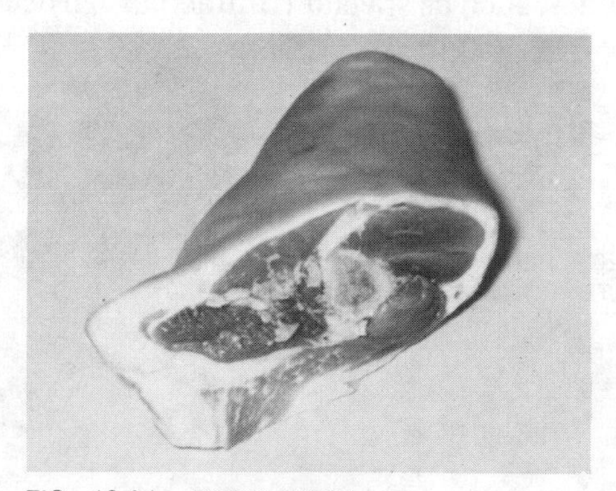

FIG. 16.21 PORK HOCK

FIG. 16.22. FAT BACK

Fat Backs

These are solid, relatively uniformly shaped, rectangular fat slabs removed from the surface of the pork loin (Fig. 16.22). They may be sliced thin for surface larding or cut into thin strips and drawn through various roasts with a larding needle. This is often done for roasting beef tenderloin to increase juiciness.

Fat

Lard is usually rendered from fat backs, clear plates, and leaf lard. Most lard is hydrogenated, making it virtually odorless. Coupled with a high smoke point, it makes an inexpensive, highly desirable shortening. Heavy fat bellies are usually dry salt cured for salt pork. Jowls are sometimes dry salt cured, or trimmed and smoked.

Pork Belly

The pork belly is mostly trimmed, cured, and smoked for bacon. Better bacons are dry cured. Pork belly is sometimes sold fresh as side meat.

Boston Butt

This is a portion of the shoulder containing the blade bone. The butt is made by a cut perpendicular to either side of the shoulder, just above the arm muscle joint through the tip of the blade bone. The cellar trim (C.T.) butt is the meaty solid portion removed from below the blade bone. When a solid piece is needed, the C.T. butt may prove most economical portionwise.

Neck Bones

These are similar to spare ribs, less desirable for institutional use, and less expensive, recommended for institutional use.

Feet

The feet are saved and sometimes sold fresh. Feet are usually sold as a pickled product, sometimes cooked and cured. The feet are cut below the shanks.

Skin

The meat industry markets all the by-products, the skin, as well as the snout, ears, and tail. The skin is sometimes fried and sold as fluffs.

Chitterlings

Chitterlings are made of the large and small intestines of the hog, which are completely emptied, and rinsed thoroughly. They are cooked in this form.

SUCKLING PIG

Suckling pigs are very young animals from 20 to 35 lbs dressed. They are usually delivered with the head on, and priced per pig rather than by the pound. Although basically ornamental, they are a delicacy, and properly prepared and garnished, make a most delightful banquet item.

BY-PRODUCTS

Because the pork products are, for the most part, young, flavorful, and tender, the lean trim or by-products of almost any cut can be interchanged. Lean pieces of pork from virtually any part of the carcass can be cut into boneless chops or pork steaks, or into cutlets which may be cubed. The cutlet may be upgraded by giving it the name of "tenderette." It also may be breaded, which will create an additional crusty flavor. The lean trim may be cut into cubes, and used for various braised dishes, or may be mixed with fat to make pork sausage.

When setting up a menu plan for pork, depending on the quality desired and the portion cost that can be afforded, almost all the cuts may be considered.

REFERENCES

ANON. 1945. Meat Handbook of the U.S. Navy. U.S. Navy Dept., Washington, D.C.

ANON. 1959. Profits with pork. National Live Stock and Meat Board. Chicago, Ill.

ANON. 1960. Institutional meat purchase specifications for fresh pork. Series *400*. U.S. Dept. Agr., Washington, D.C.

ANON. 1961. Cashing in on pork. National Live Stock and Meat Board. Chicago, Ill.

ANON. 1961. Meat Buyer's Guide to Standardized Meat Cuts. National Association of Hotel and Restaurant Meat Purveyors. Chicago, Ill.

ANON. 1962. Facts about pork. National Live Stock and Meat Board. Chicago, Ill.

BULL, S. 1951. Meat for the Table. McGraw-Hill Book Co., New York.

ROMANS, J. R. and ZIEGLER, P. T. 1977. The Meat We Eat, 11th Edition. Interstate Printers and Publishers, Danville, Illinois.

17

Processed, Smoked, and Variety Meats

CURING

The word "processed" has a very specific meaning. Definite changes are made in the product due to the application of heat and chemical processes.

Basically, the curing process is the addition of salt to retard bacterial action, and to promote preservation. In spite of the unfavorable comments of some persons relative to cured meat "not being what it used to be," the fact is that modern curing processes are faster, more uniform, more economical, and generally result in less spoilage.

The cure may be applied as a salt to the surface, which is referred to as dry-curing, a brine solution in which the product is immersed, or a brine solution injected into the product with a needle. "Vascular curing" is the injection of brine into one of the main arteries. "Stitch curing" is done by a multiple injection.

The rate of curing may be stated as the rate of diffusion of the curing brine relative to three factors: method of application, thickness of the cut, and the temperature at which the product is stored. Frequently, garlic, pepper, and spices are added to enhance the flavor of the cured product. When a brine solution is used, it is usually referred to as "sweet pickle," usually consisting of salt, sugar, and saltpeter dissolved in water. The cure room should be maintained at 30° to 40°F. At temperatures in excess of 50°F, the brine will sour, and the product will develop off-flavor and sour around the bones.

"Corning" is a variation of this curing process, which includes salt, sugar, baking soda, nitrates, and nitrites. This curing solution usually is employed for beef cuts.

The effect of the curing process is that the salt inhibits the development of bacteria, and reduces the water content of the product. Sugar is used to counteract the hardening effect of the salt, and improve the flavor. The nitrates and nitrites help retain and enhance the characteristic pink color of the meat. The cure considerably increases the color of corned beef. In some markets gray colored corned beef is desired, but this is the exception.

SMOKING

Smoking is done in addition to curing. Hams and bacon, for example, are first cured, then smoked. The process is essentially one where the product is placed in an airtight smokehouse with heat and smoke simultaneously applied. Hardwood logs and sawdust are used to generate smoke. Smoking gives the product a characteristic flavor and color, which can be varied slightly with the cure and types of smoke. Smoking retards fat oxidation. The reduction of water content as well as the deposition of smoke components have a bacteriostatic effect. A product labeled "ready-to-eat" under U. S. Department of Agriculture inspection must be maintained at an internal temperature of not less than 140°F for a period of at least 30 minutes. The smoke flavor cannot be induced by a chemical or smoke substitute.

PROCESSED BEEF ITEMS

Corned Beef is probably the most popular of the cured beef items, usually made of cuts of the beef round or the brisket. Corned rounds are usually intended for the consumers demanding very little fat. They are generally dry, lean, easier to slice, and more economical. The beef brisket tends to be juicy and flavorful.

Dried beef is usually made from defatted top round (beef inside) by dry-curing or sweet pickling. It is hung in a ventilated room for further drying. Dried beef is available by the piece or pre-sliced. Large slices are expensive, frequently used for hors d'oeuvres. Sometimes the beef is chipped, compressed into a loaf, and then sliced. This is an economical way of preparing dried beef slices.

Jerked beef was probably discovered by the Indians of the United States or South America. The more tender pieces are selected, cut into long strips, hung, and dried in the sun. The dried strips are hard and inflexible.

Pastrami is usually made from the navel portion of the beef plate, which is cured, covered with pepper and spices, and smoked after cooking.

Peppered beef is usually made from the rounds, cured and smoked, ready to eat.

Rendered beef suet is often used as a shortening. When hydrogenated, it has very little odor and a relatively high breaking point. It is a most economical shortening.

PROCESSED PORK ITEMS

Hams are marketed green (unprocessed), sweet pickled, and smoked. Pork shoulders, picnics, and butts are also available in these forms. A vast variety of finished products are available.

Smoked hams are the common variety. These are sometimes dark-cured.

Boneless hams on the market consist of the "boned and rolled ham," which is formed in a Visking casing into a relatively uniform diameter. The buffet or tavern type ham is tied back into what is roughly a ham shape. Other boneless hams include the pear shaped canned hams, the pullman canned hams (resembling a square loaf), and boiled hams, which are cured, cooked in water or steam, and not smoked.

Virginia ham is a distinct product. The ham is first cured in salt in barrels for about seven weeks. It is then rubbed with a mixture of molasses, brown sugar, black pepper, cayenne pepper, and saltpeter, and cured for two more weeks. It is then hung hock down by a string through the ham for at least 30 days and often for over a year. It is not smoked. This ham must be parboiled before it is cooked.

Smithfield type Virginia ham is rubbed with saltpeter, then salt, and shelf cured for three to five days, then given another salt rubbing, and cured one day per pound. At the end of the curing period, it is washed, rubbed with black pepper, and smoked for about 30 days. It is allowed to age for ten months to a year in a bag.

Proscuitto ham is ready-to-eat, a highly seasoned product, originated in Italy. It is prepared with bone-in or boneless. It is very dry, very hard, and has a distinct and subtle flavor. It is generally sliced very thin and stuffed and rolled into hor d'oeuvres or served with melon slices as an appetizer.

Westphalian ham is named after the German province of origin. It is smoked in a blend of juniper twigs and berries with a beechwood fire, which gives it its distinct flavor. It is first dry-cured, then sweet pickled. After a four week ripening period, it is cleaned, and smoked, ready-to-eat. It is available both boneless and with bone-in, and makes an outstanding hor d'oeuvre item, comparable to the Proscuitto ham.

Scotch ham is a boneless defatted skinned ham with a mild cure put up in a cellulose casing and not smoked. These hams are probably not available in the domestic market.

Bacon.—Bacon is produced from fat bellies, cured and smoked. It is available in slab, stick, or sliced. The size of the belly frequently relates itself to the quality. Moderately small bellies from young hogs are the most desirable. Canadian bacon is made from fresh Canadian backs, processed the same as bacon. It is sometimes smoked ready to eat, sometimes canned.

Sausages.—The word "sausage" usually associated with fresh ground pork items, applies to a broad variety of items, including bologna, wieners, loaves, liverwurst, scrapple, etc.

Sausage processing consists of grinding and blending the meat, seasoning it, stuffing, shaping or molding it, adding a preservative, and sometimes smoking it. Three natural casings are used: the hogbun, for liver sausage; the small intestines of the hog, for a variety of European type sausages; and the sheep intestines, generally for breakfast link sausage. In addition, synthetic casings are used. For example, skinless wieners are made with an artificial casing which is peeled off after smoking, before the product is marketed. More than a hundred different kinds of sausages are offered, including cold cuts and lunch meats. Some are pre-cooked, others require cooking. In Europe, sausage is most important in total meat production. Bratwurst, mettwurst, and wienerwurst, the many hard salamis of Italian, German, and Swiss origin, are very popular in our market. Over a hundred million pounds of sausage are produced in the United States annually.

VARIETY MEATS

To upgrade and create a greater market for what was formerly termed "offal," which literally means "off fall," that portion of the product which falls off the carcass as it is dressed, the term "variety meats" has been successfully substituted. As the science of nutrition developed, and as the public palate searched for a variety of taste sensations, variety meats became most desirable. The classes of variety meats are liver, tongue, sweetbread, heart, kidney, brain, tripe, and ox tail.

The brightness of color usually specifically relates itself to the quality of the product, except in the case of brains, where a pinkish-gray product is most desirable. Variety meats are high on the nutrition scale, high in protein, iron and vitamins. They have a distinct flavor with a broad appeal to gourmets. In spite of the greatly increased popularity, the cost per serving remains relatively low.

Livers

Beef and veal liver are the most acceptable. Both are used either as sautéed slices or liver steaks. The veal and calf liver, produced from a younger animal, are more delicately flavored, more tender, and higher priced (Fig. 17.1).

Beef livers under 13 lbs from high quality, young beef are the better quality. Cow and bull livers are heavier, darker, and tougher.

Pork liver has a relatively strong flavor. It is used primarily for making liverwurst. Some pork liver is used in institutional kitchens, most frequently for paté.

Lamb liver is a marginal menu item, sometimes used in a fashion comparable to beef where a low menu price warrants using it. It is more tender than beef liver, but does not have an equally satisfying flavor.

Livers are frequently merchandised frozen because of their highly perishable nature. This appears to have little effect on palatability.

Beef and veal livers are generally graded by size. Veal livers are generally 2 to 4 lbs, calf livers 4 to 7 lbs, and 7 lbs up. The heavy calf livers generally come from marginal animals, which very much resemble beef, sometimes referred to as kip calves. Beef livers, as a rule, are graded under 10, 10 to 12, and 12 up. Veal livers under 1½ lbs come from "slunk calves" and are not desirable.

Liver may be prepared by braising, pan broiling, or frying.

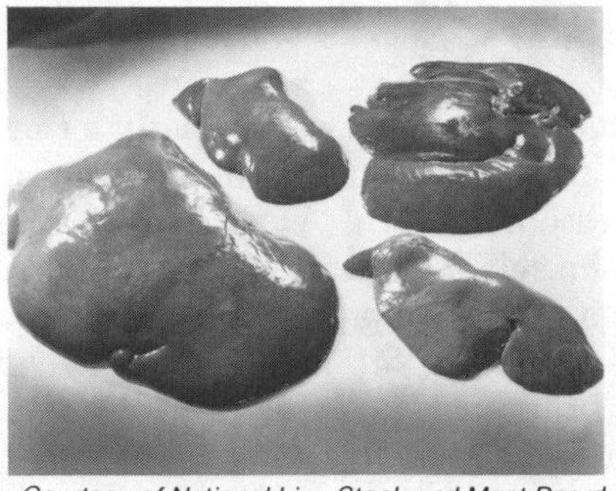

Courtesy of National Live Stock and Meat Board

FIG. 17.1. LIVERS

From left to right: Beef, Veal, Pork, and Lamb

Sweetbreads

Sweetbreads (thymus gland Fig. 17.2), are located in the area of the neck as well as the heart. Probably the most delicate of the variety meats, they are produced from beef, veal, and lamb. As the bovine matures, the sweetbreads disappear. They are frequently frozen as they are produced. Freezing has little effect on palatability.

Veal sweetbreads are generally packed in pairs and individually wrapped. Usually the larger the pair, the higher the price. The relationship between the size and price is one of supply and demand, not palatability. Veal sweetbreads that are not paired, usually sell at a considerable discount.

Sweetbreads are generally washed before cooking. The covering membrane can be removed before or after cooking. It is not necessary to precook sweetbreads before they are braised, sautéed, or fried. In general, they are precooked by simmering for some 15 minutes, which makes them more tender, permits easy removal of the outer membrane, and renders them sterile so that they will hold up considerably longer than they otherwise would after being defrosted.

Kidneys

Veal, beef, lamb, and pork kidneys (Fig. 17.3) all find their way into the market. The beef kidney is the least tender and the strongest in flavor. Kidneys may be prepared by braising, pan frying, sautéing, or broiling. Sometimes kidneys are used in a pot pie—beefsteak and kidney pie, sometimes cut into the loin chop. Where the kidney does not naturally occur in a chop, it can be easily inserted and held with a skewer.

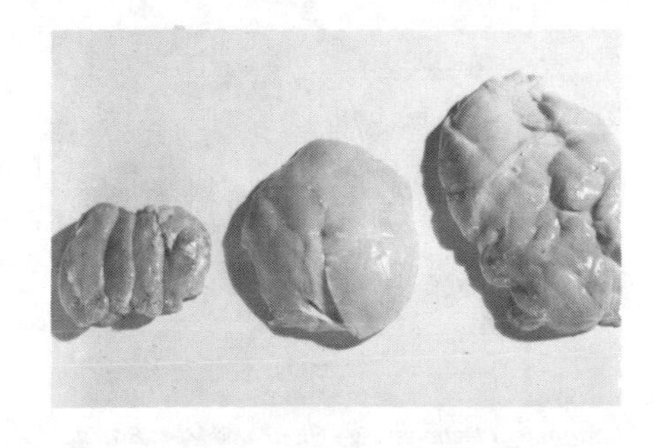

FIG. 17.2. SWEETBREADS

From left to right:
Lamb, veal, and beef

Courtesy of National Live Stock and Meat Board

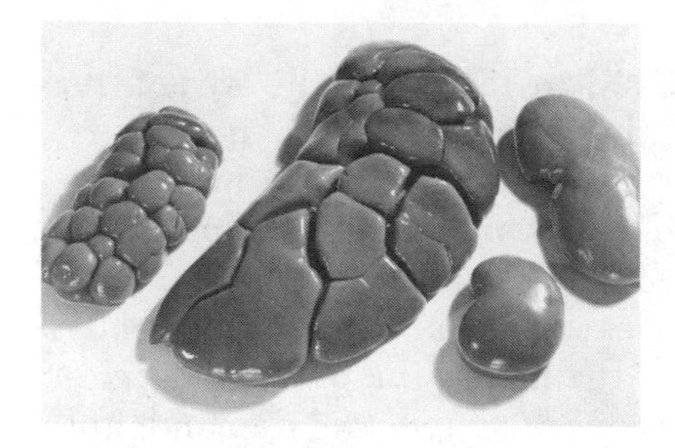

FIG. 17.3. KIDNEYS

From left to right:
Veal, beef, lamb, and pork

Courtesy of National Live Stock and Meat Board

Brains

Since the institution of humane slaughter, calf brains have been a relatively scarce item. Brains (Fig. 17.4) are extremely tender with a very delicate flavor. Among the gourmets, and particularly among the Europeans, they are a delicacy. Simmering about 20 minutes in water with a small amount of acid, as vinegar or lemon juice, will help blanch the product and make it firmer. The covering membrane should be removed. Brains may be sautéed, creamed, breaded, used in salads, or pan broiled.

Hearts

The heart (Fig. 17.5) is probably the least desirable and consequently the most economical variety meat. It may be used in very low priced menu operations. There is little to choose among the hearts: beef, veal, pork, or lamb. In general, this muscle is relatively tough. Hearts are usually braised or cooked in a liquid.

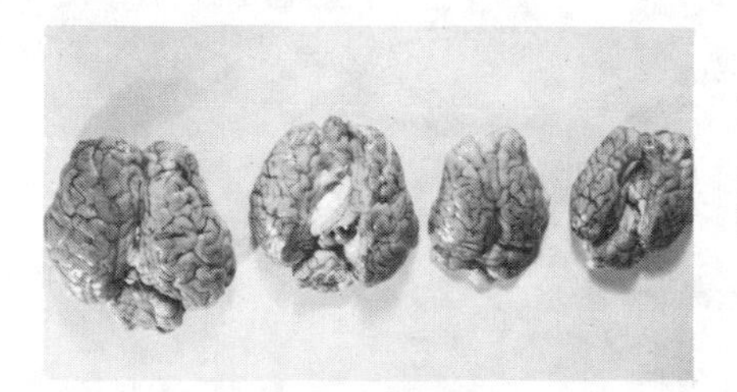

FIG. 17.4. BRAINS

From left to right:
Beef, veal, pork, and lamb

Courtesy of National Live Stock and Meat Board

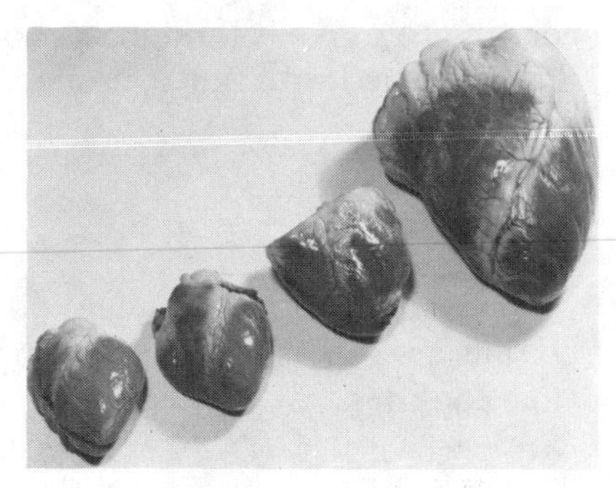

FIG. 17.5. HEARTS

From left to right:
Lamb, pork, veal, and beef

Courtesy of National Live Stock and Meat Board

Tongues

Beef and veal tongues are most desirable, primarily because of the size and the fact that they lend themselves to portions and slicing. In any case, tongues (Fig. 17.6) are a relatively tough product, and should be cooked with moist heat. Tongues are merchandised fresh, frozen, sweet pickled, and smoked. Processed tongues are considerably enhanced flavorwise. The sweet pickled tongues usually contain a large quantity of brine. Smoked tongues, if reasonably dry, although higher in price, may yield a better end product cost.

Tripe

The first and second stomachs of the beef are marketed as tripe. They are washed, soaked in lime water, and scraped to remove the inside wall, and then cooked before they leave the packing house. They are sometimes pickled or canned. "Honeycomb tripe" is the most desirable, usually used in the institutional kitchens. Most tripe is made into sausage. Although tripe is precooked, it requires considerable additional

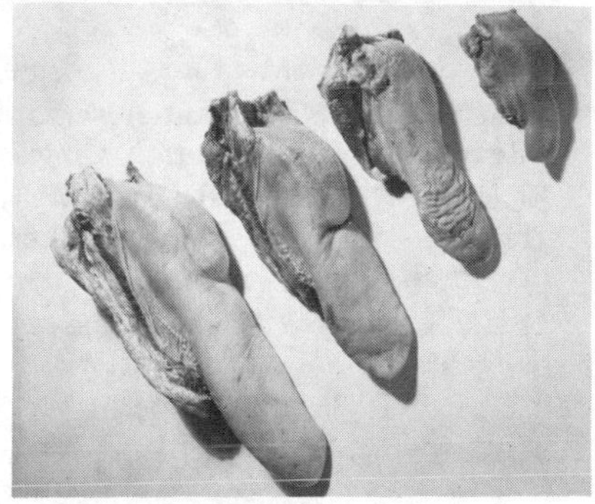

FIG. 17.6. TONGUES

From left to right:
Beef, veal, pork, and lamb

Courtesy of National Live Stock and Meat Board

cooking to make it tender. It should be braised up to two hours before it is garnished with sauce, broiled, deep fried, creamed, or used in pepper pot soup.

Oxtails

There is considerable meat on the oxtail which is very rich in flavor. Oxtails are especially adapted to braising and soups. They require long, slow cooking.

GAME MEAT

A market exists for domestically produced game slaughtered under inspection. Several different varieties are offered, as for example, buffalo, bear, elk, whale steaks, and even kangaroo fillets from Australia. These products are usually expensive, and in most cases the meat is tough and has a very gamy flavor. Though not intended for the general palate, game meats may be used to create romance for gourmet dinners. Generally, game should be prepared with moist heat.

BIBLIOGRAPHY

AMERICAN MEAT INSTITUTE FOUNDATION. 1960. Science of Meat and Meat Products. W. H. Freeman and Co., San Francisco, Calif.

ANON. 1950. Ten Lessons on Meat. 7th Edition. National Live Stock and Meat Board. Chicago, Ill.

ANON. 1955. Variety meats. American Meat Institute. Chicago, Ill.

ANON. 1957. Variety meats. National Live Stock and Meat Board. Chicago, Ill.

BULL, S. 1951. Meat for the Table. McGraw-Hill Book Co., New York.

DOTY, D. M. 1955. Meat preservation—past, present, and future. Circ. *13*, Am. Meat Institute Foundation, Chicago, Ill.

LECHOWICH, R. V., BROWN, W. L., DEIBEL, R. H. and SOMERS, I. I. 1978. The role of nitrite in the production of canned cured meat products. Food Technol. *32*, No. 5, 49–50, 52, 56, 58.

NIVEN, C. F., JR. 1960. Some prejudices and outmoded beliefs in meat curing. Circ. *56*, Am. Meat Institute Foundation, Chicago, Ill.

ZIEGLER, P. T., 1954. Meat We Eat. 4th Edition. Interstate Printers and Publishers. Danville, Ill.

18

Cooking and Palatability

Meat is usually cooked before serving. It is more attractive and more palatable when hot and aromatic. Cooking brings out the best flavor attributes; frequently contributes to tenderness. In addition, cooking renders the product nearly sterile and easily digested.

The three basic cooking methods and their variations are listed in Table 18.1.

Moist heat contributes to the tenderness of meat. The collagenous tissues in combination with heat and water change from a chewy substance to tender gelatin. There is some flavor loss in the drippings.

Dry heat methods produce maximum flavor, but the cooked products are not as tender as those obtained when moist heat is used. Dry, intense heat solidifies proteins. This increases the chewiness. On the other hand, a crust forms which is most flavorful and helps to retain the extractives. The results in frying are similar to dry heat.

ROASTING

The most controversial cooking method today is roasting. There are two diametrically opposed schools on temperature, and a half dozen areas that can arouse considerable discussion. A point of view will be presented here that is based on current investigation, which is strongly in favor of controlled, low temperature roasting, using a good oven with an accurate regulator.

Through the application of heat, some shrinkage and cooking loss will occur. Some fat, water, extractives, and food constituents are going to escape, although most of them are retained in the drippings. The application of heat also affects appearance and palatability.

There have been many controlled experiments under U.S. Department of Agriculture sponsorship at university level concerned with the temperature of roasting. The conclusions of this work may be summarized as follows:

TABLE 18.1. GENERAL COOKING METHODS

Cooking with Dry Heat	Frying	Cooking with Moist Heat
Roasting	Pan frying	Braising
Broiling	Deep fat	Stewing
Pan broiling	Sautéing	Pressure cooking

Also: Smoking, Barbecuing, Microwave Cooking

(1) Shrinkage is directly proportional to temperature: the higher the cooking temperature, the greater the shrinkage.
(2) Flavor, juiciness, and tenderness, are inversely proportional to the cooking temperature: the lower the temperature (within certain limits), the higher these qualities.
(3) Slower roasting will yield a more even "doneness."

Low Temperature Roasting

There are other advantages with lower roasting temperatures. Less supervision or attention is required as nothing happens very fast, equipment is easier to clean, and in the final analysis, the portion cost is reduced, due to less shrinkage. There is this additional factor: the meat roasted at lower temperatures tends to hold together better, which means less waste in carving or more usable slices per roast.

Some chefs bring a roast to room temperature by taking it out of the cooler and holding it at room temperature for some time before it goes into the oven. This warming prior to cooking: (1) reduces cooking time required as internal temperature is raised before roasting; and (2) permits accelerated enzymatic action in the pre-roasting period. Though the product is out of refrigeration, it is not there long enough for bacteria, molds, or any other harmful substance to taint or spoil it.

Slow cooking also permits maximum enzymatic action during the warm-up prior to cooking. The action will continue until an internal temperature of 175°F is attained.

An example of this would be a "ship round" 50 of some to 60 lbs roasted in a 250° to 300°F oven for many hours. The only apparent explanation for the tenderness of the whole round, as compared with the individual cuts which seem to be less tender when roasted separately, is the protracted roasting period.

The use of a thermometer is essential to obtain uniform results. If a meat thermometer is not used, it is necessary to use a uniform oven temperature, a uniform size roast, uniform internal starting roast temperatures, the same quality of meat, the same oven load, and by experimentation determine the oven time that will produce the desired results.

There are many slight variables that can upset the cooking time formula approach, as for example: (1) a slight variation of temperature; (2) a variation of customary shape of the cut; or a (3) change in style of preparing the roast.

Large roasts continue cooking for some time after they are removed from the oven. The roast should be removed from the oven when the thermometer is several degrees below the desired finished temperature. To determine the exact reading, trial and error will have to be employed. Removal temperatures of a large roast might be suggested at 120° to 125°F for rare, 135° to 145°F for medium, and 150° to 160°F for well done. This presupposes that the roast will be allowed time to "set up" or continue cooking at room temperature before it is sliced.

Roasting Techniques

There are seven basic steps to satisfactory roasting:

(1) Season as desired with salt and pepper, condiments, and garlic. Preseasoning with herbs would tend to create a bitterness in the finished product. Adding flour increases the possibility of scorching. Brown flour in drippings to thicken gravy. Salt brings out the natural flavor.

(2) Use a rack and shallow roasting pan. A rack is advantageous as it will hold the roast out of drippings.

(3) Attempt to place the bulb of the thermometer in the center of the largest muscle. The bulb should not rest in fat or touch bone.

(4) Add no water and *do not cover* if the flavor of dry roasting is desired.

Wrapping an oven roast in tight foil is equal to using a covered pan. The contained steam will cause the roast to be more tender as in all cases of moist heat cooking. However, the finished product will lack the desired flavor of dry roasting.

(5) Preheat oven to 375°F, then place the roast on a rack, fat side down, to the point of "caramelization" or surface crusting, then turn the roast fat side up and reduce the heat to 325° to 350°F. The roast will be self-basting due to the fat drippings starting at the top.

(6) Roast to the desired degree of doneness as indicated on the thermometer. Metal skewers may be used to reduce the cooking time as well as the cooking losses. They will leave an undesirable pattern inside the roast.

(7) Any reasonably tender cut of beef may be used. As for the more youthful type animals, veal, pork, and lamb—almost any cut may be roasted, with the possible exception of shanks, limited only by how tender the finished product must be. Pork should be well done. Internal

temperature should reach at least 165°F to be safe for consumption, all trichinae killed, and 185°F to be palatable.

Roasting Variations

There are a variety of ideas worked out in experimental kitchens that should be considered. These may be treated as variations to the basic technique.

(1) Salt Shells.—Some successful prime rib operations use a complete cover of rock salt with good results. It is entirely possible that the salt shell acts as an insulator absorbing a great amount of heat and slowing the roasting process in otherwise hot ovens. Steam may be retained in the salt shell making the roast more tender, but less flavorful. In a final test perhaps the same results could be achieved with a saving of both labor and salt by using a lower roasting temperature.

(2) Finishing in Steam Table.—By maintaining a temperature below that which further cooks the roast and keeping the lid down on the steam table or on a portable cart when the roast is not being carved, the results would be to keep the product hot and allow moist heat to work on the connective tissue, which may contribute to the tenderness.

(3) Searing.—It is an established fallacy that the coagulation of the surface of a piece of meat by searing stops the cooking loss but searing does cause a flavorful crust with a desirable brown color.

(4) Larding.—Lardoons, or strips of back fat, may be inserted into a piece of meat with a larding needle to increase the juiciness; or layers of back fat, bacon, or salt pork may be placed on roasts to provide melting fat and natural basting. As fat is a better heat conductor than lean, it will speed up the rate of cooking.

(5) Basting.—This is not necessary. Meat placed in the oven with the fat side up will be self basting as the fat melts. If the meat is lean, it may be larded.

(6) Low Temperature.—There are roasting devices on the market in which temperatures as low as 160°F can be maintained. It is also feasible to use infrared lamps, cooking at very low temperature for extended periods. The Alto-Shaam oven is heated by a radiant heat source.

Very low constant temperatures can be achieved. A variation of this oven, the Thunderbolt, has a two stage control, roasting at 250°F with a second stage as low as 135°F for holding.

(7) Bag in Water.—The meat is placed in a plastic impervious bag, and roasted in hot water below boiling. The heat exchange is very slow. Roast may be finished rare. One disadvantage, from a display point of view, is that the meat will not get the attractive brown surface color achieved by high heat roasting. Color can be imputed with various commercial products.

Low temperature roasting results in less shrink, and because of the accelerated enzymatic activity over a prolonged period, will yield a more tender, flavorful, juicy roast. Roasts are sometimes held up to 24 hours at 135°F. The only visible change is that it becomes more tender.

Care has to be exercised with low temperature roasting over protracted periods. If a boil in the bag technique is used, water should be over 160°F to ensure no bacterial growth. It is a good practice to freeze, store in the cooler, or use the roast shortly after it is taken from the water.

Product can safely be stored in the Alto-Shaam oven up to 24 hours, providing the product and the oven are sterile at the outset. The oven can be sterilized simply by running it at 200°F plus for 1 hour or more. Roasts coming from a conventional oven at 250°F plus are sterile going into the Alto-Shaam. They should be handled with sterile forks.

Carving

To master carving for maximum yield and the best presentation, a few simple rules should be observed: (1) use only the best knives, steel, fork, and slicing board; (2) the style of the knife is a matter of personal preference; (3) keep the knives sharp with stone and steel; (4) know the bone structure and the direction of the muscle fiber; a first rule of carving is to cut across the grain; (5) start with a plan, cut with a plan, and attempt to make neat slices; and (6) whenever possible, roast boneless cuts for maximum yield and uniform slices.

BROILING

In broiling, the protein on the surface coagulates, the extractives are retained, fats melt and are absorbed by the lean increasing juiciness, and a wonderful aroma is produced. Meat selected for broiling should be high grade. If the cut is very lean (low in fat), the cooked meat may be hard, dry, and tough.

Broiling is generally done on a rack with some type of intense heat, such as charcoal, gas with ceramic coals, an electric element or gas heated infrared radiation elements. The heat should be intense to get maximum flavor. This is difficult to accomplish by cooking with a simple gas element over the product, as a salamander. Charcoal is an effective heating element. Gas heated ceramic coals made of various materials such as lava ash work well and are clean and economical (Fig. 18.1). Many chefs broil over charcoal, yet charcoal itself makes no flavor contribution. Charcoal merely does what any other source of intense heat would do.

The success of broiling is based on simple chemical fact. Coagulation of the protein which occurs at high temperatures produces a distinct and desirable flavor. The broiling technique is directly contrary to oven roasting as far as temperature is concerned. Shrink is not a major factor in that the cooking time is relatively short. A maximum and distinct flavor sensation is the primary objective.

Broiling Technique

(1) Preheat broiler to cooking glow. Place the grate at the cooking position so that it too may be preheated.

(2) Preseason meat with salt and pepper, and other condiments, and the seasoning will become a part of the crust. Salt brings out the flavor of the meat, but retards caramelization. Seasoning must be done before cooking, as the flavorful crust forms during cooking. There are some who object to preseasoning for fear of loss of fluids. This loss is not significant in quantity.

(3) Cook as close to the heat as practical. Individual evaluations will have to be made relative to fuel, type of grill, amount of product handled, and the flames from the drippings.

(4) Cook half way on one side, and turn just once when the heat comes from below. As the second side is cooked the heat from below

Courtesy of the Pepper Mill, Pasadena, Calif.

FIG. 18.1. CERAMIC COAL TYPE BROILER

drives the natural juices to the top. Another turning would drop all of the juices in the fire and make the product less palatable.

(5) The test of doneness should be by sight, feel, probe, or cooking time. Experience on a particular broiler with careful observation will ultimately lead to uniform results.

(6) Products generally broiled are beef steaks, lamb chops, hamburgers and sausage. Pork and veal steaks and chops should not be broiled. Ham steaks may be broiled at temperatures under 400°F, and bacon is best pan broiled at temperatures below its own smoke point (about 300°F) browned in its own grease.

Pan Broiling

This is a variation of broiling employing the heat of conduction, rather than the heat of convection and radiation. A very heavy cast iron skillet should be preheated or literally "filled with heat" before broiling starts. Salt may be placed in the skillet, and when it turns brown the skillet is hot enough to broil. Pour off the salt, use no fat or water, place the product in the skillet, use no cover, and proceed as in broiling. Drippings should be poured off as they accumulate. There should be no discernible flavor difference between this and broiling; however, the score marks made by the grate will be missing and a great deal of smoke will be generated. The finished product from the point of view of flavor should be crusty and flavorful. This technique can be employed on a grill if it has sufficient elements to keep it hot with a normal service load.

Pan broiling thick cuts may cause an undesirable burned crust before the desired degree of doneness is achieved. Pan broiling is especially effective for a "minute" steak or thin cut steak as it creates a crust in a very short time without overcooking the center. It is very difficult to do a thin steak on a broiler to the rare or medium rare stage yet have a product adequately crusted.

Low Temperature Broiling

Many experts favor lower temperature broiling at 350° to 400°F, sometimes suggesting the use of a thermometer. The arguments in favor of this include more uniform cooking, lower shrinkage, a more tender product, more attractive appearance, less smoking, lower fuel cost, and greater control of cooking.

Temperature is frequently controlled or reduced by spraying the charcoal bed with water, which creates steam to tenderize some of the connective tissues. However, there will be a less desirable flavor from the moist heat and lower temperatures.

Logistically, airline feeding creates some unusual problems. Steaks which are most popular cannot be broiled while in the air. No equipment, as of yet, has been devised compatible with airline safety. Three techniques are presently employed:

(1) Just before departure, caramelize the steak, top and bottom, and finish in an oven while in flight.

(2) Broil the steak to the desired doneness and hold in an oven in the cabin until served.

(3) Caramelize the steak, freeze, and hold. Defrost before departure time and finish in an oven while in flight.

The searing, freezing, reheating process is the most unpalatable. This creates a definite negative reheat flavor, sometimes unpalatable. The other two processes are fair. The most acceptable seems to be to broil the steak to the desired doneness and hold at 135°F until served. Strict sanitation must be maintained.

FRYING

Frying may be defined as cooking in fat, and includes pan frying, sautéing, and deep fat frying. This variation of dry heat cooking produces a flavorful crust with some sacrifice of tenderness. Thin slices of tender cuts or cuts which have been tenderized by cubing, pounding, scoring, or grinding are best suited.

Pan Frying Technique

(1) Use an old fashioned iron skillet for best results as it will heat evenly and retain the heat until the cooking is started.

(2) Preseason with pepper, dredge in flour, meal, or crumbs, as recipes may indicate.

(3) Brown on both sides in a small amount of fat. The temperature should be below that at which the fat volatilizes.

(4) Do no cover. A cover would retain steam and braising will occur. The desirable crisp crust and its unique flavor will not be obtained if the skillet is covered.

(5) The proper temperature will cook the product thoroughly and also brown and crust it. Turning is necessary to ensure even cooking; season with salt before serving.

(6) The proper temperature and cooking time for any item can be arrived at by trial and error. It is most important to avoid high temperatures. Smoking fat, which is burning or breaking down, clearly indicates excessive heat. A thermometer should be used.

(7) Sautéing is similar to pan frying. Small tender pieces of meat are usually used, which are turned frequently and cooked quickly. Scallops

of veal, for example, are frequently sautéed and then blended with a special sauce as in veal scallopini.

(8) If hamburgers turn red and stay red in the center, this is not caused by the meat or the method. If water is added, which contains very small traces of nitrate, bacterial action may reduce the traces to nitrites. The older the patty, the greater the bacterial action, the redder the color. The patty may be fully cooked yet very red in the center.

Deep Fat Frying Technique

(1) The best equipment is a deep kettle made for deep fat frying equipped with a basket and a thermostatic device for controlling temperature, or a frying thermometer.

(2) Fat should be deep enough to immerse product and temperatures of 300° to 350°F should be maintained.

(3) To increase browning, crust and flavor, the product should be seasoned, and covered with crumbs, batter, flour, or meal. Salt should not be used as it will retard "caramelization" and substantially reduce the life of the frying fats.

(4) Lower meat into fat. It will cook simultaneously on all sides. The product load should be within the intended limits of the utensil. Proper cooking temperatures must be maintained.

(5) Drain off excessive fat before removing product from basket.

Frying Variation

Some items may be finished with moist heat to increase tenderness. When pan frying, cover for a short time; a small amount of water may be added. When grilling, cover for a short time with a pie tin or similar utensil. When pressure cooking is combined with oil, the two-fold effect of frying crispness and moist heat tenderness may be achieved. Excessive use of moist heat in any case will destroy the crisp crust.

Frying Fats

Many fats, both animal and vegetable, as well as blends of the two, are available. They should all be examined in terms of the particular intended function. Important considerations are initial cost, flavor tendencies, smoke point, and frying life. Various fats will be discussed in sequence of initial cost.

Rendered beef suet is probably the most consistent low priced fat. Labeled "Edible Beef Suet" under U.S. Department of Agriculture inspection, it will have no foreign, higher flavored fats in it, and if it is hydrogenated, it will impart a minimum amount of flavor. Although it has a relatively low smoke point, suet may be considered for many deep frying purposes.

Hydrogenated lard of good quality will range in price considerably below vegetable fats. The smoke point is above 380°F. Lard will change less in quality than other fats if it is not overheated. After each use, lard should be strained through cheese cloth to remove foreign particles, covered and stored in a cool place.

Vegetable shortenings and oils have the highest smoke point, and under most actual kitchen conditions, the longest life. The initial cost is highest. These fats require the same good care as animal fats. For very high temperature frying, and for minimum of flavor transfer, vegetable shortenings seem to be the most acceptable. The cost of any fat can only be measured relative to the amount of product produced per dollar of fat. This can only be determined by kitchen testing with usual use and conditions.

Oil life can be extended by daily filtration. Filter equipment using diatomaceous earth is especially effective.

BRAISING

Braising is generally employed for the less tender cuts, or for a cooking variation to create a different flavor sensation. The product is cooked in a covered utensil with a small amount of liquid. The moist heat gelatinizes the connective tissues, creates a distinct flavor sensation, and makes a more tender finished product. Some samples of braised dishes are beef pot roast, fricassee, Swiss steak, and casserole dishes. With slight variations, in each instance, the cooking principles are the same.

Braising Technique

(1) Brown the meat on all sides in a heavy skillet. If the cut is lean, add fat. This increases the aroma and develops flavor.

(2) The meat may be dredged in flour, or breaded and seasoned to suit before browning, or if not floured, may be seasoned after browning. The condiments, herbs, and spices, should be added before braising, as they become a part of the gravy and make a greater penetration into the product. Sometimes a product such as sauerbraten is marinated before cooking which adds flavor.

(3) Cook in a tightly covered utensil to which a small amount of liquid is added. The contained steam provides the moist heat that will produce tenderness. Various liquids may be used such as water, vegetable juices, soup stock, wines, and marinades. The amount of liquid should be

maintained at a desirable minimum. Liquid may be added as it cooks away. Many cuts will require no additional liquid to maintain an atmosphere of steam.

(4) Cook until tender and done at a low temperature of 300°F in the oven or below boiling on top of the range. Very high temperatures will soften the connective tissues and finally completely hydrolyze them, making the meat stringy, difficult to slice, unnecessarily dry, as well as causing excessive shrinkage.

(5) Drippings may be used for gravy. They are high in both flavor and food value.

(6) The less tender cuts with considerable connective tissue and meat from older animals, chops, steaks, and cutlets from veal and pork, and some offal, hearts, hog and beef liver, kidneys, and tripe are generally braised. Cuts may be scored with a mechanical cuber or with a hand tool or pounded to shorten the fibers. Consommé may be added to hamburger; this will help to tenderize the meat and also improve the flavor. This hamburger when broiled or fried will be less crusty, and when cooked may turn gray-brown inside.

STEWING

Stewing is a variation of utilizing moisture to break down connective tissues. In this case, the product is submerged in a liquid. A properly cooked stew is tender, juicy, will not fall apart, and is not stringy. It should be served with the drippings which contain so much of the flavor.

Stewing Technique

(1) If "brown stew" is desired, brown meat in a pan on all sides. For maximum flavor use the very minimum amount of liquid. This is optional.

(2) Then add liquid, cover and allow the meat to stew in its own juices. Preferably it is started with hot liquid, either stock or water.

(3) Season well. Use salt, pepper, herbs, spices, wines, and vegetables. A great variety may be used. Vegetables to be served with the meat, should be added just long enough before the meat is tender to cook properly. They should be selected for appearance and flavor, and should be cut in uniformly large pieces.

(4) Serve with the liquid or a gravy for maximum flavor and aroma. Vegetables will add color to the dish. Individual casseroles may be topped with a baking powder biscuit. Dumplings, noodles, macaroni, or spätzle are natural complements.

(5) When served cold, allow to cool in its own stew for maximum flavor and minimum shrink.

(6) A meat pie may be made by using a thickened gravy made of flour paste and two tablespoons of stock per cup. Bring to a hard boil to thicken. If peas are used, cook separately for maximum color retention.

The Stock Pot

The stewing process is essentially the process for making soup stock or meat stock. Through this process the soluble proteins, fats, minerals, and gelatins are extracted, giving the stock body. That portion of the fats not absorbed should be skimmed off. Many by-products of the kitchen that otherwise might be disposed of may be added to the stock pot. Bones and tendinous cuts of meat as the shank contribute the maximum amount of collagen, which is converted into gelatin in the cooking process. Meats that are added to the stock pot may later be seasoned and used for meat dishes as they still contain much of their food value. Soup stimulates the appetite and the flow of digestive juices.

COOKING FROZEN MEAT

Should frozen meat be thawed first or can cooking be started while it is still frozen? Many experiments demonstrate that this is a matter of personal preference and convenience. However, cooking of meat while still frozen requires some additional skill and experience. If cooking is begun before defrosting, considerable additional cooking time must be allowed to thaw the meat. One hundred forty-four B.t.u. are needed to change a pound of ice at 32°F to a pound of water at 32°F. New critical time tables may have to be developed. In broiling, the intensity of the heat determined by the type of fuel and height of the grate will vary with each operation. Trial and error procedures should be employed to develop time tables.

Because there are many variables, for example, grade of meat, weight and shape of the roast, and the type of oven, each operator should develop his own timetables for the various items. In the case of roasts that are not defrosted, they should be placed in the oven long enough to thaw them, then a thermometer should be inserted into each and noted from time to time to judge the proper doneness.

Cooking meat in a frozen state is a necessity as in the case of veal cutlets made by grinding, forming, and breading. They would otherwise fall apart. Other extenuating circumstances might require that other products be prepared while still frozen. Microwave cooking is generally started from the frozen state.

Sometimes the finished product is less desirable when started frozen. For example, it is possible when roasting to finish the product on the outside and part way through, with the interior still frozen, cold, or partially cooked. In the case of broiled steaks it is almost impossible to use the intense heat crusting technique and get the middle of the steak warm when the outside is done. Further, there is the economic consideration of the high cost of fuel in the prolonged cooking of frozen products.

Defrosting prior to cooking may be achieved in any convenient manner: slow defrosting in a refrigerator, at room temperature on a rack with a pan to catch the drippings, or in circulating cold water. For quick defrosting of meat in a Cryovac bag, leave the product in the bag intact and run cold water over it. The bleached appearance caused by the water is not significant. Refrigerator defrosting appears to be the best practice for a daily program. Surface bacterial action is minimized and the product has the best shelf life after defrosting. Room temperature provides an accelerated defrosting process and may increase the number of bacteria on the meat. The latter is generally not significant if the product is cooked within a relatively short period of time.

Defrosted meat should be cooked in the same manner as fresh meat.

The drip loss experienced during defrosting usually is not significant. Defrosted products enjoy a decrease in the losses during cooking, so that the over-all losses from frozen to cooked are about the same for products cooked frozen or defrosted.

COOKING WITH MICROWAVES

Due to the resistance of food products, high frequency electrical energy is transformed into heat. This is unlike heating by induction (due to motion of electrons) or dielectric heating (molecular motion). To evaluate this technique it is important to consider that the type of heat and the intensity of the heat will have a direct bearing on the ultimate flavor of the product. High frequency cooking should be examined not only for its various functions in the kitchen, but taste panel evaluations should be made.

The "diathermal" or "thermotronic" technique involves two electrodes or condenser plates from a high frequency oscillator. The product is placed between and touching the two points. The temperature increase in the product is rapid and uniform throughout. In a very short time with this device, it has been demonstrated that an internal temperature of over 200°F can be achieved with less than a 10°F temperature variation throughout.

Microwave equipment at a much higher frequency (about 2000 megacycles) has been commercially marketed. This equipment requires no electrodes and no direct contact with the product.

Many phenomena occur when cooking with microwaves. The waves produce no heat when they are transmitted through air. The microwaves have a maximum penetration of about 2½ to 3 in. For thin products, the cooking period is relatively short. For thicker products, from the maximum penetration point of energy to the center of the product, heat is transmitted by conduction which considerably slows the cooking process. This presents an interesting question. What flavor changes occur because of the dual cooking process, microwave to the 3-in. depth, and by conduction to the middle?

This quick cooking technique is still being explored. There are some disadvantages: the flavor it produces, the poor color and appearance of the cooked meat, and the absence of flavorful crusts. Some manufacturers have introduced a second cooking element in the cabinet such as quantz lamps or a browning grill. This dual heat system is an attempt to combine microwave speed with an energy source that will caramelize the surface.

Microwave cooking has some applications that should be carefully evaluated by each operator to determine if and where it fits a particular kitchen plan. Frozen products may be cooked without defrosting, cooked foods may be reheated, some items may be prepared, portioned, stored in the refrigerator, and quickly heated to order. Such products are now available commercially. Certain foods taste better the "second day" or when reheated. These are ideal for microwave cooking. Desserts from fresh fruits are an exceptional item. Cappuccino mix may be heated in an oven instead of using live steam.

SMOKING AND PIT BARBECUING

Smoking and pit barbecuing are essentially the same process, except for the temperature range, which is considerably higher for barbecuing. A smokehouse or pit should be well insulated, fireproof, impervious to the penetration of smoke, and complete with some device to generate smoke and heat. The usual medium is hardwood logs, or sawdust used in a controlled gas-type smoke generator.

The deposition of smoke gives a characteristic flavor which is unique. The smoke components, creosote, acetic acid and pyroligneous acid, and the drying effect of smoking inhibit bacterial growth. So called, "ready-to-eat" or "fully-cooked" smoked pork products must reach and sustain an internal temperature of 140°F. The trichinae are destroyed at 137°F. Enzymatic action is inhibited at 140°F which further contributes to product stability. In pit barbecuing, the temperature is considerably higher and the product is cooked and sterilized. In essence, pit barbecuing is oven roasting with a relatively high moisture condition and a prolonged enzymatic action, both making for a more tender end product. There are the additional flavor components of smoking and the flavor imputed by the basting on the crusty meat surface.

Resinous type woods such as pine should never be used. Hickory is the most popular. Many of the fruit woods, e.g. walnut, apple, plum, peach, along with other hardwoods, as maple, oak, and ash are used. Dried corn cobs are sometimes used as a source of smoke. Slight flavor variations can be detected with changes of the smoke medium.

The temperature of the smokehouse and the length of time required for smoking are inversely proportional. The end result will be approximately the same (Table 18.2).

Meats smoked until they are dark brown will have a better keeping quality but the increased surface deposits of pyroligneous acid may cause digestive disorders. If the surface is trimmed before using, the product is usable.

Tenderizers

Weak acids, salts, hydrolyzed plant proteins, and enzyme preparations are being used as tenderizers. As a general rule, tenderizers have a three-fold effect: (1) the product is more tender; (2) the tenderizer contributes to the flavor and can be detected; and (3) the flavor of the initial product is not enhanced.

Weak acids such as lactic acid, vinegar, and lemon juice used as marinades are of questionable value as far as tenderizing is concerned. They usually contribute to a flavor change in the final product.

Salts in concentrations of two per cent are effective. Application of this sort is what is commonly referred to as curing, corning, and sweet pickling. There is a definite flavor change and the final product is served as something different from the untreated raw product of the same cut.

Plant enzyme preparations such as papain, and bromelin effectively increase tenderness. Warming speeds up enzyme action (Table 18.3).

TABLE 18.2. TIME REQUIRED FOR SMOKING TO A CHESTNUT BROWN COLOR

Temperature	Time
80° to 90°F	3 to 4 days
90° to 100°F	30 to 40 hours
125° to 135°F	18 hours

TABLE 18.3. ACTION OF ENZYMATIC TENDERIZERS AT VARIOUS TEMPERATURE RANGES

Temperature	Rate of Action
0°F	No action
32° to 36°F	Very little
36° to 105°F	Builds up slowly
105° to 175°F	Sharp increase—maximum action
over 180°F	Stops

Effectiveness of enzymatic tenderizers is arrested as the 140°F mark is passed. Penetration studies have been made of tenderizers with dye molecules and microscopic examination. It is generally concluded that effective distribution cannot be accomplished by natural diffusion. Various reports indicate a maximum penetration of two to five millimeters. From this it can be concluded that the tenderizing effect is that of easing the initial bite experience on both sides of the product for a portion of the depth only. Microscopic examination further reveals that during the pre-cooking or holding period there is little tenderizing action at refrigerator temperatures, and that treated meats are more tender after cooking because the connective tissues are partially hydrolyzed. Enzyme tenderizers may have a definite negative taste contribution which occurs while chewing, as well as an "after taste" which fails to leave the palate with the "clean" sensation of a piece of USDA Choice beef.

In evaluating tenderizers, it is well to consider that tenderness is the simple quality of palatability most sought after, and that negative flavor changes occur to offset partially the advantage of tenderness. If tenderizers are used, consideration should be given to USDA Choice quality rough cuts as bottom sirloins or rounds instead of loin cuts from low grade beef.

Pre-slaughter injections experiments are being conducted. It is too early to make complete objective evaluations. These consist of oxytetracycline to protect against bacterial spoilage so that high temperature aging may be conducted, and a second process which is the injection of an enzyme to accelerate the hydrolysis of the connective tissues. A limited number of cattle are now being injected and offered through commercial channels as tenderized beef.

PALATABILITY

Palatability according to Webster is pleasant to the taste, savory, tasty. The subject is really much more complex. The pleasant experience involves much more than just taste. Because the experience is subjective, the same product will evoke different degrees of pleasantry, as well as widely different experiences relative to age group, environment, season, ethnic background, individual differences as described by Alfred Adler as "contentious apperception," and a host of other reasons.

Acceptability itself is relative. It is interesing to list exact opposites that make for high acceptability in different contexts:

Chewy—Taffy	versus	Tender—Steak
Hot—Soup	versus	Cold—Salad
Smelly—Cheese	versus	Fresh—Fish
Spicy—Chili	versus	Bland—Pudding

Relative to selling product, either in the retail market or a restaurant, palatability should be carefully related to acceptability. Significant marketing evaluations depend upon the range of acceptability in a given context. What are the limits? What is the maximum spectrum? What characteristics should be sought out? How can these be measured?

There are some sophisticated mechanical testing devices available. At best, they seem to be a method to test the testing panel. Hopefully, meat science will improve on the existing devices. For now, it appears, that one must look to some forms of subjective testing. Firstly, the factor of relativity of experience must be considered. It is sometimes fatal for a food service operator to rely on his private palate. This makes it imperative that some objective ground rules be established. There is a lot of poorly prepared food served by the restaurant industry by operators who rationalize highly personalized facts such as pride in the establishment, pride in the culinary skills of the chef, and slowly the business fails because the customer manifests his disagreement with management in the form of lack of patronage. The real problem is what is acceptable to the patron, rather than palatable to management.

To achieve the goal of acceptable products, it is important to understand the factors of palatability, and to set up standards for testing.

Factors of Palatability

Various systems have been devised. In essence the problem is *food object—human experience* interaction. Humans experience basically through five primary senses: sight, taste, smell, touch, and hearing. Food experiences are eclectic and for the experiencer, rarely broken down in components. Usually the food experience is hedonically and subconsciously rated from great to so-so to poor. From a marketing point of view, one must be more sophisticated and relate components of the experience to our senses.

Sight	—Appearance
Taste	—Flavor
Smell	—Odor
Touch	—Texture, juiciness
Hearing	—Texture

The visceral senses—appetite, hunger, thirst, satiety, peristalsis—have significant roles from time to time. In the laboratory evaluation, these will have to be eliminated. However, under special circumstances as airline feeding, certainly appetite coupled with anxiety has to be taken into consideration.

Appearance

The highly relative acceptability of food is demonstrated by appearance. Exact opposites, in different contexts, are palatable. Food molds reflecting decay and decomposition are almost universally rejected. Yet what is good piece of Italian hard salami without a moldy casing, or a piece of Roquefort cheeze without its mold?

Color or the lack of same is vital in food presentation. White is almost universally accepted throughout the world for basic foods including milk, white bread, rice, and potatoes. Blue, infrequently appearing in nature as blueberries, blue plums, and limited other items, is generally a poorly accepted color. Blue plates do not combine well with foods. Blue food coloring has very limited applications.

Colors also relate to quality, flavor, and kinesthetic experiences.

Quality colors	Yellow—Corn
	White—Veal
	Pink—Nova Scotia Salmon
Flavor colors	Red—Peppermint
	Green—Mint
	Yellow—Lemon
Kinesthetic colors	Red—Hot
	Green—Cool

There are four color relationships for meat.

(1) Cooked Color.—The color of a cooked piece of meat varies directly with the raw color. Very dark bull or cow meat will broil very dark. Perhaps most significant is the color of veal. Quality veal is highly regarded for its light color. The cooked color will only be as white as the raw product.

(2) Fat Color.—The color of cooked fat reflects the color of raw fat; yellow fats reflect the same color values after cooking. Yellow fat, associated with very mature cows and bulls, is generally unacceptable.

(3) Surface Color Reflects Cooking Technique.—Braising, broiling, or oven roasting produce their own color values.

Moist heat cookery, braising, stewing, steaming, boiling, cooking in a covered pot, or wrapping in foil turns meats grey. The product becomes tender, but the color is not too acceptable. By pre-browning the color is improved. Most important, the presentation appearancewise is improved with attractive gravies and colorful vegetables. A "colorful" name may improve the salability.

Dry heat cooking, broiling, frying, and oven roasting crusts or caramelizes the meat surface. The brown crust is a very attractive color contribution. When a grate is used, with heat from below, the intense heat of conduction at all points of contact create attractive deep colored "score marks." Markings from the grate are considered so important that many operators will not broil steaks with top heat in a salamander, or on a flat grill.

The appearance governs this decision. There is available a conveyorized broiler with infrared heat on top and bottom, and score marks on the bottom created by the transportation belt and score marks on the top "branded" on by a heated wheel to make the appearance acceptable.

(4) Internal Color Reflects Internal Temperatures.—Broiling and roasting produce a series of internal color changes. The heat transfer starts at the surface which is hottest and works slowly to the middle of the product which is coldest. When the middle of a roast is 140°F, the temperature range is progressively higher nearer the surface. A cross section of roasted meat will vary in color from red-rare in the center, through pink, through grey, through brown at the surface. To even up the doneness, a roast should be removed from the oven and held for about fifteen minutes at room temperature. The heat transfer will continue, but with the heat source eliminated, the heat in the roast will even up. *Heat seeks its own level.* The temperature will even, and the color will even.

Internal colors are due to the physics of heat and the chemistry of meat. At internal temperatures of 131°F to 149°F coagulation of oxyhemoglobin occurs. This chemistry supports the observable pink-rare to medium-rare meat color. From 150°F to 176°F, heme pigments occur, which account for meat's brown-gray color.

The roasting process may be summarized by saying that while there is little, if any flavor change, the various roasting stages produce dramatic color changes which vitally affect acceptability. Many operators will not use microwave for dry heat meat cookery. One of the reasons is color failure. This only emphasizes the consumer significance of appearance.

Higher internal roasting temperatures produce texture changes. Medium-rare through medium roasts are generally firmer and more tender than rare roasts.

Taste

"Taste" instead of "flavor" is a semantical approach. Flavor is frequently associated with a total experience—oral, olfactory, and kinesthetic. Taste evaluation, for the purpose of this phase of the discussion, is related specifically to the oral sensory experience.

The taste factor is not to be taken lightly. Taste or the lack of taste fills garbage disposals daily in homes and restaurants. Early in the 1960s, flavor experiences were pioneered by some frozen food manufacturers who added combinations of butter, salt, monosodium glutamate, sauces, and cheese to various fresh frozen vegetables. Cognizance was taken that vegetables may be eaten "because they are good for you," but a lot more are sold if they are tasty.

Taste blending, though a relatively new commercial practice, is no great innovation—even the humblest table is set with salt, probably the most important flavor enhancer.

Salt makes a cantaloupe sweeter.

Salt makes a lime less sour.

Salt brings out the natural meat flavors.

Salt makes a noticeable taste contribution when added to raw steak before cooking, as it becomes an integral part of the meat crust or surface caramelization. Some kind of comparative taste evaluation should be devised to evaluate steak preparation. Which is more acceptable: natural steak flavor or the flavor of preseasoned steak?

Some food technologists are stating that in the future natural foodstuffs will be merely the "building blocks," and that the technologist will create flavor experiences and condition whole markets. The idea is really not really new. For example, "Mint Jelly" is really apple jelly colored green with an artificial mint flavor. Today, textured soy protein is marketed. A bacon flavored solid is widely used in salads.

Taste also reflects the highly relative nature of acceptability. Taste sensations are acceptable in different degrees in different contexts for different foods. The food service operator must determine consumer preferences!

For example, how salty should a ham be? Bitey like a Smithfield?

Bland like a canned Danish pear ham? Somewhere in between if your market is Kansas? Obviously, the place, people, occasion, consumer, and provincial food habits are all relevant variables.

Although there are many systems of flavor categories, a basic system of five reasonably covers the various taste sensations: sweetness, saltiness, sourness, bitterness, and metallic.

When taste evaluations are made, results and corrections can be formalized in these terms.

Odor

Odor is a component of the flavor experience; it is that part we experience through the nose. Flavor and odor are almost as one. In laboratory conditions with blindfolds and nose plugs, denied our sense of sight and smell, flavor differentiation of many food products is impossible. Pleasant food odors improve dining pleasure. Some odors can be deliberately produced. A hot scented towel offered after dining is one device. Good operators will not overlook the opportunity to create supplementary pleasant odors or eliminate bad odors to improve the total experience. How many dollars does the sidewalk odor of fresh baked donuts bring into the bakery? How many dollars on a hot summer day does the smelly fish market drive away? The popularity of meat may be attributed to its very acceptable, delightful odor. Yet the odor of protein decomposition, the so-called "aged odor," is highly undesirable to Americans. This negative odor becomes even more pronounced in cooked meats. To confuse protein decomposition with aged meat could be an expensive mistake.

Touch

Probably the most important single meat-palatability factor is touch, which pertains to three related and overlapping kinesthetic experiences: oral, visual, and auditory. Arbitrarily, touch can be divided into four categories: texture, tenderness, hotness, and juiciness.

Texture.—Texture is defined sometimes as "mouth feel." This property relates to density, viscosity, surface tension, and other physical properties. Here are a few texture experiences combining auditory, visual, and oral experiences.

Firm	Velvety	Crisp
Chewy	Sticky	Lumpy
Rough	Mealy	Soggy
	Stale	

Celery is a good example of food texture being most acceptable when fresh and crisp. Hearing and touch may be even more important than taste in this particular food experience.

Texture plays a significant role in the evaluation of fish. Many people assert they prefer fresh to frozen fish, and there are facts supporting this conclusion. Fish texture which is very delicate changes when frozen. Fresh Dungeness crab is delightful; frozen it is more like straw. To overcome some texture changes, flavor may be added in preparation. Add a tasty cocktail sauce to the frozen crab and it becomes acceptable.

Texture experience can be heightened by a texture variation. Combining textures in the case of fish by sautéing with toasted almonds heightens the experience. Other examples of kinesthetic variations are chewy raisins in rice pudding, crisp toasted croutons in soup, and chestnuts in poultry dressing.

Tenderness.—Several investigators have concluded that tenderness is probably the single most important factor affecting consumer evaluation of meat quality and acceptability.

A survey made of a relatively sophisticated food service seminar at Cornell University (January 1967) indicated overwhelmingly that "meat tenderness is probably most important." Yet, a further survey of this same group showed, by equally overwhelming vote, they preferred top sirloin butt steaks or strip loin steaks over the more tender fillet or tenderloin steaks. Boiling, braising, or using ground or comminuted meats would produce more tender results. Yet tougher broiled meats are preferred.

In fact, meat tenderness is a relative term. Tough meat and meat that is too tender are equally unacceptable. For example, a relatively tough piece of meat passed through a mechanical tenderizer might be very acceptable. On the other hand, a USDA Choice strip loin processed through the tenderizer might be too tender and unacceptable as a strip loin steak.

Americans prefer broiled steaks and the texture experiences developed through broiling. *Broiling is antithetical to tenderness.* Dry heat cooking shortens the connective tissues and toughens the meat. To satisfy the demands for both tenderness and texture, it is important to select top grades of beef correctly aged and to eliminate through trimming so far as possible any connective tissues making for toughness. Finally, broiled steak tenderness is a kinesthetic experience that starts at the fingertips. The initial experience with a good sharp steak knife is important.

Juiciness.—Juiciness combines visual and kinesthetic experiences in a time pattern: initial visual experience, first bite, and sustained chewing. A few controversial points specifically related to meat juiciness should be examined. Although meat science is a long way from exact laboratory answers, a few statements can be made.

(1) Juiciness is not reduced by using a fork as a cooking utensil. Meat is not a balloon that bursts when punctured; it is multicellular. Steak meat can be mechanically tenderized with literally hundreds of small knife slits and will still be very juicy.

(2) Juiciness is not directly related to serum or water content. Choice beef averages about 50% water; grades as low as Utility run 70% water. A simple demonstration proves that Choice beef (with 20% less water) is far juicier. It is reasonable to presume that the experience of juiciness correlates with intramuscular fats.

(3) Freezing and defrosting meat, under optimum conditions, temperature, wrapping material, and storage period have little if any effect on juiciness.

(4) Concern about drip-loss, drying, "bleeding," defrosting, etc., should be analyzed. The visual experience of meat defrosting in a pool of serum is dramatic. But more important, what happens to a fresh steak when put on a broiler at 500°F to 1000°F? The serum evaporates! Check weight of product before and after cooking. The difference is largely moisture loss through cooking. Meat science indicates that the loss of moisture will be just about the same for fresh meat or frozen and defrosted meat. Top quality steak will finish juicy despite drip losses apparent during defrosting.

(5) The myth that salting meat makes meat less juicy probably originated with the concept of osmosis. Presalting does not reduce the juiciness experience.

(6) When broiling meat with heat from below, turn once. Heat drives moisture away. Heat from below deposits the natural juices on the top of the steak. If these juices are preserved on top of the steak, the initial visual experience of the patron will be one of juiciness.

FOOD TESTING

It must be concluded that to reasonably evaluate a product one must resort to empirical procedures—taste and tell. Two basic techniques may be employed: the triangle test, and the general acceptance test.

Triangle Test

Three samples, two of which are alike and one different are submitted to a taste panel. This test (Fig. 18.2) should only be used to evaluate

(1) Please try all three samples and check the appropriate box below.

- ☐ All three samples are the same.
- ☐ All three samples are different.
- ☐ One sample is different and two are the same.

Which sample is different? ______

(2) If you thought that one or more of the samples were different, which sample did you prefer? ______

(3) Why did you prefer this sample? ______

Name ______
Date ______

FIG. 18.2. TRIANGLE TEST

differences. The panel is instructed to select the "unlike" sample. Such a test might be used to compare fresh and frozen items to determine if the expert palate can discern any difference. This test can provide additional feedback relative to preference of samples and reasons for choice.

TASTE PANEL SCORE SHEET

ITEM ______ PAGE ______

You have been given three samples. Two of the samples are identical; one sample is different.

I SAMPLES ______ and ______ are identical.
II SAMPLE ______ is different.

Grade the two samples as follows using the thermometer type scale illustrated below:

Scale	Score
Excellent	10
Very Good	
Good	
Fair	5
Poor	
Very Poor	
Extremely Poor	0

	Sample I	Sample II
Appearance:		
Aroma:		
Flavor:		
Tenderness:		
Juiciness:		
Texture:		
TOTAL		

Name ______
Date ______

FIG. 18.3. GENERAL ACCEPTANCE TEST

General Acceptance Test

Various techniques may be used to have a taste panel rate different samples (See Chap. 8, **Palability Test** for basic principles of testing.)

(1) Score each sample as a "total experience" according to a predetermined descriptive scale.

(2) Rank each sample as "good," "better," "best."

(3) Employ a hedonic rating chart, ranging from negative experience as "disliked very much" through the scale to "enjoyed very much."

(4) Profile the product by employing a hedonic scale and by evaluating such characteristics as flavor, juiciness, texture, tenderness, appearance, etc. Supplementary to general acceptability, solicit additional general subjective reactions from the panel (Fig. 18.3).

BIBLIOGRAPHY

AMERICAN MEAT INSTITUTE FOUNDATION. 1960. Science of Meat and Meat Products. W. H. Freeman and Co., San Francisco.

ANON. 1942. Meat and Meat Cookery. National Live Stock and Meat Board. Chicago.

ANON. 1945. Meat Handbook of the U.S. Navy. U.S. Navy Dept., Washington, D.C.

ANON. 1950. Ten Lessons on Meat, 7th Edition. National Live Stock and Meat Board, Chicago.

ANON. 1958. Principles of Microwave Cooking. Raytheon Corp., Waltham, Mass.

ANON. 1959A. Cooking Meat in Quantity, 2nd Edition. National Live Stock and Meat Board. Chicago.

ANON. 1959B. Better Homes and Gardens Meat Cook Book. Meredith Publishing Co., Des Moines, Iowa.

ANON. 1960. Meat Manual, 6th Edition. National Live Stock and Meat Board. Chicago.

ANON. 1962A. Fun with Meat Outdoors. National Live Stock and Meat Board. Chicago.

ANON. 1962B. Food Buyers Guide, A circular. Food Publications, Los Angeles.

ANON. 1962C. Facts about Beef. National Live Stock and Meat Board, Chicago.

ANON. 1962D. Proteases, Circ. *SP-40*. Rohm and Haas Co., Philadelphia, Pa.

BRELAND, J. H. 1947. Chef's Guide to Quantity Cooking. Harper Bros. New York.

BULL, S. 1951. Meat for the Table. McGraw-Hill Book Co., New York.

CHANG, S. S. *et al.* 1966. Methodology of Flavor Evaluation. Publications Dept., Packaging Institute, New York.

CULLEN, M. O. 1941. How to Carve Meat, Game, and Poultry. Grosset and Dunlap, New York.

DECAREAU, R. V. Microwave oven in hospital food service. Hosp. Management *87,* No. 3, 96.

DOTY, D. M. 1955. Meat Preservation—Past, Present, and Future, Circ. *13.* Am. Meat Inst. Foundation, Chicago.

DOTY, D. M., and PIERCE, J. C. 1961. Beef muscle characteristics as related to carcass grade, carcass weight, and degree of aging. Tech. Bull. *1231.* U.S. Dept. Agr., Washington, D.C.

EVANS, N.R. 1959. Food Preparation Manual. Harper Bros., New York.

JELLINEK, G. 1964. Introduction to and critical review of modern methods of sensory analysis. J. Nutr. Dietet. *1,* 219–260.

LEVIE, A. 1967. Convenience meats. Cornell Hotel Restaurant Admin. Quart. *8,* No. 1, 80–85.

MATZ, S. A. 1962. Food Texture. Avi Publishing Co., Westport, Conn.

MCLEAN, B. B., and CAMPBELL, T. H. 1953. The Complete Meat Cookbook. Charles A. Bennett Co., Peoria, Ill.

MINOR, L. J. 1966. Food flavor. Cornell Hotel Restaurant Admin. Quart. *7,* No. 3, 69–81.

PLOTKIN, S. 1976. Keep Your Roast Beef Bacteria Safe. Journal of Environmental Health 38, No. 4, 230–233.

RIETZ, C. A. 1961. A Guide to the Selection, Combination, and Cooking of Foods, Vols. 1 and 2. Avi Publishing Co., Westport, Conn.

ROHANS, J. R., and ZIEGLER, P. T. 1977. The Meat We Eat, 11th Edition. Interstate Printers and Publishers, Danville, Ill.

SCHULTZ, H. W. 1960. Food Enzymes. Avi Publishing Co., Westport, Conn.

STOKES, J. W. 1960. Food Service in Industry and Institutions. Wm. C. Brown Co., Dubuque, Iowa.

TAPPEL, A. L., MIYADA, D. S., STERLING, C., and MAIER, V. P. 1956. Application of meat tenderizer. Calif. Agr. *10,* No. 10, 10–15.

WEBER, E. S. 1962. Frozen, Prepared Foods. Cornell Hotel Restaurant Admin. Quart. *3,* No. 2, 3–8.

WICK, EMILY L. *et al.* 1966. Flavor: reflection and directions. Food Technol. *20,* No. 12, 43–52.

WOLGAMOT, I. H. 1957. Beef facts for consumer education, *AIB-84.* U.S. Dept. Agr., Washington, D.C.

WOOD, A. 1957. Quantity Buying Guides, Rev. Ahrens Publishing Co., New York.

Selected References

ANON. 1975. Official United States Standards For Grades of Carcass Beef, Code of Federal Regulations.

DESROSIER, N., and DESROSIER, J. N. 1977. The Technology of Food Preservation, 4th Edition. AVI Publishing Company, Westport, CT.

FABBRICANTE, T., and SULTAN, W. J. 1978. Practical Meat Cutting and Merchandising, Volume 1, Beef, 2nd Edition. AVI Publishing Company, Westport, CT.

FABBRICANTE, T., and SULTAN, W. J. 1974. Practical Meat Cutting and Merchandising, Volume 2, Pork, Lamb and Veal. AVI Publishing Company, Westport, CT.

FABBRICANTE, T. 1977. Training Manual for Meat Cutting and Merchandising. AVI Publishing Company, Westport, CT.

FORREST, J. C., ABERLE, E. D., HEDRICK, H. B., JUDGE, M.D., and MERKEL, R. A. (1975). Principles of Meat Science. W. H. Freeman & Co., San Francisco.

HENRICKSON, R. L. 1978. Meat, Poultry and Seafood Technology. Prentice-Hall, Englewood Cliffs, N.J.

KRAMLICH, W. E. PEARSON, A. M., and TAUBER, F. W. 1973. Processed Meats. AVI Publishing Company, Westport, CT.

LAWRIE, R. A. (1974). Meat Science, 2nd Edition. Pergamon Press, New York. Reprinted 1975.

POTTER, N. 1978. Food Science, 3rd Edition. AVI Publishing Company, Westport, CT.

WEISER, H. H., MOUNTNEY, G. J., and GOULD, W. A. 1971. Practical Food Microbiology and Technology, 2nd Edition. AVI Publishing Company, Westport, CT.

Index